Basic Cell Culture Protocols

METHODS IN MOLECULAR BIOLOGY™

John M. Walker, SERIES EDITOR

Basic Cell Culture Protocols

Third Edition

Edited by

Cheryl D. Helgason

*Department of Cancer Endocrinology, British Columbia Cancer
Agency, Vancouver, British Columbia, Canada*

Cindy L. Miller

StemCell Technologies Inc., Vancouver, British Columbia, Canada

HUMANA PRESS ✳ TOTOWA, NEW JERSEY

© 2005 Humana Press Inc.
999 Riverview Drive, Suite 208
Totowa, New Jersey 07512

www.humanapress.com

This publication is printed on acid-free paper. ∞
ANSI Z39.48-1984 (American Standards Institute)

Permanence of Paper for Printed Library Materials.

Production Editor: Nicole E. Furia
Cover design by Patricia F. Cleary
Cover Illustration: Figure 2 from Chapter 18, "Generation and Differentiation of Neurospheres From Murine Embryonic Day 14 Central Nervous System Tissue," by Sharon A. Louis and Brent A. Reynolds.

For additional copies, pricing for bulk purchases, and/or information about other Humana titles, contact Humana at the above address or at any of the following numbers: Tel.: 973-256-1699; Fax: 973-256-8341; E-mail: humana@humanapr.com; or visit our Website: www.humanapress.com

Printed in the United States of America. 10 9 8 7 6 5 4 3 2

eISBN 1-59259-838-2

ISSN 1064-3745

Library of Congress Cataloging-in-Publication Data

Basic cell culture protocols.-- 3rd ed. / edited by Cheryl D. Helgason,
Cindy L. Miller.
    p. ; cm. -- (Methods in molecular biology ; 290)
  Includes bibliographical references and index.
  ISBN 1-58829-284-3 (hardcover : alk. paper) -- ISBN 1-58829-545-1 (pbk. : alk. paper)
  1. Cell culture--Laboratory manuals. 2. Tissue culture--Laboratory manuals.
  [DNLM: 1. Cell Culture--methods--Laboratory Manuals. 2. Tissue
Culture--methods--Laboratory Manuals. QS 525 B297 2005] I. Helgason, Cheryl D.
II. Miller, Cindy L. III. Series: Methods in molecular biology
(Clifton, N.J.) ; v. 290.
  QH585.2.B375 2005
  571.6'38--dc22
                                                                    2004011235

Preface

Tissue culture techniques were first developed at the beginning of the 20th century and have undergone dramatic changes and improvements since that time. They are invaluable tools for the exploration of numerous biological questions related both to cellular processes and to the signaling mechanisms that regulate them. At some point in their careers, virtually every scientist, technician, and many medical professionals have a need to utilize cell culture systems, regardless of their area of specialization.

Our objective in preparing this book was to provide the novice cell culturist with sufficient information to perform the basic techniques, to ensure the health and identity of their cell lines, and to be able to isolate and culture specialized primary cell types. It is not the intent to educate cell culturists on any specific cell type or organ system, but rather to offer clear methodologies pertinent to current areas of investigation, as well as provide a valuable resource book for years to come. It is anticipated that many readers will have a solid background in the fundamentals of anatomy, histology, and biochemistry, but little or no experience in cell culture. We anticipate that this book will prove a useful resource for technicians, graduate students, postdoctoral fellows, as well as to the research leaders (both basic scientists and clinicians) and those cell culture experts who are moving toward the use of new model systems.

The chapters that follow provide step-by-step instructions for the isolation and growth of various primary cell types. In addition, they illustrate the techniques required for defining the properties of various types of cells, as well as for cell differentiation and expansion of cultures for large-scale experimentation.

Finally, we wish to extend our sincerest appreciation to all the contributors who willingly took their time to share their expertise and knowledge, and to the many individuals who assisted us in the preparation of this book. Special thanks go to Christine Kelly who was instrumental in maintaining organization amidst the chaos.

Cheryl D. Helgason
Cindy L. Miller

Contents

Contributors

ELIANE ALEXANDRE • *Laboratoire de Chirurgie Expérimentale, Fondation Transplantation, Strasbourg, France*

LUDMILA I. BERNSTAM • *Department of Environmental Health Sciences, University of Michigan School of Public Health, Ann Arbor, MI*

FERNANDO D. CAMARGO • *Center for Cell and Gene Therapy, Cell and Molecular Biology Program, Baylor College of Medicine, Houston, TX*

SARAH K. CHO • *Division of Biological Science, University of California, San Diego, La Jolla, CA*

EMER CLARKE • *StemCell Technologies Inc., Vancouver, British Columbia, Canada*

D. JAMES COON • *CellzDirect Inc., Pittsboro, NC*

SUSAN L. CUVELIER • *Immunology Research Group, Department of Physiology and Biophysics, University of Calgary, Calgary, Alberta, Canada*

STEPHEN M. DANG • *Institute of Biomaterials and Biomedical Engineering, Department of Chemical Engineering and Applied Chemistry, University of Toronto, Toronto, Ontario, Canada*

JOHN Q. DAVIES • *Sir William Dunn School of Pathology, University of Oxford, Oxford, UK*

RENÉE F. DE POOTER • *Department of Immunology, University of Toronto, Sunnybrook & Women's College Health Sciences Centre, Toronto, Ontario, Canada*

WILHELM G. DIRKS • *DSMZ - German Collection of Microorganisms and Cell Cultures, Braunschweig, Germany*

HANS G. DREXLER • *DSMZ - German Collection of Microorganisms and Cell Cultures, Braunschweig, Germany*

ELAINE A. DZIERZAK • *Department of Cell Biology and Genetics, Erasmus University Medical Center, Rotterdam, the Netherlands*

CONNIE J. EAVES • *Terry Fox Laboratory, British Columbia Cancer Agency, Vancouver, British Columbia, Canada*

JOANNE T. EMERMAN • *Department of Anatomy, University of British Columbia, Vancouver, British Columbia, Canada*

MARGARET A. GOODELL • *Center for Cell and Gene Therapy, Department of Immunology and Cell and Molecular Biology Program, Baylor College of Medicine, Houston, TX*

SIAMON GORDON • *Sir William Dunn School of Pathology, University of Oxford, Oxford, UK*

GERALDINE A. HAMILTON • *CellzDirect Inc., Pittsboro, NC*

KIRSTY HARVEY • *Department of Cell Biology and Genetics, Erasmus University Medical Center, Rotterdam, the Netherlands*

CHERYL D. HELGASON • *Department of Cancer Endocrinology, British Columbia Cancer Agency, Vancouver, British Columbia, Canada*

SUMMER JOLLEY • *Department of Drug Delivery and Disposition, University of North Carolina, School of Pharmacy, Chapel Hill, NC*

ALY KARSAN • *Department of Pathology and Laboratory Medicine, University of British Columbia and Departments of Medical Biophysics and Pathology and Laboratory Medicine, British Columbia Cancer Agency, Vancouver, British Columbia, Canada*

DAN S. KAUFMAN • *Stem Cell Institute and Department of Medicine, University of Minnesota, Minneapolis, MN*

BECKY LAI • *StemCell Technologies Inc., Vancouver, British Columbia, Canada*

BRUNO LARRIVÉE • *Department of Medicine, University of British Columbia, and Department of Medical Biophysics, British Columbia Cancer Agency, Vancouver, British Columbia, Canada*

EDWARD L. LECLUYSE • *Department of Drug Delivery and Disposition, University of North Carolina, School of Pharmacy, Chapel Hill, NC*

CUNLAN LIU • *Division of Experimental Immunology, Institute for Genome Research, University of Tokushima, Tokushima, Japan*

SHARON A. LOUIS • *StemCell Technologies Inc., Vancouver, British Columbia, Canada*

RODERICK A. F. MACLEOD • *DSMZ - Department of Human and Animal Cell Cultures, German Collection of Microorganisms and Cell Cultures, Braunschweig, Germany*

SHANNON MCKINNEY-FREEMAN • *Center for Cell and Gene Therapy, Department of Immunology, Baylor College of Medicine, Houston, TX*

CINDY L. MILLER • *StemCell Technologies Inc., Vancouver, British Columbia, Canada*

TAKESHI NITTA • *Division of Experimental Immunology, Institute for Genome Research, University of Tokushima, Tokushima, Japan*

ROBERT A. J. OOSTENDORP • *The Stem Cell Physiology Laboratory, III, Medizinische Klinik, Klinikum Rechts der Isar, Technical University, Munich, Germany*

KAMALA D. PATEL • *Immunology Research Group, Department of Physiology and Biophysics, University of Calgary, Calgary, Alberta, Canada*

BRENT A. REYNOLDS • *StemCell Technologies Inc., Vancouver, British Columbia, Canada*

LYSIANE RICHERT • *Laboratoire de Chirurgie Expérimentale, Fondation Transplantation, Strasbourg, France and Laboratoire de Biologie Cellulaire, Faculté de Médecine et de Pharmacie, Besançon, France*

KLAUS-DIETER SCHLÜTER • *Physiologisches Institut, Justus-Liebig-Universität Giessen, Giessen, Germany*

DANIELA SCHREIBER • *Physiologisches Institut, Justus-Liebig-Universität Giessen, Giessen, Germany*

GABI SHEFER • *Department of Biological Structure, School of Medicine, University of Washington, Seattle, WA*

JOHN STINGL • *Terry Fox Laboratory, British Columbia Cancer Agency and StemCell Technologies Inc., Vancouver, British Columbia, Canada*

YOUSUKE TAKAHAMA • *Division of Experimental Immunology, Institute for Genome Research, University of Tokushima, Tokushima, Japan*

MARY TAUB • *Biochemistry Department, State University of New York at Buffalo, Buffalo, NY*

XINGHUI TIAN • *Stem Cell Institute and Department of Medicine, University of Minnesota, Minneapolis, MN*

TOMOO UENO • *Division of Experimental Immunology, Institute for Genome Research, University of Tokushima, Tokushima, Japan*

CORD C. UPHOFF • *DSMZ - German Collection of Microorganisms and Cell Cultures, Braunschweig, Germany*

FRIZELL L. VAUGHAN • *Department of Environmental Health Sciences, University of Michigan School of Public Health, Ann Arbor, MI*

CATHERINE VIOLLON-ABADIE • *Laboratoire de Biologie Cellulaire, Faculté de Médecine et de Pharmacie, Besançon, France*

ZIPORA YABLONKA-REUVENI • *Department of Biological Structure, School of Medicine, University of Washington, Seattle, WA*

PETER W. ZANDSTRA • *Institute of Biomaterials and Biomedical Engineering, Department of Chemical Engineering and Applied Chemistry, University of Toronto, Toronto, Ontario, Canada*

JUAN CARLOS ZÚÑIGA-PFLÜCKER • *Department of Immunology, University of Toronto, Sunnybrook & Women's College Health Sciences Centre, Toronto, Ontario, Canada*

Value-Added eBook/PDA

This book is accompanied by a value-added CD-ROM that contains an eBook version of the volume you have just purchased. This eBook can be viewed on your computer, and you can synchronize it to your PDA for viewing on your handheld device. The eBook enables you to view this volume on only one computer and PDA. Once the eBook is installed on your computer, you cannot download, install, or e-mail it to another computer; it resides solely with the computer to which it is installed. The license provided is for only one computer. The eBook can only be read using Adobe® Reader® 6.0 software, which is available free from Adobe Systems Incorporated at www.Adobe.com. You may also view the eBook on your PDA using the Adobe® PDA Reader® software that is also available free from Adobe.com.

You must follow a simple procedure when you install the eBook/PDA that will require you to connect to the Humana Press website in order to receive your license. Please read and follow the instructions below:

1. Download and install Adobe® Reader® 6.0 software
   You can obtain a free copy of the Adobe® Reader® 6.0 software at www.adobe.com
   Note: If you already have the Adobe® Reader® 6.0 software installed, you do not need to reinstall it.
2. Launch Adobe® Reader® 6.0 software
3. Install eBook: Insert your eBook CD into your CD-ROM drive
   PC: Click on the "Start" button, then click on "Run"
   At the prompt, type "d:\ebookinstall.pdf" and click "OK"
   Note: If your CD-ROM drive letter is something other than d: change the above command accordingly.
   MAC: Double click on the "eBook CD" that you will see mounted on your desktop. Double click "ebookinstall.pdf"
4. Adobe® Reader® 6.0 software will open and you will receive the message "This document is protected by Adobe DRM" Click "OK"
   Note: If you have not already activated the Adobe® Reader® 6.0 software, you will be prompted to do so. Simply follow the directions to activate and continue installation.

Your web browser will open and you will be taken to the Humana Press eBook registration page. Follow the instructions on that page to complete installation. You will need the serial number located on the sticker sealing the envelope containing the CD-ROM.

If you require assistance during the installation, or you would like more information regarding your eBook and PDA installation, please refer to the eBookManual.pdf located on your CD. If you need further assistance, contact Humana Press eBook Support by e-mail at ebooksupport@humanapr.com or by phone at 973-256-1699.

*Adobe and Reader are either registered trademarks or trademarks of Adobe Systems Incorporated in the United States and/or other countries.

1

Culture of Primary Adherent Cells and a Continuously Growing Nonadherent Cell Line

Cheryl D. Helgason

Summary

Cell culture is an invaluable tool for investigators in numerous fields. It facilitates analysis of biological properties and processes that are not readily accessible at the level of the intact organism. Successful maintenance of cells in culture, whether primary or immortalized, requires knowledge and practice of a few essential techniques. The purpose of this chapter is to explain the basic principles of cell culture using the maintenance of a nonadherent cell line, the P815 mouse mastocytoma cell line, and the isolation and culture of adherent primary mouse embryonic fibroblasts (MEFs) as examples. Procedures for thawing, culture, determination of cell numbers and viability, and cryopreservation are described.

Key Words: Cell culture; nonadherent cell line; adherent cells; P815; primary mouse embryonic fibroblasts; MEF; hemocytometer; viability; subculturing; cryopreservation.

1. Introduction

There are four basic requirements for successful cell culture. Each of these will be briefly reviewed in this introduction. However, a more detailed description is beyond the scope of this chapter. Instead, the reader is referred to one of a number of valuable resources that provide the information necessary to establish a tissue culture laboratory, as well as describe the basic principles of sterile technique (*1–4*).

The first necessity is a well-established and properly equipped cell culture facility. The level of biocontainment required (Levels 1–4) is dependent on the type of cells cultured and the risk that these cells might contain, and transmit, infectious agents. For example, culture of primate cells, transformed human cell lines, mycoplasma-contaminated cell lines, and nontested human cells require a minimum of a Level 2 containment facility. All facilities should be

From: *Methods in Molecular Biology, vol. 290: Basic Cell Culture Protocols, Third Edition*
Edited by: C. D. Helgason and C. L. Miller © Humana Press Inc., Totowa, NJ

equipped with the following: a certified biological safety cabinet that protects both the cells in culture and the worker from biological contaminants; a centrifuge, preferably capable of refrigeration and equipped with appropriate containment holders that is dedicated for cell culture use; a microscope for examination of cell cultures and for counting cells; and a humidified incubator set at 37°C with 5% CO_2 in air. A 37°C water bath filled with water containing inhibitors of bacterial and fungal growth can also be useful if warming of media prior to use is desired. Although these are the basic requirements, there are numerous considerations regarding location of the facility, airflow, and other design features that will facilitate contamination-free culture. If a new cell culture facility is being established, the reader should consult facility requirements and laboratory safety guidelines that are available from your institution's biosafety department or the appropriate government agencies.

The second requirement for successful cell culture is the practice of sterile technique. Prior to beginning any work, the biological safety cabinet should be turned on and allowed to run for at least 15 min to purge the contaminated air. All work surfaces within the cabinet should be decontaminated with an appropriate solution; 70% ethanol or isopropanol are routinely used for this purpose. Any materials required for the procedure should be similarly decontaminated and placed in or near the cabinet. This is especially important if solutions have been warmed in a water bath prior to use. The worker should don appropriate personnel protective equipment for the cell type in question. Typically, this consists of a lab coat with the cuffs of the sleeves secured with masking tape to prevent the travel of biological contaminants and Latex or vinyl gloves that cover all exposed skin that enters the biosafety cabinet. Gloved hands should be sprayed with decontaminant prior to putting them into the cabinet and gloves should be changed regularly if something outside the cabinet is touched. Care should be taken to ensure that anything coming in contact with the cells of interest, or the reagents needed to culture and passage them, is sterile (either autoclaved or filter-sterilized). The biosafety office associated with your institution is a valuable resource for providing references related to the discussion of required and appropriate techniques required for the types of cells you intend to use.

A third necessity for successful cell culture is appropriate, quality controlled reagents and supplies. There are numerous suppliers of tissue culture media (both basic and specialized) and supplements. Examples include Invitrogen (www.invitrogen.com), Sigma–Aldrich (www.sigmaaldrich.com), BioWhittaker (www.cambrex.com), and StemCell Technologies Inc. (www.stemcell.com). Unless otherwise specified in the protocols accompanying your cells of interest, any source of tissue-culture-grade reagents should be acceptable for most cell culture purposes. Similarly, there are numerous suppliers of the plasticware needed for most cell culture applications (i.e., culture dishes and/or flasks, tubes, disposable pipets). Sources for these supplies include Corning (www. corning.com/lifesciences/), Nunc (www.nuncbrand.com), and Falcon (www. bdbiosciences.com/discovery_labware). Two cautionary notes are essential. First, sterile culture dishes can be purchased as either tissue culture treated or Petri style. Although either can be used for the growth of nonadherent cells, adherent cells require tissue-culture-treated dishes for proper adherence and growth. Second, it is possible to use glassware rather than disposable plastic for cell culture purposes. However, it is essential that all residual cleaning detergent is removed and that appropriate sterilization (i.e., 121°C for at least 15 min in an autoclave) is carried out prior to use.

If the three above-listed requirements have been satisfied, the final necessity for successful cell culture is the knowledge and practice of the fundamental techniques involved in the growth of the cell type of interest. The majority of cell culture carried out by investigators involves the use of various nonadherent (i.e., P815, EL-4) or adherent (i.e., STO, NIH 3T3) continuously growing cell lines. These cell lines can be obtained from reputable suppliers such as the American Tissue Type Collection (ATCC; www.atcc.org) or DSMZ (the German Collection of Microorganisms and Cell Cultures) (www.dsmz.de/ mutz/mutzhome.html). Alternatively, they can be obtained from collaborators. Regardless of the source of the cells, it is advisable to verify the identity of the cell line (refer to Chapters 4 and 5) and to ensure that it is free of mycoplasma contamination (refer to Chapters 2 and 3). In addition to working with immortalized cell lines, many investigators eventually need or want to work with various types of primary cells (refer to Chapters 6–21 for examples). Bacterial contaminations, as a consequence of the isolation procedure, and cell senescence are two of the major challenges confronted with these types of cell.

The purpose of this chapter is to explain the basic principles of cell culture using the maintenance of a nonadherent cell line, the P815 mouse mastocytoma cell line, and adherent primary mouse embryonic fibroblasts (MEF) as examples. Procedures for thawing, subculture, determination of cell numbers and viability, and cryopreservation are described.

2. Materials

2.1. Culture of a Continuously Growing Nonadherent Cell Line (see Note 1)

1. P815 mastocytoma cell line (ATCC, cat. no. TIB-64).
2. High-glucose (4.5 g/L) Dulbecco's Modified Essential Medium (DMEM). Store at 4°C.
3. Fetal bovine serum (FBS) (*see* **Note 2**). Sera should be aliquoted and stored at –20°C.

4. Penicillin–streptomycin solution. 100X stock solution. Aliquot and store at –20°C (*see* **Note 3**).
5. L-Glutamine, 200 m*M* stock solution. Aliquot and store at –20°C.
6. DMEM+ growth medium: high-glucose DMEM (**item 2**) supplemented with 10% FBS, 4 m*M* glutamine, 100 IU penicillin, and 100 µg/mL streptomycin. Prepare a 500-mL bottle under sterile conditions and store at 4°C for up to 1 mo (*see* **Note 4**).
7. Trypan blue stain (0.4% w/v trypan blue in phosphate-buffered saline [PBS] filtered to remove particulate matter) or eosin stain (0.14% w/v in PBS; filtered) for determination of cell viability.
8. Tissue-culture-grade dimethyl sulfoxide (DMSO) (i.e., Sigma) stored at room temperature.
9. Freezing medium, freshly prepared and chilled on ice, consisting of 90% FBS and 10% DMSO (*see* **Note 5**).

2.2. Culture of Primary Mouse Embryonic Fibroblasts

1. High-glucose (4.5 g/L) DMEM (*see* **Subheading 2.1.**).
2. FBS (*see* **Subheading 2.1.**).
3. Penicillin–streptomycin solution (100X) (*see* **Subheading 2.1.**).
4. MEF culture medium. DMEM supplemented with 10% FBS and 1X (100 IU penicillin and 100 µg/mL streptomycin) antibiotics.
5. Dulbecco's Ca^{2+}- and Mg^{2+}-free PBS (D-PBS). D-PBS can be purchased as 1X or 10X stocks from numerous suppliers or a 1X solution can be prepared in the lab as follows: Dissolve the following in high-quality water (*see* **Note 6**): 8 g/L NaCl, 0.2 g/L KCl, 0.2 g/L KH_2PO_4, 2.16 g/L $Na_2HPO_4 \cdot 7H_2O$; adjust pH to 7.2. Filter-sterilize using a 0.22-µm filter and store at 4°C.
6. 0.25% Trypsin–0.5 m*M* EDTA (T/E) solution (*see* **Note 7**). Store working stocks at 4°C.
7. Freezing medium (*see* **Subheading 2.1.**).
8. Timed pregnant female mouse (*see* **Note 8**).
9. 70% Ethanol solution or isopropanol.
10. Two sets of forceps and scissors; one set sterilized by autoclaving at 121°C for 15 min.
11. Fine forceps (sterile) (Fine Science Tools, cat. no. 11272-30).
12. Small fine scissors (sterile).
13. 18-Gage blunt-end needles (sterile) (StemCell Technologies Inc.).

3. Methods

Prior to the initiation of any cell culture work, it is essential to ensure that all equipment is in optimal working condition. Moreover, if cell culture is to become a routine technique utilized in the laboratory, scheduled checks and regular maintenance of the equipment are required. A partial checklist of things to consider includes the following: check to ensure that the temperature and CO_2 levels in the incubator are at the desired levels; check to be sure that the water pan in the incubator is full of clean water and that it contains copper sulfate to inhibit bacterial growth; check to ensure that the water bath is at the required temperature and contains adequate amounts of clean water; check to ensure that the biological safety cabinet to be used is certified and operating correctly; ascertain that the centrifuge is cleaned and decontaminated.

3.1. Culture of a Continuously Growing Nonadherent Cell Line

3.1.1. Thawing Cryopreserved P815 Cells

1. In the biological safety cabinet, prepare one tube containing 9 mL of DMEM+ growth medium warmed to at least room temperature.
2. Remove one vial of cells from the storage container (liquid nitrogen or ultralow temperature freezer) (*see* **Note 9**).
3. Transfer the vial of cells to a 37°C water bath until the suspension is just thawed (*see* **Note 10**).
4. In the cell culture hood, use a sterile glass or plastic pipet to transfer the contents of the vial slowly into the tube containing the growth medium.
5. Centrifuge the cells at 1200 rpm (300*g*) for 7 min to obtain a pellet.
6. Aspirate the supernatant containing DMSO and suspend the cell pellet in 10 mL of DMEM+ growth medium (*see* **Note 11**).
7. Transfer the cells to a tissue culture dish (100 mm) and incubate at 37°C, 5% CO_2.
8. Examine cultures daily using an inverted microscope to ensure that the culture was not contaminated during the freeze–thaw process and that the cells are growing.

3.1.2. Determination of Cell Number and Cell Viability

Every cell line has an optimal concentration for maintaining growth and viability. Until sufficient experience is gained with a new cell line, it is recommended to check cell densities and viability every day or two to ensure that optimal health of the cultures is maintained.

1. Gently swirl the culture dish to evenly distribute the cell suspension.
2. Under sterile conditions, remove an aliquot (100–200 µL) of the evenly distributed cell suspension.
3. Mix equal volumes of cells and viability stain (eosin or trypan blue); this will give a dilution factor of 2.
4. Clean the hemocytometer using a nonabrasive tissue.
5. Slide the cover slip over the chamber so that it covers both sides.
6. Fill the chamber with the well-mixed cell dilution and view under the light microscope.
7. Each 1-mm^2 square should contain between 30 and 200 cells to obtain accurate results (*see* **Note 12**).

8. Count the numbers of bright clear (viable) and nonviable (red or blue depending on the stain used) cells in at least two of the 1-mm^2 squares, ensuring that two numbers are similar (i.e., within 5% of one another). Count all five of the 1-mm^2 squares if necessary to ensure accuracy (*see* **Note 13**).

9. Calculate the numbers of viable and nonviable cells, as well as the percentage of viable cells, using the following formulas where *A* is the mean number of viable cells counted, *B* is the mean number of nonviable cells counted, *C* is the dilution factor (in this case, it is 2), *D* is the correction factor supplied by the hemocytometer manufacturer (this is the number required to convert 0.1 mm^3 into milliliters; it is usually 10^4).

Concentration of viable cells (per mL) = $A \times C \times D$
Concentration of nonviable cells (per mL) = $B \times C \times D$
Total number of viable cells = concentration of viable cells $\times$ volume
Total number of cells = number of viable + number of dead cells
Percentage viability = (number of viable cells $\times$ 100)/total cell number

3.1.3. Subculture of Continuously Growing Nonadherent Cells

Maintenance of healthy, viable cells requires routine medium exchanges or passage of the cells to ensure that the nutrients in the medium do not become depleted and/or that the pH of the medium does not become acidic (i.e., turn yellow) as a result of the presence of large amounts of cellular waste.

1. View cultures under an inverted phase-contrast microscope. Cells growing in exponential growth phase should be round, bright, and refractile. If necessary, determine the cell density as indicated in **Subheading 3.1.2.**

2. There is no need to centrifuge the cells unless the medium has become too acidic (phenol red = yellow), which indicates the cells have overgrown, or if low viability is observed.

3. Transfer a small aliquot of the well-mixed cell suspension into a fresh dish containing prewarmed DMEM+ growth medium (*see* **Note 14**), ensuring that the resulting cell density is in the optimal range for the particular cell line.

4. Repeat this subculture step every 2–3 d to maintain cells in an exponential growth phase.

3.1.4. Cryopreservation of Continuously Growing Nonadherent Cells

Continuous culture of cell lines can lead to the accumulation of unwanted karyotype alterations or the outgrowth of clones within the population. In addition, continuous growth increases the possibility of cell line contamination by bacteria or other unwanted organisms. The only insurance against loss of the cell line is to ensure that adequate numbers of vials (i.e., at least 10) are cryopreserved for future use. For newly acquired cell lines, cryopreservation of stock (master cell bank) vials should be done as soon as possible after the cell line has been confirmed to be free of mycoplasma (*see* Chapters 2 and 3).

1. View the cultures under a phase-contrast inverted microscope to assess cell density and confirm the absence of bacterial or fungal contamination.
2. Remove a small aliquot of the cells for determination of cell numbers as outlined in **Subheading 3.1.2.** Cells for cryopreservation should be in log growth phase with greater than 90% viability.
3. Prepare the cryopreservation vials by indicating the name of the cell line, the number of cells per vial, the passage number, and the date on the surface of the vial using a permanent marker (*see* **Note 15**).
4. Prepare the required volume of freezing medium as outlined in **Subheading 2.1.** and chill on ice.
5. Centrifuge the desired number of cells at 1200 rpm (300g) for 5–7 min and aspirate the supernatant from the tube.
6. Suspend the cells to a density of (1–2) × 10^6 cells/mL in the freezing medium.
7. Quickly aliquot 1 mL into each of the prepared cryovials using a pipet. Care is required to ensure that sterility is maintained throughout the procedure.
8. Place cryovials on dry ice until cells are frozen and then transfer to an appropriate ultralow temperature storage vessel (freezer or liquid-nitrogen tank) for long-term storage (*see* **Notes 16** and **17**).

3.2. Culture of Primary Mouse Embryonic Fibroblasts

3.2.1. Isolation of MEF

1. In order to obtain embryos at the desired stage of development set up female and male mice 14 d prior to the anticipated harvest date. On the following morning check for copulation plugs and remove the mated females to a separate cage. The day the plug is found is designated d 1.
2. On d 13 of pregnancy, sacrifice the females according to institutional guidelines. Spray or wipe the fur on the abdominal cavity of the dead mouse with 70% ethanol or isopropanol to reduce contamination risk and prevent fur from flying about.
3. Expose the skin of the abdominal cavity by cutting through the fur using a pair of scissors and forceps (sterility is not critical at this step).
4. Using the sterile scissors and forceps, cut through the abdominal wall and remove the uteri containing the embryos into a dish containing D-PBS.
5. In a biosafety cabinet, place the uteri into a sterile 100-mm dish. Dissect the embryos away from the yolk sac, amnion, and placenta using the sterile scissors and forceps.
6. Transfer the embryos to a clean dish and wash thoroughly to remove any blood.
7. Transfer the embryos to another sterile dish and use a pair of sterile fine forceps to pinch off the head and remove the liver from each embryo.
8. Transfer the remainder of the carcass into a fresh culture dish and gently mince the tissue using the fine sterile scissors into pieces small enough to be drawn into a 10-mL disposable pipet.
9. Add 0.5 mL of MEF culture medium per embryo to the minced tissue and draw the slurry up into a syringe of the appropriate volume through a sterile 18-gage

blunt needle. Expel and draw up the minced tissue through the needle four to five times to generate small clumps of cells.

10. Add 10 mL of MEF culture medium per two embryos and culture in a 100-mm tissue-culture-treated (*not* Petri style) cell culture dishes. This is considered passage 1 (P1).
11. Incubate overnight at 37°C, 5% CO_2 in a humidified cell culture incubator. Clusters of adherent cells should be visible, attached to the surface of the dish. Aspirate the medium containing floating cell debris and add an equal volume of fresh MEF culture medium.
12. Cultures should become confluent in 2–3 d. The expected yield is 1×10^7 cells per confluent 100-mm dish.

3.2.2. Subculture of MEF

Mouse embryonic fibroblasts should be subcultured when they reach 80–90% confluence. If the MEF are allowed to reach 100% confluence, growth arrest can result with a decrease in the subsequent proliferative potential of the cells.

1. Aspirate the MEF medium from the dishes that have achieved the desired level of confluence and wash the monolayer of cells with 2–3 mL of room-temperature D-PBS to remove any residual growth medium.
2. Aspirate the D-PBS and add 3–4 mL of room-temperature trypsin–EDTA (T/E). Incubate the dishes at 37°C for 3–5 min. Progress should be monitored by examining the cultures using an inverted phase-contrast microscope.
3. Once the cells have begun to detach, transfer them to a centrifugation tube containing 6–7 mL MEF medium (which contains sufficient FBS to inhibit the trypsin activity) for centrifugation. Residual cells can be collected by rinsing the dish once or twice with 5 mL of the cell/medium mixture.
4. Centrifuge at 1200 rpm (300g) for 5–7 min.
5. Aspirate the T/E containing medium and add fresh MEF culture medium (3 mL per initial input dish).
6. Split the cells at no more than a 1:3 ratio to expand their numbers. Dishes should be labeled as "P2" to indicate that this is the second plating of these cells.
7. After 2–3 d, the cells should again reach confluence and are ready to use or to cryopreserve.

3.2.3. Cryopreservation of MEF

The protocol for freezing MEF is the same as that described in **Subheading 3.1.4.** (*see* **Notes 16–18**).

3.2.4. Thawing MEF

The thawing of MEF follows **steps 1–5** outlined for thawing the P815 cell line (*see* **Subheading 3.1.1.**). Once the thawed cells have been pelleted by centrifugation, the protocols diverge. The following steps are required to obtain healthy MEF cultures.

1. Resuspend the thawed MEF cell pellet in MEF culture medium supplemented with 30% FBS instead of the normal 10%. The additional FBS facilitates cell attachment to the tissue culture treated dishes. Culture $(1–2) \times 10^6$ thawed MEF cells per 100-mm tissue-culture-treated dish.
2. Allow the cells to adhere by overnight culture in a humidified incubator at 37°C, 5% CO_2.
3. The following morning (or at least 6 h after plating), remove the high FBS medium containing dead and nonadherent cells and replace it with regular MEF culture medium.
4. Subculture of the MEF can typically be carried out for 5–10 passages using the procedures described in **Subheading 3.2.2.** (*see* **Note 18**).

4. Notes

1. One of the primary sources of contamination arising during cell culture is the use of shared stock solutions that are accessed repeatedly by several lab workers. It is advisable to store all stock solutions in aliquots of a size that is typically used, thus eliminating this concern.
2. Any FBS selected for cell culture applications should be specified by the manufacturer as mycoplasma-free and endotoxin low/negative. In addition, for sensitive cell types, it might be necessary to pretest lots of FBS to ensure that it supports optimal growth. FBS can be heat inactivated by incubation at 56°C for 30 min, with frequent swirling, to inactivate complement if this is a concern. Heat-inactivated FBS should be cooled overnight at 4°C and then aliquoted under sterile conditions for long-term storage at –20°C.
3. Antibiotics are not essential for the culture of mammalian cells. However, they do help to protect against inadvertent bacterial contamination of the cultures arising through the use of inappropriate sterile technique and are thus recommended for use by novice culturists. It is recommended that once you become more competent with the required techniques, the antibiotics be omitted from the media formulation to reduce the emergence of antibiotic-resistant bacterial strains. Antibiotics are routinely used for the culture of primary cells because of the increased risk of bacterial contamination associated with the isolation procedures. For primary cells and newly acquired cell lines, it is advisable to culture cells with and without antibiotics or antimycotics to exclude the possibility of biological effects of these agents on the cells.
4. Most cell culture media contain phenol red as a pH indicator. Repeated entry into the medium bottle can result in a shift in the pH and, thus, a change in the color from red to a more purple color. Most cells (both primary and immortalized) display optimal growth within a defined physiological pH range. If the pH of the media does change, the media should be discarded and fresh media prepared. If this happens regularly, it is advisable to make smaller volumes of the growth media that can be used completely before the pH changes.

5. Addition of DMSO to the FBS results in an exothermic reaction that can denature the proteins in the serum. To prevent this occurrence, the FBS should be aliquoted into a tube and chilled on ice. The room-temperature DMSO should be added slowly dropwise. Do not put the bottle of DMSO on ice because it will freeze. As an alternative, the freezing medium can consist of the cell culture medium supplemented with 10% DMSO. However, higher concentrations of FBS ($\geq 30\%$) tend to increase the recovery of viable cells.

6. The water used to prepare any tissue culture reagents should be of high quality. Water (18 Megohm) prepared using ion-exchange and reverse-osmosis apparatus is recommended. Routine testing for bacterial, fungal, and endotoxin contaminants in the water supply is also suggested.

7. Trypsin is an enzyme that is active at 37°C. If large bottles of T/E are purchased (i.e., 500 mL), it is advisable to thaw the solution overnight at 4°C and then aliquot into convenient sizes (i.e., 40 mL/tube) for storage at –20°C. Avoid repeatedly warming and cooling the solution, as it will reduce the activity of the enzyme.

8. Mouse embryonic fibroblasts can be isolated from all strains of mice. However, if a specific strain is not required, it is advisable to use one that generally produces large litters (i.e., CD1) so that fewer female mice are needed to yield large numbers of MEFs.

9. Extreme caution must be used when removing vials that have been stored in the liquid phase of liquid nitrogen because the possibility exists that liquid nitrogen might have seeped into the vial and the pressure generated as the vial warms might cause it to explode. Always wear a face shield and insulated gloves when removing frozen vials of cells.

10. Be careful to immerse only the bottom half of the vial into the water bath to prevent seepage of water into the vial. Once the cells have almost completely thawed, remove the vial from the water bath. Note the information recorded on the vial and then rinse the outside of the vial with 70% ethanol or isopropanol to decontaminate it prior to proceeding with the thawing procedures.

11. The volume in which the cells are suspended and the amount of time required to reach confluence in the culture is dependent on the number of viable cells recovered from the freezer. If the vial has been frozen for a long period of time so that viability is questionable or if the number of cells frozen was low, it is better to err on the side of caution and suspend the cells in a smaller volume; you can always add more medium after a day or two.

12. The central area of the counting chamber is 1 mm^2 and is divided into 25 smaller units surrounded by a triple line. This central square is surrounded diagonally by 4 other 1-mm^2 squares each subdivided into 16 smaller units. The depth of a hemocytometer is 0.1 mm. Every hemocytometer manufacturer provides a diagram and counting instructions that should be consulted prior to carrying out cell counts for the first time.

13. There are several sources of inaccuracy that should be avoided when doing cell counts: the presence of air bubbles and debris in the counting chamber; overfill-

ing or underfilling the chamber; cells not evenly distributed in the chamber; too few or too many cells in the chamber. If problems are encountered, clean the chamber well, fill properly, and ensure that a well-mixed cell suspension is used. Decrease the cell volume or increase the dilution factor if too few or too many cells, respectively, are present in the chamber.

14. A seeding density of approx 1×10^5 cells/mL works well for P815 cells. To ensure continued exponential growth, the cell density should be maintained between 1×10^5 and 1×10^6 cells/mL. Refer to the data information sheet provided with each cell line, as this density can vary from one cell line to another.

15. Although some cell lines are not affected by the temperature of the vials, other cells (i.e., MEF) are more sensitive. To avoid further shock to the cells, the cryovials can be chilled in a –80°C freezer prior to use. Before chilling the vials, it is important that all pertinent information be noted on the vials. In addition, the same information should be noted in the freezer log book that indicates the position of the cells in the freezing vessel.

16. Some cell lines (i.e., P815) can be rapidly frozen on dry ice without loss of viability. Other cell lines (i.e., MEF) exhibit a significant loss in viability if frozen rapidly. Cryovials containing these types of cell should be placed inside a passive freezing container (i.e., Nalgene "Mr. Frosty") and stored at –80°C overnight before transfer to the long-term storage vessel. If no freezing containers are available, cells can be placed in a Styrofoam rack inside a Styrofoam box for overnight storage.

17. It is highly recommended that the cell line be maintained in culture and frozen cells tested to ensure that viable uncontaminated cells can be recovered following the freezing process before the cell line in discarded. One to two weeks after the cryopreservation of the cells, one or two vials should be thawed and placed into culture. If cells recover well and no signs of contamination are observed immediately or within 1 wk after thawing, it should be safe to discard the original cultures.

18. It is advisable to freeze MEF at higher densities (i.e., $[2–5] \times 10^6$ cells per vial) than is typically used for most cell lines. All primary cell types, including MEF, have a finite life-span in culture because of cell senescence. Senescent changes in the MEF culture are characterized by a decrease in the growth rate and a change in cell morphology to a more elongated and stringy looking cell rather than a rounded cell. It is critical to record the passage number of all primary cells and to ensure that aliquots are frozen for future use as soon as possible if future experiments are anticipated.

Acknowledgments

The Michael Smith Foundation for Health Research and the Canadian Institutes of Health Research are acknowledged for salary support. Special thanks to the members of my lab for their patience and understanding while I worked on this chapter.

References

1. Freshney, R. I. (ed.) (2000) *Culture of Animal Cells. A Manual of Basic Techniques*, 2nd ed., Wiley, New York.
2. Celis, J. F. (ed.) (1998) *Cell Biology: A Laboratory Handbook*, 2nd ed., Academic, New York.
3. Davis, J. M. (ed.) (2002) *Basic Cell Culture*, 2nd ed., IRL, Oxford.
4. Bonifacino, J. S., Dasso, M., Harford, J. B., Lippincott-Schwartz, J., and Yamada, K. M. (eds.) (2000) *Current Protocols in Cell Biology*, Wiley, New York.

2

Detection of Mycoplasma Contaminations

Cord C. Uphoff and Hans G. Drexler

Summary

Mycoplasma contamination of cell lines is one of the major problems in cell culture technology. The specific, sensitive, and reliable detection of mycoplasma contamination is an important part of mycoplasma control and should be an established method in every cell culture laboratory. New cell lines as well as cell lines in continuous culture must be tested in regular intervals. The polymerase chain reaction (PCR) methodology offers a fast and sensitive technique to monitor all cultures in a laboratory. The technique can also be used to determine the contaminating mycoplasma species.

The described assay can be performed within 3 h, including sample preparation, DNA extraction, performing the PCR reaction, and analysis of the PCR products. Special precautions necessary to avoid false-negative results resulting from inhibitors of the *Taq* polymerase present in the crude samples and the interpretation of the results are also described.

Key Words: Bacteria; cell lines; contamination; mycoplasma; PCR.

1. Introduction
1.1. Mycoplasma Contaminations of Cell Lines

Acute contaminations of cell lines are frequently observed in routine cell culture and can often be attributed to improper handling of the growing culture. These contaminations can usually be detected by the turbidity evolving after a short incubation time or by routine observation of the culture under the inverted microscope. In addition to these obvious contaminations, other hidden infections can occur consisting of mycoplasmas, viruses, or cross-contaminations with other cell lines. Although known for many years and despite the multitude of publications dealing with mycoplasma infections of cell cultures, a high proportion of scientists are not aware of the potential contamination of cell cultures with mycoplasmas. As seen in our cell repository, more than 25%

From: *Methods in Molecular Biology, vol. 290: Basic Cell Culture Protocols, Third Edition*
Edited by: C. D. Helgason and C. L. Miller © Humana Press Inc., Totowa, NJ

of the incoming cell lines are infected with mycoplasmas, and in most cases, the depositor was not aware of this. Whereas in the early years of cell culture, bovine serum was one of the major sources of infections, nowadays mycoplasmas seem to be mainly transferred from one infected culture to another by using laboratory equipment, media, or reagents that came into contact with infected cultures. This culture hopping is concordant with the occurrence of cross-contaminations with a proved incidence of 16% plus an estimated number of unknown cases *(1)*. Thus, methods for the detection, elimination (*see* Chapter 3), and prevention of mycoplasma contaminations should belong to the basic panel of cell culture techniques applied.

The term "Mycoplasma" is usually used as a synonym for the class of Mollicutes that represents a large group of highly specialized bacteria and are all characterized by their lack of a rigid cell wall. Mycoplasma is the largest genus within this class. Because of their small size and flexibility, these bacteria are able to pass through conventional microbiological filters. Mycoplasmas can be seen as commensales, because their reduced metabolic abilities cause a relatively long generation time, which is in the range of that of cell lines, and they do usually not overgrow or kill the eukaryotic cells. However, their influence on the biological characteristics of the eukaryotic cells is manifold and almost every experimental or production setting can be influenced. The identification of infecting mycoplasmas shows that only a limited number of about seven *Mycoplasma* and *Acholeplasma* species from human, swine, and bovine hosts occur predominantly in cell cultures, and no species specificity can be observed. Additionally, a couple of mycoplasma species were shown to enter the eukaryotic cells actively and to exist intracytoplasmic *(2)*. Hence, sensitive methods need to be established and frequently employed in every cell culture laboratory to detect mycoplasma contaminations.

1.2. Mycoplasma Detection

The biological diversity of mycoplasmas and their close adaptation to cell cultures renders it very difficult to detect all contaminations in one general assay. A large spectrum of approaches have been described to detect mycoplasma in cell cultures. Many of these methods are lengthy, complex, and not applicable in routine cell culture (e.g., electron microscopy, biochemical and radioactive incorporation assays, etc.) or are restricted to specific groups of mycoplasmas. Molecular biological methods were the first to be able to detect all the different mycoplasma types in cell cultures, regardless of their biological properties, with a relatively low effort in terms of time and labor *(3)*.

Polymerase chain reaction (PCR) provides a very sensitive and specific option for the direct detection of mycoplasmas in cell cultures. PCR combines many of the features that were covered earlier by different assays: sensitivity, specificity, low expenditure of labor, time, and costs, simplicity of the assay, objectivity of interpretation, reproducibility, and documentation of the results. On the other hand, a number of indispensable control reactions must be included in the PCR assay to avoid false-negative or false-positive results. A comparison of the PCR method with other well-established assays (DNA/RNA hybridization, microbiological culture) showed that the PCR assay is a very robust, efficient, and reliable method for the detection of mycoplasmas *(4)*.

The choice of the primer sequences is one of the most crucial decisions. Several primer sequences are published for both single and nested PCR (*see* **Note 1**) and with narrow or broad specificity for mycoplasma or eubacteria species. In most cases, the 16S rDNA sequences are used as target sequences, because this gene contains regions with more and less conserved sequences. This gene also offers the opportunity to perform a PCR with the 16S rDNA or an RT-PCR (reverse transcriptase–PCR) with the cDNA of the 16S rRNA (*see* **Note 2**) *(5)*. Here, we describe the use of a mixture of oligonucleotides for the specific detection of mycoplasmas. This approach reduces significantly the generation of false-positive results resulting from possible contamination of the solutions used for sample preparation and the PCR run and from other materials with airborne bacteria. Nevertheless, major emphasis should be placed on the preparation of the template DNA, the amplification of positive and negative control reactions, and the observance of general rules for the preparation of PCR reactions. One of the main problems concerning PCR reactions with samples from cell cultures is the inhibition of the *Taq* polymerase by unspecified substances. To eliminate those inhibitors, we strictly recommend that the sample DNA be extracted and purified by conventional phenol–chloroform extraction or by the more convenient column or matrix-binding extraction methods. To confirm the error-free preparation of the sample and PCR run, appropriate control reactions have to be included in the PCR. These comprise internal control DNA for every sample reaction and, in parallel, positive and negative as well as water control reactions. The internal control consists of a DNA fragment with the same primer sequences for amplification, but it is of a different size than the amplicon of mycoplasma-contaminated samples. This control DNA is added to the PCR mixture in a previously determined limiting dilution to demonstrate the sensitivity of the PCR reaction. In this chapter, detailed protocols are provided to establish the PCR method for the monitoring of mycoplasma contaminations in any laboratory.

2. Materials

1. PBS (phosphate-buffered saline): 140 mM NaCl, 27 mM KCl, 7.2 mM Na$_2$HPO$_4$, 14.7 mM KH$_2$PO$_4$, pH 7.2. Autoclave 20 min at 121°C to sterilize the solution.

2. 50X TAE (Tris–acetic acid–EDTA): 2 *M* Tris base, 5.71% glacial acetic acid (v/v), 100 m*M* EDTA. Adjust to pH of approx 8.5.

3. DNA extraction and purification system (e.g., phenol–chloroform extraction and ethanol precipitation, or DNA extraction kits applying DNA binding matrices).

4. GeneAmp 9600 thermal cycler (Applied Biosystems, Weiterstadt, Germany).

5. *Taq* DNA polymerase (Qiagen, Hilden, Germany).

6. 6X Loading buffer: 0.09% (w/v) bromophenol blue, 0.09% (w/v) xylene cyanol FF, 60% glycerol (v/v), 60 m*M* EDTA.

7. Primers (any supplier) (*see* **Note 3**):

> 5' primers (Myco-5'):
> cgc ctg agt agt acg t**w**c gc
> tgc ctg **r**gt agt aca ttc gc
> cgc ctg agt agt atg ctc gc
> cgc ctg ggt agt aca ttc gc
>
> 3' primers (Myco-3'):
> gcg gtg tgt aca a**r**a ccc ga
> gcg gtg tgt aca aac ccc ga
> (**r** = mixture of g and a; **w** = mixture of t and a)

Primer stock solutions: 100 µ*M* in dH$_2$O, stored frozen at –20°C. Working solutions: mix of forward primers at 5 µ*M* each (Myco-5') and mix of reverse primers at 5 µ*M* each (Myco-3') in distilled water (dH$_2$O), aliquoted in small amounts (i.e., 25 to 50-µL aliquots), and stored frozen at –20°C.

8. Internal control DNA: can be obtained from the DSMZ (German Collection of Microorganisms and Cell Cultures, Braunschweig, Germany) *(4)*. A limiting dilution should be determined experimentally by performing a PCR with a dilution series of the internal control DNA.

9. Positive control DNA: a 10-fold dilution of any mycoplasma-positive sample prepared as described in **Subheading 3.1.** or obtained from the DSMZ.

10. Deoxy-nucleotide triphosphate mixture (dNTP mix): mixture contains 5 m*M* each of deoxyadenosine triphosphate (dATP), deoxycytidine triphosphate (dCTP), deoxyguanosine triphosphate (dGTP), and deoxythymidine triphosphate (dTTP) (Peqlab, Erlangen, Germany) in H$_2$O and stored as 50-µL aliquots at –20°C.

11. 1.3% Agarose–TAE gel *(6)*.

3. Methods

The following subsections describe the sample collection, extraction of the DNA, setting up and performing the PCR reaction, the interpretation of the results, and, in addition, the identification of the mycoplasma species. These techniques can also be used to detect mycoplasma contamination in culture media or other supplements (*see* **Note 4**).

Every incoming cell culture should be kept in quarantine until mycoplasma detection assays are completed and the infection status is clearly determined.

Positive cultures should either be discarded and replaced by clean cultures or cured with specific antibiotics (*see* Chapter 3). Only definitely clean cultures should be used for research experiments and for the production of biologically active pharmaceuticals. Additionally, stringent rules for the prevention of further mycoplasma contamination of cell cultures should be strictly followed *(1)*.

3.1. Sample Collection and Preparation of DNA

1. Prior to collecting the samples, the cell line to be tested for mycoplasma contamination should be in continuous culture for several days and without any antibiotics (even penicillin and streptomycin) or after thawing for at least 2 wk. This should assure that the titer of the mycoplasmas in the supernatant is within the detection limits of the PCR assay.
2. One milliliter of the supernatant of adherently growing cells or of cultures with settled suspension cells are taken for the analysis. Collecting the samples in this way, some viable or dead eukaryotic cells are included in the test. This is of advantage, as some mycoplasma strains predominantly adhere to the eukaryotic cells or even invade them. Thus, it is also not necessary to centrifuge the sample to eliminate the eukaryotic cells. The crude cell culture supernatants can be stored at 4°C for a few days or frozen at –20°C for several weeks. After thawing, the samples should be further processed immediately.
3. The cell culture suspension is centrifuged at 13,000*g* for 5 min. The pellet is resuspended in 1 mL PBS by vortexing.
4. The suspension is centrifuged again and washed one more time with PBS as described in **step 3**.
5. After centrifugation, the pellet is resuspended in 100 µL PBS by vortexing and then heated to 95°C for 15 min.
6. Immediately after lysing the cells, the DNA is extracted and purified by standard phenol–chloroform extraction and ethanol precipitation *(6)* or other DNA isolation methods (*see* **Note 5**).

3.2. PCR Reaction

The amplification procedure and the parameters described here are optimized for the use in thin-walled 0.2-mL reaction tubes in an Applied Biosystems GeneAmp 9600 thermal cycler. An adjustment to any other equipment might be necessary (*see* **Note 6**). Amplified positive samples contain high amounts of target DNA. Thus, established rules to avoid DNA carryover should be strictly followed: (1) The places where the DNA is extracted, the PCR reaction is set up, and the gel is run after the PCR should be separated from each other; (2) all reagents should be stored in small aliquots to provide a constant source of uncontaminated reagents; (3) avoid reamplifications; (4) reserve pipets, tips, and tubes for their use in the PCR only and irradiate the pipets frequently by ultraviolet (UV) light; (5) the succession of the PCR setup described below should be followed strictly; (6) wear gloves during the whole sample preparation and PCR setup; (7) include the appropriate control reactions, such as internal, positive, negative, and the water control reaction.

1. Per sample to be tested, two reactions are set up with the following solutions. Sample only: 1 μL dNTPs, 1 μL Myco-5', 1 μL Myco-3', 1.5 μL of 10X PCR buffer, 9.5 μL dH$_2$O; sample and DNA internal standard: 1 μL dNTPs, 1 μL Myco-5', 1 μL Myco-3', 1.5 μL of 10X PCR buffer, 8.5 μL dH$_2$O, 1 μL internal control DNA.

   For several samples, premaster mixtures can be performed. For the reaction without internal control DNA, three reactions have to be added (for the positive, negative, and the water control reactions), and for the reactions with the internal control DNA, two reactions have to be added for the positive and the negative control reaction (*see* **Notes 7** and **8**). For both premaster mixtures, add also the amounts for an additional reaction to have a surplus for pipetting variations.

2. Transfer 14 μL of each of the pre-master mixtures to 0.2 mL PCR reaction tubes and add 1 μL dH$_2$O to the water control reaction.

3. Prepare the *Taq* DNA polymerase mix (10 μL per reaction, plus one additional reaction for pipetting variations) containing 1X PCR buffer and 1 U *Taq* polymerase per reaction.

4. Set aside all reagents used for the preparation of the master mix. Take out the samples of DNA to be tested and the positive control DNA. Do not handle the reagents and samples simultaneously. Add 1 μL per DNA preparation to one reaction tube that contains no internal control DNA and to one tube containing the internal control DNA.

5. To perform a hot-start PCR, transfer the reaction mixtures (without *Taq* polymerase) to the thermal cycler and start one thermo cycle with the following parameters: step 1, 7 min at 95°C; step 2, 3 min at 72°C; step 3, 2 min at 65°C; step 4, 5 min at 72°C.

   During step 2, open the thermal lid and add 10 μL of the *Taq* polymerase mix to each tube. For many samples, the duration of this step can be prolonged. Open and close each reaction tube separately to prevent evaporation of the samples. Allow at least 30 s after adding the *Taq* polymerase to the last tube and closing the lid of the thermal cycler for equilibration of the temperature within the tubes and removal of condensate from the lid before continuing to the next cycle step.

6. After this initial cycle, perform 32 thermal cycles with the following parameters: step 1, 4 s at 95°C; step 2, 8 s at 65°C; step 3, 16 s at 72°C plus 1 s of extension time during each cycle.

7. The reaction is finished by a final amplification step at 72°C for 10 min and the samples are then cooled down to room temperature.

8. Prepare a 1.3% agarose–TAE gel containing 0.3 μg of ethidium bromide per milliliter (*6*). Submerge the gel in 1X TAE and add 12 μL of the amplification product (10 μL reaction mixtures plus 2 μL of 6X loading buffer) to each well and run the gel at 10 V/cm.

9. Visualize the specific products on a suitable UV light screen and document the results.

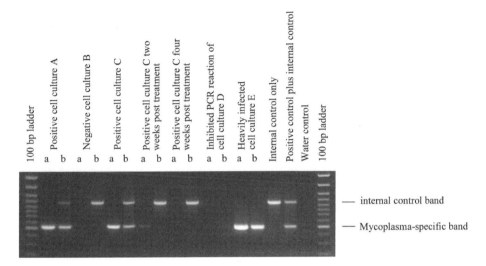

Fig. 1. The PCR analysis of mycoplasma status in cell lines. Shown is an ethidium bromide-stained gel containing the reaction products following PCR amplification with the primer mix listed in the Materials section. Products of about 510 bp were obtained; the differences in length reflect the sequence variation between different mycoplasma species. Shown are various examples of mycoplasma-negative and mycoplasma-positive cell lines. Two paired PCR reactions were performed: one PCR reaction contained an aliquot of the sample only (a) and the second reaction contained the sample under study plus the control DNA as internal standard (b). Cell cultures A, C, and E are mycoplasma positive; cell culture B is mycoplasma negative. The analysis of cell culture D is not evaluable because the internal control was not amplified and no other mycoplasma-specific band appeared in the gel. In this case, the analysis needs to be repeated. Cell line C 2 wk after antibiotic treatment shows a weak but distinctive band in the reaction without internal control. This band results from residual DNA in the medium, because after a further 2 wk of culture, no contamination was detected.

3.3. Interpretation of Results

Figure 1 shows a representative ethidium bromide-stained gel with some samples that produce the following results:

- Ideally, all samples containing the internal control DNA show a band at 986 bp. This band might be more or less bright, but the band has to be visible if no other bands are amplified (*see* **Note 9**). Otherwise, the reaction might have been contaminated with *Taq* polymerase inhibitors from the sample preparation. In this case, it is usually sufficient to repeat the PCR run with the same DNA solution as previously. It is not necessary to collect a new sample from the cell culture. Even if the second run also shows no band for sample and the internal control, the whole procedure should be repeated.

- Mycoplasma-positive samples show a band at 502–520 bp, depending on the mycoplasma species. In the case of *Acholeplasma laidlawii* contamination and applying the DSMZ internal control DNA, a third band might be visible between the internal control band and the mycoplasma-specific band. This is formed by cross-hybridization of the complementary sequences of the single-stranded long internal control DNA and the shorter single-stranded mycoplasma DNA form.
- Contaminations of reagents with mycoplasma-specific DNA or PCR product are revealed by a band in the water control and/or in the negative control sample.
- Weak mycoplasma-specific bands can occur after treatment of infected cell cultures with antimycoplasma reagents for the elimination of mycoplasma or when other antibiotics such as penicillin–streptomycin are applied routinely. In these cases, the positive reaction might either be the result of residual DNA in the culture medium derived from dead mycoplasma cells or from viable mycoplasma cells present at a very low titer. Therefore, special caution should be taken when cell cultures are tested that were treated with antibiotics. Prior to PCR testing, cell cultures should be cultured for at least 2–3 wk without antibiotics or retested at frequent intervals to demonstrate either a decrease or increase of mycoplasma infection.

3.4. Identification of Mycoplasma Species

Although the method described is sufficient to detect mycoplasma contaminations, it might be of advantage to know the infecting mycoplasma species (e.g., in efforts to determine the source of a contamination). This PCR method allows the identification of the mycoplasma species most commonly infecting cell cultures by modified restriction fragment length polymorphism analysis. In case of a contamination detected by PCR, the PCR reaction is repeated in a 50-µL volume without the internal control DNA to amplify only the mycoplasma-specific PCR fragment. Per reaction, 8 µL of the amplified DNA is directly taken from the PCR reaction and is digested in parallel reactions with the restriction endonucleases *Asp*I, *Hae*III, *Hpa*II, and *Xba*I by the addition of 1 µL of the appropriate 10X restriction enzyme buffer and 1 µL of the restriction enzyme. The mycoplasma species can be determined directly by the restriction pattern (*see* **Fig. 2**). This analysis allows only the determination of those mycoplasma species that most often (>98%) occur in cell cultures and is not suitable for the global identification of all types of mycoplasma species. Cell culture infections are commonly restricted to about a half dozen mycoplasma species listed in **Fig. 2**.

4. Notes

1. Originally, the described method was also designed as nested PCR *(7)*. Here, the second round of PCR was omitted, because in standard applications, no significant differences in the results were observed between one round of PCR only and

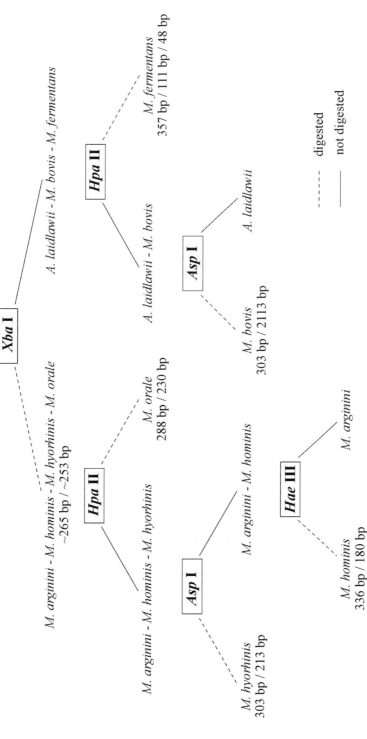

Fig. 2. Flowchart for the identification of the mycoplasma species. Digesting aliquots of the amplified PCR product with the indicated restriction enzymes will result in undigested (solid lines) or digested (dashed lines) fragments of the sizes mentioned below the species names.

nested PCR. Mycoplasma-positive cell cultures were detected as positive in the first round of PCR and negative samples were consistently negative employing nested PCR. Furthermore, applying a nested PCR increases the risk of transmission of first-round PCR products to the reagents used in the second amplification and potentially to those shared with the first round.

2. In this protocol, genomic DNA is used for the PCR reaction. As the primers hybridize to the 16S rRNA, an RT-PCR can also be performed after extracting RNA and preparation of cDNA. RT-PCR might increase the sensitivity of the assay, because the number of rRNA molecules per organism is much higher than the coding gene. Nevertheless, we find that the sensitivity of the described method is high enough for routine applications, and the excess of labor, time, and costs required for RT-PCR protocols is not warranted.

3. The primers can be designed using the degenerated code to incorporate two different nucleotides to form a mixture of two primers. When the forward or reverse primers are mixed and aliquoted for use in the PCR reaction, it must be taken into account that the molarities of the oligonucleotides with mixed bases are reduced by 50%. The primer solutions should be aliquoted into small portions (i.e., 25-μL aliquots) and stored frozen at –20°C to avoid multiple freeze–thawing cycles and to minimize contamination risks.

4. To use this PCR method for the testing of cell culture media or supplements (e.g., fetal bovine serum [FBS]), the sample sizes can be increased and centrifugation performed in an ultracentrifuge.

5. We do not recommend using the crude lysate of the sample for the PCR reaction as described in some publications, because it often contains inhibitors of the *Taq* polymerase and could lead to false-negative results. For convenience and speed of the assay, we apply commercially available DNA extraction/purification kits based on binding of the DNA to matrices and subsequent elution of the DNA. We tested normal phenol–chloroform extraction and subsequent ethanol precipitation, the High Pure PCR Template Preparation Kit from Roche (Mannheim, Germany), the Invisorb Spin DNA MicroKit III from Invitek (Berlin, Germany), and the Wizard DNA Clean-Up System from Promega (Mannheim, Germany). Following the recommendations of the manufacturers, the amplification of the mycoplasma sequences were all similar when the same amounts were used for the elution or resuspension. For screening many samples, the Wizard system works very well with the vacuum manifold.

6. The use of thermal cyclers other than the GeneAmp 9600 might require some modifications in the amplification parameters (e.g., duration of the cycling steps, which are short in comparison to other applications). Also, magnesium, primer, or dNTP concentrations might need to be altered. The same is true if another *Taq* polymerase is used, either polymerases from different suppliers or different kinds of *Taq* polymerase; for example, we found that the parameters described were not transferable to HotStarTaq with a prolonged denaturation step (Qiagen).

7. The limiting dilution of the internal control DNA can be used maximally for 2 or 3 mo when stored at 4°C. After this time, the amplification of the internal control

DNA might fail even when no inhibitors are present in the reaction, because the DNA concentration might be reduced because of degradation or attachment to the plastic tube.

8. Applying the internal control DNA, the described PCR method is competitive only for the group of mycoplasma species that carries primer sequences identical to the one from which the internal control DNA was prepared. The other primer sequences are not used up in the PCR reaction because of mismatches. Usually, one reaction per sample is sufficient to detect mycoplasma in long-term infected cell cultures. However, to avoid the possibility of performing a competitive reaction and of decreasing the sensitivity of the PCR reaction (e.g., after antimycoplasma treatment or for the testing of cell culture reagents), two separate reactions are performed: (1) without internal control DNA to make all reagents available for the amplification of the specific product and (2) including the internal control DNA to demonstrate the integrity of the PCR reaction (*see* **Fig. 1**).

9. Heavily infected cell cultures might show the mycoplasma specific band, whereas the internal control is not visible. In this case, the mycoplasma target DNA suppresses the internal control, which is present in the reaction mixture at much lower concentrations. The reaction is classified mycoplasma positive (*see* **Fig. 1**).

References

1. Uphoff, C. C. and Drexler, H. G. (2001) Prevention of mycoplasma contamination in leukemia–lymphoma cell lines. *Hum. Cell* **14,** 244–247.
2. Drexler, H. G. and Uphoff, C. C. (2002) Mycoplasma contamination of cell cultures: incidence, sources, effects, detection, elimination, prevention. *Cytotechnology* **39,** 23–38.
3. Drexler, H. G. and Uphoff, C. C. (2000) Contamination of cell cultures, mycoplasma, in *The Encyclopedia of Cell Technology* (Spier, E., Griffiths, B., and Scragg, A. H., eds.), Wiley, New York, pp. 609–627.
4. Uphoff, C. C. and Drexler, H. G. (2002) Comparative PCR analysis for detection of mycoplasma infections in continuous cell lines. *In Vitro Cell. Dev. Biol. Anim.* **38,** 79–85.
5. Uphoff, C. C. and Drexler, H. G. (1999) Detection of mycoplasma contamination in cell cultures by PCR analysis. *Hum. Cell* **12,** 229–236.
6. Sambrook, J., Fritsch, E. F., and Maniatis, T. (eds.) (1989) *Molecular Cloning, A Laboratory Manual*, 2nd ed., Cold Spring Harbor Laboratory Press, Cold Spring Harbor, NY.
7. Hopert, A., Uphoff, C. C., Wirth, M., Hauser, H., and Drexler, H. G. (1993) Specificity and sensitivity of polymerase chain reaction (PCR) in comparison with other methods for the detection of mycoplasma contamination in cell lines. *J. Immunol. Methods* **164,** 91–100.

3

Eradication of Mycoplasma Contaminations

Cord C. Uphoff and Hans G. Drexler

Summary

Mycoplasma contaminations have a multitude of effects on cultured cell lines that can potentially influence the results of experiments and pollute bioactive substances used in human medicine. The elimination of mycoplasma contaminations in cell cultures has become a practical alternative to discarding and re-establishing important or irreplaceable cell lines. Different quinolones, tetracyclins, and macrolides shown to have strong antimycoplasma properties are employed for the decontamination. We provide detailed descriptions to assure eradication of mycoplasma, to prevent formation of resistant mycoplasma strains, and to cure heavily contaminated and damaged cells. To date, no consistent and permanent alterations that affect the eukaryotic cells during or after the treatment have been detected.

Key Words: Antibiotic elimination; cell lines; mycoplasma.

1. Introduction

The use of human and animal cell lines for the examination of biological functions and for the production of bioactive substances requires rigorous quality control to exclude contamination with organisms (i.e., other eukaryotic cells, bacteria, and viruses). In this respect, mycoplasmas play an important but undesirable role, because a high portion (approx 25%) of the cell cultures arriving at our cell lines collection are contaminated with these wall-less bacteria. Mycoplasma can have a multitude of effects on eukaryotic cells and can alter almost every cellular parameter from proliferation via signaling pathways to virus susceptibility and production. Most strikingly are the effects regarding the competition in nutrition consumption that lead to the depletion of a number of essential nutrients. Consequentially, many downstream effects can be detected such as altered levels of protein, DNA and RNA synthesis, and

From: *Methods in Molecular Biology, vol. 290: Basic Cell Culture Protocols, Third Edition*
Edited by: C. D. Helgason and C. L. Miller © Humana Press Inc., Totowa, NJ

alterations of cellular metabolism and cell morphology. Mycoplasmas do not gain energy by oxidative phosphorylation, but from fermentative metabolism of diverse nutrients. This can lead to an alteration of the pH value and to the production of metabolites that are toxic to the eukaryotic cells (e.g., NH_3). The dependence of many mycoplasmas on cholesterols, sterols, and lipids can result in an alteration of the membrane composition. Other activation and suppression processes have also been described (e.g., lymphocyte activation, cytokine expression, induction of chromosomal aberrations, etc.). It has been noted that many experimentally analyzed parameters that were at first attributed to the eukaryotic cells were later ascribed to the contaminating mycoplasmas or were caused by them. For example, mycoplasmas carry a uridine phosphorylase that can inactivate the artificial deoxynucleotide, bromodeoxyuridine (BrdU). Cells with a thymidine kinase defect are commonly used for cell fusions and selected by the addition of BrdU. If mycoplasmas inactivate BrdU, the growing eukaryotic cells might appear to carry the enzyme deficiency and are misleadingly selected for cell fusions. Cell lines for virus propagation are also often affected by mycoplasma infections, leading to higher or lower titers of viruses *(1)*.

When an infected cell culture is detected, it should be autoclaved and discarded immediately and replaced by a mycoplasma-free culture. However, some cell lines are not replaceable because of unique characteristics of the cells or all the work that has been invested to manipulate these particular cells.

A number of methods have been described to eradicate mycoplasmas from cell cultures. They comprise physical, chemical, immunological, and chemotherapeutic treatment. Some of these treatments are restricted to surfaces only (e.g., exposure to detergents), to eukaryotic-cell-free solutions, such as fetal bovine serum (FBS) (e.g., filtration through microfilters), and to specific mycoplasma species (culture with antimycoplasma antisera), are not practicable for a standard cell culture laboratory (in vivo passage of continuous cell lines through nude mice cell cloning), or are ineffective in eliminating the mycoplasmas quantitatively (heat treatment, exposure to complement) *(2)*. It also has to be taken into account that some mycoplasma species are competent to penetrate the eukaryotic cell. *Mycoplasma fermentans* is one of the main infecting mycoplasma species that could also enter the cells. Thus, eliminating agents also have to be active intracytoplasmically.

Chemotherapeutic treatment can be efficiently employed using specific antibiotics. Because mycoplasmas possess no rigid cell walls and have a highly reduced metabolism, many of the common antibiotics exhibit no effect on the viability of the mycoplasmas. They are naturally resistant to antibiotics targeting cell wall biosynthesis (e.g., penicillins) or have an acquired resistance against other antibiotics that are often prophylactically used in cell culture (e.g., streptomycin), or the antibiotics are effective only at concentrations that have detrimental effects on the eukaryotic cells as well. Hence, the general use of antibiotics in cell culture is not recommended except under special circumstances and then only for short durations. General use of antibiotics could lead to selection of drug-resistant organisms, to lapses in aseptic technique, and to delayed detection of low-level infection with either mycoplasmas or other bacteria *(3)*.

Three classes of antibiotic have been shown to be highly effective against mycoplasmas, both in human/veterinary medicine and in cell culture: tetracyclines, macrolides, and quinolones. These antibiotics can be applied at relatively low concentrations, with a negligible likelihood of resistance development, and, finally, with low or no effects on the eukaryotic cells. Tetracyclines and macrolides inhibit protein synthesis by binding to the 30S and 50S ribosomal subunits, respectively *(4)*. Quinolones inhibit the bacterial DNA gyrase, which is essential for the replication of the DNA. The risk of development of resistant clones is minimized by the application of antibiotics with different mechanisms of action, by sufficient treatment durations, and by constant concentrations of the antibiotics in the medium *(5)*. Here, we describe the use of several antibiotics for the treatment of mycoplasma-contaminated cells, the rescue of heavily infected cultures, salvage treatment of resistant cultures, and some pitfalls during and after the treatment.

2. Materials (*see* Note 1)

1. BM-Cyclin (Roche, Mannheim, Germany) contains the macrolide tiamulin (BM-Cyclin 1) and the tetracycline minocycline (BM-Cyclin 2), both in lyophilized states. Dissolve the antibiotics in 10 mL sterile distilled water (dH$_2$O), aliquot in 1-mL fractions and store at –20°C. These stock solutions have concentrations of 2.5 mg/mL and 1.25 mg/mL, respectively. Repeated freezing and thawing of the solutions is not detrimental for the activity of the antibiotics. The dissolved solutions can be used at 1:250 dilutions in cell culture (at 10 μg/mL and 5 μg/mL final concentration, respectively).
2. Plasmocin (InvivoGen, San Diego, CA) contains two antibiotics, one is active against protein synthesis of the bacteria and one inhibits the DNA replication (gyrase inhibitor). The mixture is a ready-to-use solution and applied 1:1000 in the cell culture (at 25 μg/mL final concentration).
3. Ciprobay 100 (Bayer, Leverkusen, Germany) is a ready-to-use solution and contains 2 mg/mL ciprofloxacin. It can be used 1:200 in cell culture (at 10 μg/mL final concentration). One-milliliter aliquots should be taken sterile from the bottle and stored at 4°C. Crystals form at 4°C and can be redissolved at room temperature.
4. Baytril (Bayer) contains 100 mg/mL of enrofloxacin and is diluted 1:100 with RPMI 1640 medium immediately prior to the treatment. The dilution should be

prepared freshly for every antimycoplasma treatment. This solution is used as
1:40 final dilution in cell culture (at 25 µg/mL).

5. Zagam (Aventis-Pharma, Ireland) contains the antibiotic sparfloxacin as powder
   and the stock solution is prepared by dissolving the antibiotic in freshly prepared
   0.1 *N* NaOH to a concentration of 20 mg/mL. This solution can be stored at 4°C.
   Before treatment, the stock solution is diluted 1:1 with RPMI 1640 medium and
   used in cell culture at a 1:1000 final dilution (at 10 µg/mL).
6. MRA (Mycoplasma Removal Agent, ICN, Eschwege, Germany) is a ready-
   to-use dilution and contains 50 µg/mL of a 4-oxo-quinolone-3-carboxylic acid
   derivative. It is used in the treatment of cell cultures at a 1:100 dilution (at 0.5 µg/mL).
7. PBS: 140 m*M* NaCl, 27 m*M* KCl, 7.2 m*M* Na_2HPO_4, 14.7 m*M* KH_2PO_4. Adjust
   to pH 7.2 and autoclave for 20 min at 121°C.
8. Cell culture media and supplements as appropriate and recommended for the particular cultured cell lines.

3. Methods
3.1. Pretreatment Procedures

1. If no frozen reserve ampoules of the cell line are available, aliquots of the contaminated cell line should be stored frozen before treatment. Whenever possible,
   the ampoules should be kept isolated from noninfected cultures, either at –80°C
   for short time (over the complete curation time of 1–2 mo) or, preferably, in
   liquid nitrogen in separate tanks (*see* **Note 2**). The ampoules have to be marked
   properly as "mycoplasma positive" to prevent a mix up of ampoules containing
   cured or infected cells. After successful cure, these mycoplasma-positive
   ampoules should be removed and the cells destroyed by autoclaving.
2. Prepare the antibiotic working solutions freshly for every treatment and add the
   solution directly to the cell culture, not to the stored medium.
3. The FBS concentration should be increased to 20% before, during, and for at
   least 2 wk after the treatment to ensure optimal growth conditions, even if the
   cells grow well at lower concentrations.

3.2. Antibiotic Treatment

Mycoplasma infection often impairs the growth and viability of eukaryotic
cells. After addition of the antibiotic, heavily infected cells might recover significantly and the viability of the culture might increase rapidly. However, in
several other cases, the delicate health of the cells is further aggravated by the
exposure to the antibiotics. One reason might be the partial inhibition of mito-
chondrial respiration by the antibiotic(s). Even though optimal concentrations
of the antibiotics were determined in many trials, different cell types and infection conditions might behave differently upon treatment. Thus, in some
instances, the cultures might be killed by the treatment *(5)*. In these events, the
treatment has to be repeated with another culture that was stored frozen prior to

the treatment. Even when no antibiotics are added to the medium, the cells might reach a crisis and die. To counteract the treatment-associated harm, a few general rules should be followed to improve the culture conditions and to reduce the stress of infection and treatment on the eukaryotic cells (these rules are suitable for most cell lines, but some cell lines require special care which has to be determined by the user):

- Keep the concentration of the antibiotic constant during the treatment period; degradation of the antibiotic can be avoided by frequent complete medium exchanges noting the following caveats:
- Culture the cells at a medium or higher cell density and keep this density almost constant during the treatment and a few weeks after; a higher density of the cells demands a more frequent change of medium, which is commonly more favorable than a relatively low cell density and long intervals between medium changes; however, some cell lines reportedly produce their own growth factors and, therefore, the medium should not be fully exchanged, depending on the cell line.
- Observe the culture daily under the inverted microscope to recognize quickly any alteration in general appearance, growth, morphology, decrease in cell viability, detachment of cells, formation of granules, vacuoles, and so forth.
- In the case of deterioration of the cell culture, interrupt the treatment for a few days and let the cells recover (but this should only be the last resort); culture conditions should be changed immediately after recognition of the alterations, because if the cells are already beyond a certain degree of damage, it is usually difficult to reverse the progression of apoptosis.
- If possible, frequently detach slowly growing adherent cells in order to facilitate the exposure of all mycoplasmas to the antibiotic; the contaminants should not have the opportunity to survive in sanctuaries such as cell membrane pockets (it is similarly helpful to break up clumps of suspension cells by vigorous pipetting or using other reagents [e.g., trypsin or Accutase]).
- As antibiotics are light sensitive, protect cultures from the light, as much as possible.

Generally, three different methods are applied for the treatment of cell cultures: (1) the use of a single antibiotic compound (e.g., the quinolones), which is basically the same procedure for each antibiotic of that group; (2) the simultaneous application of two different antibiotics in the case of Plasmocin; and (3) the use of a combination therapy applying the two antibiotics minocycline (tetracycline) and tiamulin (macrolide) in alternating cycles (BM-Cyclin) *(4)* (*see* **Fig. 1** and **Note 3**). The latter method is more time-consuming, but also highly effective. We recommend applying two of the three types of treatment in parallel or subsequently, if one method fails.

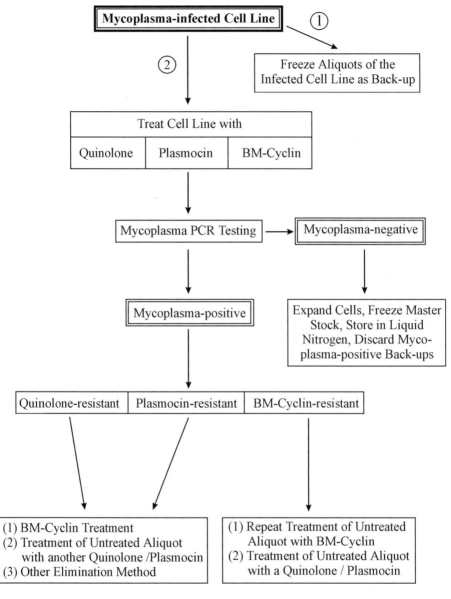

Fig. 1. Scheme for mycoplasma eradication. Different antibiotics can be used to treat mycoplasma-contaminated cell lines with a high rate of expected success. We recommend (1) cryopreservation of original mycoplasma-positive cells as backups and (2) splitting of the growing cells into different aliquots. These aliquots should be exposed singly to the various antibiotics. Posttreatment mycoplasma analysis and routine monitoring with a sensitive and reliable method (e.g., by polymerase chain reaction [PCR]) are of utmost importance.

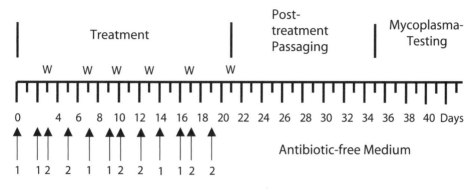

Fig. 2. Treatment protocol for BM-Cyclin. Antibiotics are given on the days indicated by arrows. Cells are washed (indicated by w) with PBS prior to the cyclical change of antibiotics to avoid formation of resistant mycoplasmas resulting from low concentrations of the antibiotics. At the end of the decontamination period, cells are washed with PBS and suspended in antibiotic-free medium. After a minimum of 2 wk posttreatment, the mycoplasma status of the cells is examined with sensitive and robust methods (e.g., by PCR).

A schematic overview of the procedure is given in **Fig. 1**; an exemplary representation of the treatment with BM-Cyclin is shown in **Fig. 2**.

3.2.1. Treatment With BM-Cyclin

1. Prepare a cell suspension (detach adherent cells, break up clumps by pipetting or using other methods) (*see* **Note 4**); determine the cell density and viability by trypan blue exclusion staining. Seed out the cells at a medium density (*see* **Note 5**) in a 25-cm² flask or one well of a 6- or 24-well-culture plate with the appropriate fresh and rich culture medium (10 mL for the flask, and 4 mL and 2 mL for the wells, respectively). Add 4 µL of a 2.5-mg/mL solution of BM-Cyclin 1 (tiamulin) per milliliter of medium. Incubate the cell culture for 2 d.
2. Remove all cell culture medium in flasks or wells containing adherent cells or after centrifugation of suspension cells. If applicable, dilute the cell cultures to a medium cell density. Add fresh medium and the same concentration of BM-Cyclin 1 as used in **step 1**. Incubate for another day. This procedure will keep the concentration of the antibiotic approximately constant over the 3-d applying tiamulin.
3. Remove the medium and wash the cells once with PBS to remove the residual antibiotic agent completely from the cells and loosely attached mycoplasmas. Seed out the cells at the appropriate density (as described in **step 1**; *see* **Note 5**)

and add 4 µL of the 1.25-mg/mL solution BM-Cyclin 2 per milliliter of medium. Incubate the culture for 2 d.

4. Remove the culture medium and substitute with fresh medium. Add the same concentration of BM-Cyclin 2 as used in **step 3**. Washing with PBS is not necessary at this step. Incubate the cell culture for 2 d to complete the 4-d of minocycline treatment.

5. After washing the cells with PBS, repeat **steps 1–4** twice (three cycles of BM-Cyclin 1 and BM-Cyclin 2 altogether). Proceed with **Subheading 3.3.**

3.2.2. Treatment With Quinolones and Plasmocin

1. Prepare a cell suspension (detach adherent cells, break up clumps by pipetting or using other methods) (*see* **Note 4**); determine the cell density and viability by trypan blue exclusion staining. Seed out the cells at a medium density (*see* **Note 5**) in a 25-cm^2 flask or one well of a 6- or 24-well-culture plate with the appropriate fresh and rich culture medium (10 mL for the flask, and 4 mL and 2 mL for the wells, respectively). Add *one* of the following antibiotics to the cell culture and incubate for 2 d.

   • 25 µL of a 1-mg/mL solution of enrofloxacin (Baytril) per milliliter of medium;
   • 10 µL of a 50-µg/mL solution of MRA per milliliter of medium;
   • 1 µL of a 10-mg/mL solution of sparfloxacin (Zagam) per milliliter of medium;
   • 5 µL of a 2-mg/mL solution of ciprofloxacin (Ciprobay) per milliliter of medium;
   • 1 µL of a 25-mg/mL solution of Plasmocin per milliliter of medium.

2. Remove all cell culture medium in flasks or wells containing adherent cells or after centrifugation of suspension cells. If applicable, dilute the cell cultures to a medium cell density. Add fresh medium and the same concentration of the respective antibiotic as used in **step 1**. Incubate for another 2 d.

3. Applying enrofloxacin, MRA, or sparfloxacin, repeat **step 2** another two times (altogether an 8-d treatment). Employing ciprofloxacin or Plasmocin, repeat **step 2** five times (altogether 14-d treatment). Proceed with **Subheading 3.3.**

3.3. Culture and Testing Posttreatment

1. After completion of the treatment, the antibiotics are removed by washing the cells with PBS. The cells are then further cultured in the same manner (enriched medium, higher cell concentration, etc.) as during the treatment period except that no antibiotics are added. Even penicillin and streptomycin should not be added to the medium. The cells should be cultured for at least another 2 wk. Even if initially the cells appear to be in good health after the treatment, we found that the cells might go into a crisis after the treatment, especially following treatment with BM-Cyclin. The reason for this posttreatment crisis is not clear, but it might also be a result of a reduced activity of the mitochondria. Thus, the cell status should be frequently examined under the inverted microscope.

2. After passaging, test the cultures for mycoplasma contamination. If the cells are clean, freeze and store the aliquots in liquid nitrogen. The cells in active culture

have to be retested periodically to ensure continued freedom from mycoplasma contamination (*see* **Note 6**).

3. After complete decontamination, expand the cells and freeze master stocks of the mycoplasma-free cell line and store them in liquid nitrogen to provide a continuous supply of clean cells. Discard the ampoules of mycoplasma-infected cells.

4. Notes

1. Store the antibiotics at the recommended concentrations, temperatures, and usually in the dark, and do not use them after the expiration date. Upon formation of precipitates, completely dissolve the crystals at room temperature in the dark before use. As the antibiotics are light sensitive, protect both the stock and working solutions from light.

2. Storage in liquid nitrogen might be one of the potential contamination sources of cell cultures with mycoplasmas. Mycoplasmas were shown to survive in liquid nitrogen even without cryopreservation. Once introduced into the nitrogen, mycoplasmas could persist in the tank for an indefinite time, not proliferating, but being able to contaminate cell cultures stored in the liquid phase of the nitrogen. The infection might happen when the ampoules are inserted into the tank, cooled down to –196°C, and the unfilled part of the ampoule is filled with liquid nitrogen because of leaks in the screw caps and the low pressure inside the vials. Thus, we strongly recommend storing the ampoules in the gaseous phase of the nitrogen to prevent contamination. Additionally, contaminated cell cultures and those of unknown status should be stored separately from noninfected cells, preferably in separate tanks. If this is not possible, be sure to store the ampoules at different locations of one tank and in the gaseous phase (high positions in the tank). Do not fill with liquid nitrogen above a certain level.

3. In our experience, it is of advantage to employ two types of treatment (BM-Cyclin and one of the quinolones or Plasmocin) in parallel, as usually at least one of the treatments is successful. In the rare event of resistance, cells of the untreated frozen backup aliquots can be thawed and treated again with another antibiotic. As MRA, ciprofloxacin, enrofloxacin, and sparfloxacin all belong to the group of quinolones, it is likely that the use of an alternative compound from the same group will produce the same end result (cure, resistance, or culture death). In the case of loss of the culture during or after the treatment, aliquots can be treated with quinolones, as these are usually better tolerated by the eukaryotic cells. We recommend using MRA, which shows almost no effect on the growth parameters during the treatment of 1 wk. The use of 5 µg/mL sparfloxacin might be an alternative to the treatment procedure described, as this concentration was also shown to be effective against mycoplasma in most cases. One of the latter two treatments is also recommended when the cells are already in very poor condition prior to treatment and the number of available cells would suffice only for one single treatment. Sometimes, the cells recover rapidly after starting the treatment because of the immediate reduction of the mycoplasmas.

4. Adherent cells are detached by methods appropriate for the cell line being treated. It is important to break up all clumps and clusters and to detach cells from the surface of the culture vessels. Although the antibiotics are in solution and should be accessible to all parts of the cells, the membranes might be barriers that cannot be passed by the antibiotics. Mycoplasmas trapped within clumps of eukaryotic cells or even in cavities formed by the cell membrane of a single cell might be protected from the antibiotic. This is also the reason for the advice to keep the concentration of the antibiotic constantly high by frequently exchanging the medium. Some mycoplasma species were shown to penetrate the eukaryotic cells. This might also be a possible source of resistance, when the eukaryotic cell membrane would be a barrier for the antibiotics. On the other hand, it was shown that specific antibiotics (e.g., ciprofloxacin) are accumulated in the eukaryotic cells so that the concentration is higher inside the cells compared to the extracellular environment.

5. Depending on the growth rate of the cell line, which might be severely altered by the antibiotic, the cell density should be diluted, kept constant, or even concentrated. If no data are available at all for a given cell culture or if the cell culture is in very poor condition, the cell density, growth rate, and viability should be recorded frequently to improve the condition of the culture.

6. Applying the overly sensitive polymerase chain reaction for the detection of mycoplasma, we found that the treated cell cultures might show a weak false-positive signal even after 2 wk of post-treatment passaging. This is not necessarily the result of a resistance of the mycoplasma, but might result from residual DNA in the culture medium. These cell cultures should not be discarded after being tested positive, but retested after further culturing (*see* Chapter 2).

References

1. Barile, M. F. and Rottem, S. (1993) Mycoplasmas in cell culture, in *Rapid Diagnosis of Mycoplasmas* (Kahane, I. and Adoni, A., eds.), Plenum, New York, pp. 155–193.
2. Drexler, H. G. and Uphoff, C. C. (2000) Contamination of cell cultures, mycoplasma, in *The Encyclopedia of Cell Technology* (Spier, E., Griffiths, B., and Scragg, A. H., eds.), Wiley, New York, pp. 609–627.
3. Uphoff, C. C. and Drexler, H. G. (2001) Prevention of mycoplasma contamination in leukemia–lymphoma cell lines. *Hum. Cell* **14**, 244–247.
4. Schmidt, J. and Erfle, V. (1984) Elimination of mycoplasmas from cell cultures and establishment of mycoplasma-free cell lines. *Exp. Cell Res.* **152**, 565–570.
5. Uphoff, C. C. and Drexler, H. G. (2002) Comparative antibiotic eradication of mycoplasma infections from continuous cell lines. *In Vitro Cell. Dev. Biol. Anim.* **38**, 86–89.

4

Authentication of Scientific Human Cell Lines

Easy-to-Use DNA Fingerprinting

Wilhelm G. Dirks and Hans G. Drexler

Summary

Human cell lines are an important resource for research and most often used in reverse genetic approaches or as in vitro model systems of human diseases. In this regard, it is crucial that the cells faithfully correspond to the purported objects of study. A number of recent publications have shown an unacceptable level of cell lines to be false, in part as a result of the nonavailability of a simple and easy DNA profiling technique. We have validated different single- and multiple-locus variable numbers of tandem repeats (VNTRs) enabling the establishment of a noncommercial, but good laboratory practice, method for authentication of cell lines by DNA fingerprinting. Polymerase chain reaction amplification fragment length polymorphism (AmpFLP) of six prominent and highly polymorphic minisatellite VNTR loci, requiring only a thermal cycler and an electrophoretic system, was proven as the most reliable tool. Furthermore, the generated banding pattern and the determination of gender allows for verifying the authenticity of a given human cell line by simple agarose gel electrophoresis. The combination of rapidly generated DNA profiles based on single-locus VNTR loci and information on banding patterns of cell lines of interest by official cell banks (detailed information at the website www.dsmz.de) constitute a low-cost but highly reliable and robust method, enabling every researcher using human cell lines to easily verify cell line identity.

Key Words: Authentication; cross-contamination; PCR; DNA fingerprinting; false cell lines; VNTR; AmpFLP.

1. Introduction
1.1. The Neglected Problem of False Cell lines

Most facilities culturing cells use multiple cell lines simultaneously. Because of the complexity of experimental designs today and because of the fact that the broad use of cell lines in science and biotechnology continues to increase, the possibility of inadvertent mixture of cell lines during the course of

From: *Methods in Molecular Biology, vol. 290: Basic Cell Culture Protocols, Third Edition*
Edited by: C. D. Helgason and C. L. Miller © Humana Press Inc., Totowa, NJ

day-to-day cell culture is always present. Based on the reputation of a laboratory, the information on an exchanged cell line within a scientific cooperation is normally thought to be correct. A number of studies have shown an unacceptable level of leukemia–lymphoma cell lines to be false *(1)*. Results from authentication studies of a comprehensively large sample of cell lines using DNA fingerprinting and cytogenetic evaluation have shown a high incidence (approx 15%) of false cell lines observed among cell lines obtained directly from original investigators or from secondary sources *(2)*. Routine identification and early detection of contamination of a given cell line with another is necessary to prevent mistaken interpretation of experimental results.

1.2. History of Cell Line Discrimination

The requirement for authentication of cell lines has a history almost as long as cell culturing itself, presumably beginning when more then one cell line could be cultured continuously. In the early 1960s, the application of specific species markers, including cell surface antigens and characteristic chromosomes, showed that interspecies misidentification was a widespread problem *(3,4)*. Compared to historical analyses of polymorphic isoenzymes, a much higher resolution in discrimination among human cell lines was achieved using restriction fragment length polymorphism (RFLP) of simple repetitive sequences *(5)*, which lead subsequently to the concept of "DNA fingerprinting" (better termed DNA profiling) *(6)*. The principle of the method is based on the phenomenon that genomes of higher organisms harbor many variable numbers of tandem repeats (VNTR) regions, which show multiallelic variation among individuals *(7)*. Sequence analysis demonstrated that the structural basis for polymorphism of these regions is the presence of tandem-repetitive and nearly identical DNA elements, which are inherited in a Mendelian way. Depending on the length of the repeats, VNTRs are classified into minisatellites consisting of 9- to > 70-bp core sequences and microsatellites, which include all short tandem repeats (STRs) with core sizes from 1 to 6 bp (*see* **Table 1**). Both categories of repeats can be governed by one definite locus or are spread all over the genome and belong to the single-locus system (SLS) or multiple-locus system (MLS), respectively. Using SLS fingerprinting, a few loci used in sequential combination can distinguish between two individuals who are not identical twins. MLS fingerprints using hundreds of relevant polymorphic loci in a single step generally have a higher resolution potential but also are restricted to classical RFLP techniques and analyses.

1.3. Amplification-Based DNA Fingerprinting in Scientific Cell Culture

The innovation of the polymerase chain reaction (PCR) technology and availability of complete sequence information of the human genome have

Table 1
Amplifiable Human Fragment Length Polymorphism Loci

Designation	Status	Synonym	Chromosomal location	Repeat length (bp)	Product length (bp)	Percent heterozygosity (%)
Minisatellites						
Apo-B1	SLS	—	2p23-p24	15	522–909	80[a]
Col2A1	SLS	—	12q12-q13.1	31–34	600–850	81[b]
D1S80	SLS	MCT118	1	16	400–940	85[c]
D17S5	SLS	YNZ22	17p13.3	70	168–1080	78[d]
D2S44	SLS	YNH24	2pter	31	600 – >5000	97[e]
PAH	SLS	—	12q22-q24.2	30	370–760	78[f]
Microsatellite						
(GTG)$_N$	MLS	—	Chromosome ends	3	10 – >15000	>99.9[g]

SLS, single-locus system; MLS, multiple-locus system.
Note: Footnotes *a–g* with regard to heterozygosity are cited in **ref. 8**.

revolutionized DNA fingerprinting technology. Several hundred accurately mapped minisatellite and microsatellite markers are available for each chromosome. The primer sequences for amplification of specific STRs or VNTRs, as well as the information on the PCR product sizes and estimated heterozygosity, are available from a number of genome databases within the World Wide Web (*see* **Note 1**). A modern average laboratory applying molecular biology and cell culture techniques is normally not equipped with expensive robots and kits for (forensic) DNA profiling. Therefore, the purpose of this chapter is to propose a rapid, practical, inexpensive, robust, and reliable method with a high discrimination potential available to students, technicians, and scientists. DNA fingerprinting should be carried out if one of the following necessities or problems arises:

1. Confirmation and identity control of a newly generated and immortalized cell line (*see* **Note 2**)
2. Characterization of somatic hybrid cell lines involving human cells (e.g., species-specific monochromosomal cell lines)
3. Confirmation of cell line identity between different passages of an intensively used cell line (e.g., the human embryonic kidney cell line 293 [HEK 293]) (*see* **Note 3**)
4. Mapping of loss of heterozygosity (LOH) of chromosomal regions (e.g., for detection of tumor suppressor genes)

   In the following chapter, the technique of pool-plexed PCR amplification fragment length polymorphism (AmpFLP) of six prominent and highly polymorphic minisatellite VNTR loci (for detailed information, *see* **Table 2**) and one additional locus for sex determination using the detection of the *SRY* gene on the Y chromosome is presented. The combination of six VNTRs increases the exclusion rate to a sufficient extent and allows discrimination of one human cell line from another at the level of 10^6. In order to definitely rule out any false positivity, it is highly recommended to test suspicious cell lines further using the multilocus fingerprint system if they reveal identical or similar DNA profiles based on AmpFLP VNTR. The combination of rapidly generated DNA profiles based on single-locus VNTR loci and confirmation of duplicate AmpFLP banding patterns using a multilocus fingerprint constitute a highly reliable and robust method independent of the quantity of individual cell lines examined *(8)*.

2. Materials
2.1. Preparation of High-Molecular-Weight DNA

1. Phosphate-buffered saline (PBS): 140 mM NaCl, 27 mM KCl, 7.2 mM Na$_2$HPO$_4$ × 12 H$_2$O, 14.7 mM KH$_2$PO$_4$, pH 7.2; autoclave at 121°C for 20 min.
2. Absolute isopropanol and absolute ethanol.

Table 2
Primer Sequences of Highly Polymorphic Human VNTR Loci

Primer designation	Primer sequences
ApoB1-F	5'-ATGGAAACGGAGAAATTATGGAGGG-3'
ApoB1-R[a]	5'-CCTTCTCACTTGGCAAATACAATTCC-3'
D1S80-F	5'-GAAACTGGCCTCCAAACACTGCCCGCCG-3'
D1S80-R[c]	5'-GTCTTGTTGGAGATGCACGTGCCCCTTGC-3'
D17S5-F	5'-AAACTGCAGAGAGAAAGGTCGAAGAGTGAAGTG-3'
D17S5-R[d]	5'-AAAGGATCCCCCACATCCGCTCCCCAAGTT-3'
D2S44-F	5'-AGCAGTGAGGGAGGGGTGAGTTCAAGAG-3'
D2S44-R[e]	5'-GAAAACACTTCAGTGTATCTCCTACTCC-3'
COL2A1-F	5'-CCAGGTTAAGGTTGACAGCT-3'
COL2A1-R[b]	5'-GTCATGAACTAGCTCTGGTG-3'
PAH-F	5'-GTTATGTGATGGATATGCTAATTAC-3'
PAH-R[f]	5'-GTGGTGTATATATATGTGTGCAATAC-3'
SRY-F	5'-CTCTTCCTTCCTTTGCACTG-3'
SRY-R[h]	5'-CCACTGGTATCCCAGCTGC-3'
(GTG)-MLS[g]	5'-GTGGTGGTGGTGGTG-3'

Note: Footnotes *a–g* with regard to sequence information are cited in **ref. 8**. *h* are unpublished primer sequences spanning exact 200 bp of the *SRY* gene of the Y chromosome.

3. TE 10/1: 10 m*M* Tris-HCl, 1 m*M* EDTA, pH 8.0; prewarmed to 50°C.
4. High Pure PCR Template Preparation Kit (Roche) (*see* **Note 4**).
5. Water bath prewarmed to 72°C.
6. Standard tabletop microcentrifuge capable of 13,000*g* centrifugal force.

2.2. Hot-Start PCR

For a highly standardized procedure, prepare a premaster mix calculated for 40 µL per reaction of each sample, plus one additional reaction according to **Table 3**. We recommend using colored tubes for aliquots of primer stocks as well as for the premaster mixtures and PCR reactions for the individual loci.

1. Thermal cycler: Perkin-Elmer Cetus 480 (*see* **Note 5**).
2. *Taq* DNA polymerase (Qiagen); 10X PCR reaction buffer (Qiagen); 5X Q-solution (Qiagen); SureStart *Taq* DNA Polymerase (Stratagene); 10X PCR reaction buffer (Stratagene).
3. 6X Loading buffer: 0.01% (w/v) bromophenol blue, 0.01% (w/v) xylene cyanol, 60% glycerol (v/v), 60 m*M* EDTA in bidistilled water.
4. Primers (any supplier): *See* **Table 2**. The primers should be concentrated at 100 µ*M* in TE (10/1) as stock solution and stored at –20°C, whereas working solutions should be aliquoted at 10 µ*M* in small amounts (approx 25- to 50-µL aliquots) and stored frozen at –20°C.

Table 3
Preparation of Premaster Mixtures for Printing Individual Single-Locus Spots

Stock solution	Apo-B (blue)	D17S5 (red)	D1S80 (yellow)	D2S44 (green)	Col2A1 (white)	PAH (orange)	SRY (pink)
10X PCR buffer	4	4	4	4	4	4	4
5X Q-solution	—	10	10	10	10	—	10
dNTP (2 μM)	1	1	1	1	1	1	1
Forward/reverse primer	1	1	0.5	2	1	1	1
H$_2$O	33	23	24	18	23	33	23
DNA (10–20 ng/μL)	1	1	0.5	5	1	1	1

2.3. Agarose Gel Electrophoresis

1. Erlenmeyer flask, 500 mL.
2. 40X TAE stock solution: 1.6 M Trizma base, 0.8 M Na–acetate, 40 mM EDTA; adjust pH to 7.2 with glacial acetic acid.
3. Ultra Pure agarose (Invitrogen)
4. Microwave oven (any supplier).
5. Electrophoresis system consisting of gel tray and comb, electrophoresis chamber, and power supply.
6. Digoxigenin-labeled molecular-weight DNA marker II (Roche).
7. Ethidium bromide solution (5 mg/mL in bidistilled water).
8. Ultraviolet (UV) transillumination screen.

2.4. High-Resolution Multilocus DNA Fingerprinting

2.4.1. Restriction Endonuclease Digestion

1. 2-mL Reaction tubes.
2. Multiblock heater.
3. Microcentrifuge.
4. Disposable pipets.
5. Restriction endonuclease *Hin*fI (high concentration 50 U/μL).
6. 10X *Hin*fI restriction buffer.
7. 6X Gel loading buffer, bidistilled water
8. Absolute isopropanol (–20°C).
9. 70% (v/v) Ethanol (–20°C).

2.4.2. Southern Blotting and DNA Fixation

1. Whatman paper.
2. Nylon membrane positively charged (Roche).
3. Parafilm.
4. Two glass plates (30 cm × 30 cm).
5. Paper towels.
6. 500-g Weight.

7. 0.4 *M* NaOH.
8. Oven capable of 120°C temperature.

2.4.3. Preblocking, Prehybridization, and Hybridization

1. Hybridization oven with rotating bottles.
2. Plastic wrap.
3. 6X SSC: 0.9 *M* NaCl, Na citrate, pH 7.0.
4. 0.4% (v/w) Blocking Reagent (Roche) in 6X SSC, pH 7.0.
5. Prehybridization solution (50 mL): 28 mL bidistilled water, 16 mL 25% (v/w) dextran sulfate, 4 mL 10% (v/w) sodium dodecyl sulfate (SDS), 2.32 g NaCl, 100–200 μg/mL of heat-denatured *Escherichia coli* DNA.
6. Hybridization solution: 10 mL of prehybridization solution complemented with 100–130 pmol of the digoxigenin-labeled oligonucleotide $(GTG)_5$ per 150 cm^2 of nylon membrane.
7. Wash buffer I: 3X SSC, pH 7.5.
8. Wash buffer II: 3X SSC, pH 7.5, 0.1% SDS.

2.4.4. Chemiluminescent Detection of Digoxigenin-Labeled DNA

1. Maleic acid buffer: 0.1 *M* maleic acid, 0.9 *M* NaCl, adjust pH to 7.5 with concentrated NaOH, autoclave at 120°C for 20 min.
2. 10% (v/w) Blocking Reagent in maleic acid buffer.
3. Wash buffer A: 0.3% (v/v) Tween-20 in maleic acid buffer.
4. Activation buffer: 0.1 *M* Tris-HCl, 0.9 *M* NaCl, 50 m*M* MgCl$_2$; adjust pH to 9.5 with HCl.
5. Substrate solution: CSPD (Roche) 1:100 dilution in activation buffer.
6. X-ray supplies including X-ray developer, X-ray fixer, X-ray film, X-ray film cassettes.

3. Methods

3.1. Preparation of High-Molecular-Weight DNA

The principle of this assay is that cells are lysed during a short incubation time with proteinase K in the presence of a chaotropic salt (guanidinium hydrochloride), which immediately inactivates all nucleases. Nucleic acids bind selectively to glass fibers prepacked in the filter tube. Bound genomic DNA is purified in a series of rapid washing and spinning steps to remove inhibiting cellular components. Finally, low-salt elution releases the DNA from the glass fiber cushion (*see* **Note 4**). The cell lines to be tested for identity should be taken from cell cultures with viabilities over 80% in order to prevent isolation of DNA fragments from apoptotic cells.

1. The cell culture suspension containing $(3–5) \times 10^6$ diploid cells is centrifuged in an Eppendorf tube at 2000*g* for 2 min; the supernatant is removed with a disposable pipet and discarded; the remaining pellet is carefully resuspended in 1 mL PBS and centrifuged again.

2. After the washing step, the pellet is resuspended in 200 µL PBS by vortexing; make sure that even tiny clumps of cells are carefully resuspended. Prewarm the water bath to 72°C.
3. For isolation of the genomic DNA, the commercially available DNA extraction kit from Roche is applied; 200 µL of solution I (guanidinium hydrochloride; well mixed) is added to the sample solution and mixed by pipetting.
4. Add immediately 40 µL proteinase K, mix well using a vortex, and incubate at 72°C for 10 min.
5. Add 100 µL of isopropanol to the sample, mix well, and apply the whole mixture to a filter tube; centrifuge for 1 min at 8000 rpm (5900g).
6. Discard the flowthrough, add 500 µL of inhibitor removal buffer, and centrifuge again for 1 min at 8000 rpm (5900g).
7. Discard the flowthrough, add 500 µL of wash buffer, and centrifuge again for 1 min at 8000 rpm (5900g).
8. Repeat **step 7**.
9. For elution of the DNA solution, place a new tube under the column and add 200 µL of elution buffer preheated to 72°C. Centrifuge for 1 min at 8000 rpm (5900g). For maximum yield, the elution step should be repeated using 100 µL elution buffer. The purified genomic DNA concentration should be approx 10 ng/µL per sample depending on the ploidy status of the cell line used. The genomic DNA should be stored at 4°C temperature.

3.2. (Manual) Hot-Start PCR (see Note 5)

The amplification procedure and the parameters described here provide a hot-start PCR protocol that can be carried out manually or by applying an inactivated hot-start *Taq* polymerase. Furthermore, this protocol is optimized for the application in 0.5-mL reaction tubes in a Perkin-Elmer DNA Thermal Cycler 480 or Bio-Rad I-Cycler (*see* **Note 5**). An adjustment to any other equipment might be necessary. General rules to avoid DNA carryover contaminations should be strictly followed:

1. DNA extraction should be carried out using equipment (pipets, microcentrifuge, etc.) that is independent of the PCR setup; optimally, this laboratory is separated from those rooms where the PCR reaction is set up or the PCR products are analyzed.
2. All reagents should be stored in small aliquots to provide a constant source of uncontaminated reagents; new aliquot batches should be tested and compared for quality prior to any use.
3. Reamplifications should never be conducted.
4. If possible, the place of setting up the reactions should be a PCR working station or a hood capable of UV irradiating the required pipets, tips and tubes.
5. It is highly recommended that gloves be worn during the whole procedure.
6. Finally, it is also fundamental to integrate the appropriate positive and negative controls (e.g., HeLa DNA and H_2O, respectively).

If the above prerequisitions for PCR setup are fulfilled, the reaction mixture should be carried out as follows:

1. Transfer 40 μL of the premaster mix to 0.5-mL PCR reaction tubes according to the sample number; add 1 μL distilled water (dH$_2$O) to the water control reaction.
2. Prepare the *Taq* DNA polymerase mix (10 μL per reaction plus one additional reaction) containing 1X PCR buffer and 1 U *Taq* polymerase per reaction; keep the mixture on ice.
3. Store all reagents used for the preparation of the master mix and take out the samples of DNA to be tested; do not handle the reagents and samples simultaneously or with the same gloves; add 1 μL of the DNA preparation (approx 10–20 ng) to the reaction solutions.
4. Transfer the reaction mixtures of **step 1** without *Taq* polymerase (if the enzyme is not a hot-start *Taq* polymerase) to the thermal cycler, carefully add a drop of oil (if necessary), and start one cycle with the following parameters:

    Cycle step 1: 5 min at 96°C
    Cycle step 2: 3 min at 75°C
    Cycle step 3: 2 min at 55°C (or 62°C for D2S44 only, *see* **step 5**)
    Cycle step 4: 5 min at 72°C
    During cycle step 2, open the thermal lid and add 10 μL of the *Taq* polymerase mix to each tube to perform a hot start PCR (*see* **Note 6**).

5. After this initial cycle, perform 35 thermal cycles with the following parameters:

Common program	D2S44 only
Cycle step 1: 4 s at 95°C	4 s at 95°C
Cycle step 2: 30 s at 55°C	1 min at 62°C
Cycle step 3*: 1 min at 72°C	2 min at 72°C

    (*plus 1 s of extension time during each cycle).

6. The reaction is finished by a final amplification step at 72°C for 7 min and the samples are then cooled to room temperature.

3.3. Agarose Gel Electrophoresis

1. Prepare a 1.2% agarose gel for analyzing PCR products (as shown in **Fig. 1**) or a 0.7% agarose gel for analyzing DNA fragments using the multilocus DNA profiling procedure, respectively (*see* **Subheading 2.3.**).
2. While the agarose is cooling down to approx 60°C, prepare the gel mold by applying two pieces of autoclave tape across the open ends, place the gel mold on a level surface, pour the cooled agarose solution over the entire gel mold, and break any bubbles that may be present and immediately insert the tooth comb (1-mm thickness recommended).
3. Allow the agarose to solidify for 30 min at room temperature and another 30 min at 4°C.

    Remove the tape from each end of the gel mold and place it into the electrophoresis apparatus (*see* **Note 6**). Fill the electrophoresis chamber with 1X TAE

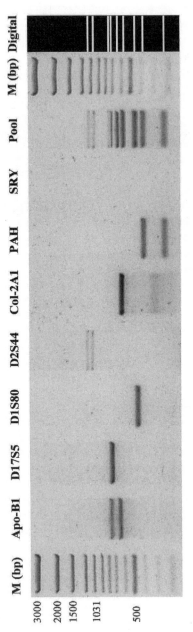

Fig. 1. Individual and pooled AmpFLP DNA fragments of KASUMI-1. Genomic DNA of the KASUMI-1 cell line was used to amplify alleles from Apo-B1, Col2A1, D17S5, D1S80, D2S44, PAH, and SRY as indicated. Y chromosomal sequences are absent (lane SRY) because of the establishment of the line from a female patient. After size determination of the amplicons from each loci on a 1% agarose gel using gel-analyzing software, the data are entered in a database, creating a specific DNA profile for each cell line as shown to the left.

44

so that the buffer is about 5 mm above the surface of the gel (*see* **Note 7**). Prepare 10 µL of each sample (8 µL reaction mix plus 2 µL of 6X loading buffer) and load samples and markers (digoxigenin-labeled analytical markers or 1-kb ladder, respectively) into the designated wells.

4. For PCR product analysis, electrophorese at 100 V (constant voltage) for 1 h and proceed with **step 5**. In the case of multilocus DNA fingerprinting, electrophorese at 40 V for 16 h or until the bromophenol blue dye has migrated to the bottom edge of the gel and proceed with **Subheading 3.4.3.**

5. Photo documentation. Carefully slide the gel onto the transilluminator and take a photograph of the gel with a ruler placed adjacent to the marker lane (1-kb ladder) for orientation. With regard to precise fingerprinting, we recommend using a gel-analyzing software (any supplier) capable of fragment length determination and saving the DNA profiles in a database.

3.4. High-Resolution Multilocus DNA Fingerprinting

DNA profiles generated by the use of multilocus probes result in a banding pattern harboring 30–50 bands in each lane. It is important to place suspicious cell lines in neighboring lanes, especially if they revealed identical or similar DNA profiles in the AmpFLP VNTRs analysis.

3.4.1. Restriction Endonuclease Digestion of Genomic DNA

1. Adjust every DNA sample containing 25 µg genomic DNA to 800 µL of volume using TE (10/1) buffer (do not vortex the DNA solution prior to restriction endonuclease digestion as this may cause random shearing); set up the following reaction in a 2.0-mL sterile reaction tube:

   10X Enzyme buffer: 100 µL
   Restriction enzyme: 150 U
   25 µg DNA: 800 µL
   Adjust to a final volume of 1000 µL with TE buffer (10/1)

2. After adding all components, flick tubes briefly and spin them in a microcentrifuge at full speed for a few seconds; incubate in a 37°C multiblock heater or water bath for 4 h on a slow-shaking platform (*see* **Note 8**).

3. Add 800 µL of ice-cold isopropanol and flick tubes to mix.

4. Immediately centrifuge tubes at 13,000g for 30 min and remove the isopropanol using a stretched glass pipet.

5. Add 200 µL of ice-cold 70% ethanol, vortex the tubes to wash DNA (may be stored overnight or longer at –70°C), remove as much as possible of the ethanol, and allow to air-dry for about 10 min (*see* **Note 8**).

6. Resuspend DNA pellet in 20 µL of TE buffer (pipetting up and down a few times accelerates the resuspension process). Add 10 µL of gel-loading buffer, vortex briefly, and microcentrifuge for a few seconds before loading the gel according to **Subheading 3.3.**

3.4.2. Southern Blotting and DNA Fixation

The objective is to set up a flow of buffer from the reservoir through the gel and the membrane so that the DNA fragments are eluted from the gel and transferred on the membrane where they will be fixed.

1. A box (22 × 22 cm) should be wrapped with 3MM Whatman paper (serving as a wick) and placed in the middle of the buffer tray or dish.
2. Apply a plastic frame with a window (e.g., Parafilm stripes) that is 1 cm shorter in length and width than the size of the gel and place it onto the wick; the Parafilm serves as a barrier to prevent transfer buffer from bypassing the gel.
3. Fill the buffer tray with 0.4 M NaOH (at least 500 mL) until the solution is 3 cm underneath the gel. Carefully place the gel, wells facing down, on the wick in the middle of the plastic frame and ensure that the edges of the gel and the frame have a 0.5-cm overlap at each site; avoid scratching the back side of the gel!
4. Center the nylon membrane on top of the gel and wet the membrane with 0.4 M NaOH. Remove all bubbles between the membrane, gel, and wick by rolling smoothly over the surface of the membrane using a 10-mL pipet; do not apply too much pressure to the gel; as this might cause distortion.
5. Wet two pieces of 3MM Whatman paper (cut to the size of the gel) in 0.4 M NaOH and place them on top of the membrane; remove air bubbles between the membrane and the 3MM Whatman paper.
6. Layer a 3-in. stack of paper towels onto the 3 MM paper. Next add the glass plate and, finally, place the 500-g weight on top of the stack. Ensure that the weight and paper towels cannot tilt.
7. Allow the transfer to proceed for 8–20 h.
8. Remove paper towels and mark the position of the comb slots and the outline of the gel onto the membrane with a fine-tip marker.
9. Place the damp membrane on a piece of 3MM Whatman paper and bake in an oven at 120°C for 30 min.
10. Remove the membrane from the oven. The membrane can be used immediately for prehybridization/hybridization or stored dry at 4°C.

3.4.3. Preblocking, Prehybridization, and Hybridization

Prehybridization prepares the membrane for probe hybridization by blocking nonspecific nucleic acid-binding sites to reduce background. Although the use of a rotating hybridization oven is recommended, it is possible to carry out the following procedure using a plastic bag and a water bath.

1. Place the membrane in a plastic bag with 10 mL 0.5% blocking reagent buffer and shake the membrane slowly for 1 h on a rotating platform at room temperature.
2. Incubate the membrane in 5–10 mL prehybridization solution prewarmed to 42°C in a rotating hybridization oven for 4 h following manufacturer's instructions.

3. Discard the prehybridization solution and replace it with 5–10 mL of prewarmed hybridization solution containing 10–20 pmol/mL of digoxigenin-labeled oligonucleotide GTG$_5$ (*see* **Note 9**).

4. After incubation overnight with rotating, the hybridization solution is discarded and replaced with prewarmed wash buffer I; wash the membrane for 25 min at 42°C and repeat this step.

5. Next, wash the membrane twice (10 min each time) using prewarmed wash buffer II. Proceed to the detection procedure or store the membrane in wash buffer I at 4°C.

3.4.4. Chemiluminescent Detection of Digoxigenin-Labeled DNA

Chemiluminescent detection is a three-step process whereby the membrane is first treated with a milk powder solution to prevent nonspecific attraction of the antibody to the membrane, then incubated with a dilution of antidigoxigenin Fab fragments, and, finally, incubated with CSPD solution, which is the substrate for generation of photons recorded on the X-ray film. Keep the membrane wet throughout the following procedure.

1. Block the membrane in 150 mL of 2% blocking buffer for 30 min at room temperature using a clean dish.

2. Centrifuge the antibody at 13,000 rpm (9600*g*) for 1 min in order to reduce background. Dilute the antidigoxigenin alkaline phosphatase Fab fragments 1:10,000 in a volume of 20 mL (approx 150 mU/mL).

3. Place the membrane in a heat-sealable plastic bag that has one or two unsealed sites. Pipet the antibody solution into the bag and remove all air bubbles.

4. Heat-seal the plastic bag and incubate the membrane for 30 min on a rotating platform at room temperature.

5. Discard the antibody solution. Wash the membrane twice for 15 min at room temperature using 200 mL of wash buffer A in a clean dish.

6. Pour off the wash buffer, remove the membrane from the plastic bag, and incubate for 2 min in a dish in 100 mL of activation buffer.

7. Place the membrane in a new plastic bag and add 10 mL of 1/100 dilution of CSPD chemiluminescent for 2 min at room temperature, ensuring that the solution is completely distributed over the membrane.

8. Place the membrane between two sheets of 3MM Whatman filter paper and wipe over it in order to remove excess CSPD solution.

9. Heat-seal the wet membrane into a new plastic bag or protect the membrane with a cling film and expose the membrane to an X-ray film for various time periods (5, 15, and 30 min). Select the exposure time for optimal signal intensities.

10. Take a picture of the autoradiograph and save the image for documentation (*see* **Fig. 2**).

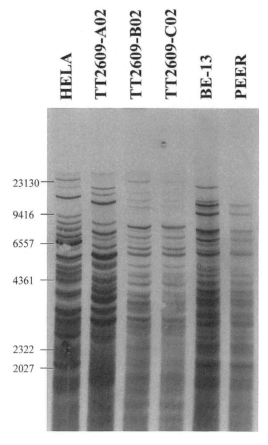

Fig. 2. High-resolution multilocus (GTG)$_5$ DNA profiles of human cell lines. The autoradiograph shows a multilocus DNA fingerprint using (GTG)$_5$ oligomers as a probe. Ten micrograms of DNA of each cell line indicated was digested to completion using *Hin*fI, size-separated on a 0.7% agarose gel, and blotted onto a nylon membrane. After hybridization with digoxigenin-labeled (GTG)$_5$ and washing procedures, a chemiluminescent detection of the probe was carried out as described in **Subheading 3.4.4.** The blot was exposed for 30 min to an X-ray film and shows the misidentification of TT2609-A02, supposed to be a sister line of TT2609-B02 and TT2609-C02, respectively, and the genetic identity of BE-13 and PEER. Lanes 1–6: HeLa, human cervix carcinoma; TT2609-A02, human follicular thyroid carcinoma (misidentified); TT2609-B02, human follicular thyroid carcinoma (authentic); TT2609-C02, human follicular thyroid carcinoma (authentic sister line); BE-13, human T cell leukaemia; PEER, human T-cell leukemia.

4. Notes

1. A good starting point for searching for microsatellite markers on a specific chromosome or for information on given markers with regard to PCR product sizes, allelic frequencies, or heterozygosity is the home page of the National Human Genome Research Institute at the National Institutes of Health, Bethesda, MD (http://nhgri.nih.gov/).
2. The establishment of cell lines from specific tumors should be carried out according to the guidelines of Drexler et al. *(9)*. With regard to an unequivocal authentication procedure, nonpathogenic tissue should be taken at the same timepoint when the tumor material is taken from the donor. It is convenient to prepare genomic DNA from blood lymphocytes. In general, 10 mL of blood will provide sufficient DNA for thousands of PCR assays.
3. The main reason for the still increasing frequency of cross-contaminated cell cultures is the uncontrolled and blind-faithed exchange of materials between scientists. It is imperative that scientists obtain the relevant cell cultures from reputable sources, like cell banks, which routinely verify the quality and authenticity of the material (DSMZ in Europe: www.dsmz.de; ATCC in USA: www.atcc.org).
4. The use of other kits capable of isolating genomic DNA is possible, but should be optimized in order to avoid the presence of inhibitory substances in the DNA preparations. Generally, DNA can be safely stored at 4°C for several months. We recommend freezing aliquots of genomic DNA for long-term storage at −20°C. However, repeated freeze–thawing will cause shearing of the high-molecular-weight DNA.
5. The use of thermal cyclers other than the Perkin-Elmer Cetus 480 or Bio-Rad I-Cycler might require some modifications in the amplification parameters (e.g., duration of the cycling steps, which are short in comparison to other applications). Similarly, Mg^{2+}, primer, or dNTP concentrations might need to be altered. The same is true if other *Taq* polymerases are used—either polymerases from different suppliers or different kinds of *Taq* polymerase.
6. The use of hot-start *Taq* polymerases, which are inactivated by chemical modifications or antibody binding at nonpermissive temperatures, prevents any activity of the polymerase at room temperature. This prevents the generation of unspecific PCR products. The use of hot-start *Taq* polymerases is recommended because it minimizes the danger of DNA carryover contamination because of opening and closing the lids of the reaction tubes. Using a chemically modified or antibody-inactivated polymerase, a single activation step (e.g., 12 min at 96°C for SureStart Taq Polymerase, Stratagene) should replace the described initiation cycle steps 1–4. For many samples, the duration of this step can be prolonged. Open and close each reaction tube separately to prevent evaporation of the samples when no oil is used. Allow at least 30 s after closing the lid of the thermal cycler to equilibrate the temperature within the tubes and to remove condensate from the lid before continuing to the next cycle step.
7. If possible, the gel run should be carried out at 4°C. If this is not possible, cool down the running buffer on ice before use.

8. Because of the viscosity of high-molecular-weight DNA, we recommend placing the DNA samples on a shaking platform (80 rpm) during restriction endonuclease digestion. In order to avoid problems with resuspension of the DNA pellets, do not overdry them (i.e., in a speed-vac).
9. Nearly all suppliers of oligonucleotides offer the possibility of chemically labeling the primers with digoxigenin. We recommend placing a single digoxigenin label at the 5' position of the primer.

References

1. MacLeod, R. A. F., Dirks, W. G., Matsuo, Y., Kaufmann, M., Milch, H., and Drexler, H. G. (1999) Widespread intraspecies cross-contamination of human tumor cell lines arising at source. *Int. J. Cancer* **83,** 555–563.
2. Drexler, H. G., Dirks, W. G., Matsuo, Y., and MacLeod, R. A. F. (2003) False leukemia–lymphoma cell lines: an update on over 500 cell lines. *Leukemia* **17,** 416–426.
3. Rothfels, F. H., Axelrad, A. A., and Simonovitch, L. (1959) The origin of altered cell lines from mouse, monkey, and man as indicated by chromosomes and transplantation studies. *Proc. Can. Cancer Res. Conf.* **3,** 189–214.
4. Simpson, W. F. and Stuhlberg, C. S. (1963) Species identification of animal cell strains by immunofluorescence. *Nature* **199,** 616–617.
5. Epplen, J. T., McCarrey, J. R., Sutou, S., and Ohno, S. (1982) Base sequence of a cloned snake chromosome DNA fragment and identification of a male-specific putative mRNA in the mouse *Proc. Natl. Acad. Sci USA* **79,** 3798–3802.
6. Jeffreys, A. J., Wilson, V., and Thein, S. L. (1985) Hypervariable minisatellite regions in human DNA. *Nature* **314,** 67–73.
7. Nakamura, Y., Leppert, M., O'Connell, P., et al. (1987) Variable number of tandem repeat (VNTR) markers for human gene mapping. *Science* **235,** 1616–1622.
8. Dirks, W. G., MacLeod, R. A., Jaeger, K., Milch, H., and Drexler, H. G. (1999) First searchable database for DNA profiles of human cell lines: sequential use of fingerprint techniques for authentication. *Cell. Mol. Biol.* **5,** 841–853.
9. Drexler, H. G., Matsuo, Y., and Minowada, J. (1998) Proposals for the characterization and description of new human leukemia–lymphoma cell lines. *Hum. Cell* **11,** 51–60.

5

Cytogenetic Analysis of Cell Lines

Roderick A. F. MacLeod and Hans G. Drexler

Summary

Cytogenetic analysis forms an essential part of characterizing and identifying cell lines, in particular those established from tumors. In addition, karyotypic analysis can be used to distinguish individual subclones and to monitor stability. This chapter describes basic cytogenetic procedures suited to cells in continuous culture. The provision of unlimited material by cell lines encourages an heuristic approach to harvesting and hypotonic treatments to yield metaphase chromosome slide preparations of improved quality suitable for subsequent banding and fluorescence *in situ* hybridization (FISH) analysis. The experience of the writers with more than 500 different cell lines has shown that no single hypotonic harvesting protocol is adequate to consistently deliver satisfactory chromosome preparations. Thus, evidence-based protocols are described for hypotonic harvesting, rapid G-banding, and FISH analysis of cell cultures to allow troubleshooting and fine-tuning to suit the requirements of individual cell lines.

Key Words: Cytogenetic methods; chromosome analysis; hypotonic treatment; G-banding; FISH.

1. Introduction

1.1. Background: The Utility of Cytogenetic Characterization

Countless cell lines have been established—more than 1000 from human hematopoietic tumors alone *(1)*—and the novelty and utility of each new example should be proved prior to publication. For several reasons, karyotypic analysis has become a core element for characterizing cell lines, mainly because of the unique key cytogenetics provides for classifying cancer cells *(2)*. Recurrent chromosome changes provide a portal to underlying mutations at the DNA level in cancer, and cell lines are rich territory for mining them. Cancer changes might reflect developmentally programmed patterns of gene expression and responsiveness within diverse cell lineages *(3)*. Dysregulation

From: *Methods in Molecular Biology, vol. 290: Basic Cell Culture Protocols, Third Edition*
Edited by: C. D. Helgason and C. L. Miller © Humana Press Inc., Totowa, NJ

of certain genes facilitates evasion of existing antineoplastic controls, including those mediated by cell cycle checkpoints or apoptosis. The tendency of cells to produce neoplastic mutations via chromosomal mechanisms, principally translocations, duplications, and deletions, renders these changes microscopically visible, facilitating cancer diagnosis by chromosome analysis. Arguably, of all neoplastic changes, those affecting chromosomal structure combine the greatest informational content with the least likelihood of reversal. This is particularly true of the primary cytogenetic changes that play key roles in neoplastic transformation and upon the presence of which the neoplastic phenotype and cell proliferation ultimately depend. Thus, all cell lines established from patients with chronic myeloid leukemia (CML) with t(9;22)(q34;q11) causing fusion of *BCR* (at chromosome 22q11) with ABL (at chromosome 9q34), which is known to be the primary change in this disease, retain this change in vitro *(4)*. Nevertheless, the usefulness of karyotype analysis for the characterization of cell lines lies principally among those derived from tumors with stronger associations with specific chromosome rearrangements [i.e., hematopoietic *(5)*, mesenchymal, and neuronal *(6)*, rather than epithelial tumors].

Cytogenetic methods facilitate observations performed at the single-cell level, thus allowing detection of intercellular differences. Accordingly, a second virtue of cytogenetic data lies in the detection of distinct subclones and the monitoring of stability therein. With the exception of doublings in their modal chromosome number from 2n to 4n "tetraploidization," cell lines appear to be rather more stable than is commonly supposed *(6–9)*. Indeed, chromosomal rearrangement in cells of the immune system could reach peak intensity in vivo during the various phases of lymphocyte development in vivo *(10,11)*.

A further application of cytogenetic data is to minimize the risk of using false or misidentified cell lines. At least 18% of new human tumor cell lines have been cross-contaminated by older, mainly "classic," cell lines, which tend to be widely circulated *(12,13)*. This problem, first publicized over 30 yr ago *(14)* but neglected of late *(15,16)*, poses an insidious threat to research using cell lines *(17)*. Ideally, authentication should be documented at the time of first publication by demonstrating concordant DNA profiling of tumor and derived cell line alike (*see* Chapter 4). Regrettably, the establishment of few cell lines has been thus documented. Hence, the vast majority of users wishing to authenticate tumor cell lines *a posteriori* are presently forced to relinquish DNA profiling in favor of cytogenetics.

In the event of cross-contamination with cells of other species, cytogenetic analysis provides a ready means of detection. Although modal chromosome numbers were formerly used to identify cell lines, their virtue as descriptors has declined along with the remorseless increase in the numbers of different cell lines in circulation. Thus, species identification necessarily rests on the ability to distinguish the chromosome banding patterns of diverse species. Fortunately, cells of the most prolific mammalian species represented in cell lines (primate, rodent, simian, as well as those of domestic animals) are distinguishable by experienced operators.

1.2. Cytogenetic Methodology

Over the last three decades, tumor cytogenetics has steadily gained in stature because of a series of advances, both technical and informational. It first became routine to distinguish and identify each of the 24 different human chromosomes (referred to as numbers 1 to 22, X, and Y) when methods for recognizing their substructures (bands) were described in the early 1970s, principally Q(uinacrine)-banding *(18)* and G(iemsa)-banding *(19)*. A further modification, trypsin G-banding *(20)*, has gained wide currency since its introduction in 1973 because of its relative speed and simplicity. Soon thereafter, banding techniques were instrumental in the identification of the "Philadelphia chromosome" (Ph) marker and its origin via a reciprocal translocation, t(9;22) (q34;q11) *(21)*, a mechanism not guessed when the Ph was first observed more than a decade earlier *(22)*. This observation marked the birth of our current picture of neoplasia as a disease of gene alteration. Improvements in speed, sensitivity, and accuracy accompanied the advent of computer-aided image analysis in the early 1990s, which enabled G-banding to handling complex tumor karyotypes.

The advent of fluorescence *in situ* hybridization (FISH) during the late 1980s *(23,24)* represented the next advance in cytogenetics. Like conventional (isotopic)-ISH, which then remained an established, though troublesome and time-consuming technique, FISH exploits the stability and specificity of DNA–DNA hybrids formed after exposure of nuclei to homologous DNA under renaturating conditions. Isotopic-ISH was superceded by FISH following the availability of nonisotopically labeled deoxynucleotides combined with a straightforward method for their efficient incorporation into DNA by nick translation. This, in turn, led to suitable probes becoming commercially available. FISH serves to bridge the gap between classical cytogenetics and molecular biology. The range of FISH is particularly impressive, enabling analysis of entire chromosomes or segments thereof ("chromosome painting") down to single genes, using probes comprising several megabases, or several kilobases or less of DNA, respectively.

Even when augmented by FISH, tumor karyotypes are often simply too complex for straightforward analysis. Complex karyotypes can be tackled using multicolor FISH (M-FISH) probes, whereby each of the 24 human chromosomes is represented by a unique mixture of 5 or more differently colored probes (reviewed in **ref. *25***). All FISH systems require broad-spectrum illumination (by ultraviolet [UV] or xenon light) and sensitive cameras to detect weaker signals, particularly those generated by short probes. All systems require special software to merge the different color channels, to improve signal-to-noise ratios and contrast, and so forth, and to generate images suitable for documentation.

The most recent advances are informational and come from sequence/mapping data of the various genome mapping and sequencing projects. Accurately mapped and sequenced bacterial/P1 artificial chromosome (BAC/PAC) clones made available as a result of these efforts allow suitably equipped investigators to map chromosome rearrangements at the level of single genes and beyond.

In this chapter, we describe basic cytogenetic procedures that have been adapted in our laboratory for use with cell cultures. For those planning *de novo* cytogenetic analysis of tumor cell lines, it is convenient to split the task into the following steps: harvesting (*see* **Subheadings 2.1.** and **3.1.**), G-banding (*see* **Subheadings 2.2.**, **2.3.**, and **3.2.**), and FISH (*see* **Subheadings 2.3.**, **2.4.**, and **3.3.**).

2. Materials

Unless otherwise indicated, reagents may be stored up to 4 wk at 4°C.

2.1. Harvesting

1. Cell culture(s) maintained in logarithmic growth phase.
2. *N*-Deacetyl-*N*-methylcolchicine (colcemid) 100X solution (Invitrogen): 4 µg/mL stock solution; store refrigerated for up to 1 yr.
3. FUDR/uridine100X stock solution. Mix 1 part 5-fluoro-2'-deoxyuridine (FUDR) (Sigma) (25 µg/mL) and 3 parts 1-β-D-Ribofuranosyluracil (uridine) (Sigma; 1 mg/mL); store refrigerated for up to 1 yr.
4. Thymidine 100X stock solution: 1-(2-deoxy-β-D-ribofuranosyl)-5-methyluracil (thymidine) (Sigma). Dissolve 50 mg in 100 mL autoclaved TE buffer (10 m*M* Tris-HCl pH 7.5, 1 m*M* EDTA). Filter-sterilize through 0.22-µm filter.
5. Trypsin 0.5 g/L–EDTA 0.2 g/L (Invitrogen) for removal and dispersal of adherent cells; store at (–20°C) for up to 6 mo.
6. Stock hypotonic solutions: KCl 5.59 g/L; or Na–citrate 9.0 g/L. Working hypotonic solutions: mix KCl and Na–citrate (e.g., 20:1, 10:1, 1:1, 1:10, 1:20, etc.) shortly before use, allowing time to reach desired temperature.
7. Fixative. Mix absolute methanol and glacial acetic acid at 3:1. Use fresh but can be stored up to 4 h at 4°C.

2.2. G-Banding Only

1. Slides (frosted ends for annotation). Wash mechanically overnight in warm ion-free detergent, rinse twice in deionized water, oven-dry, and leave overnight in

ethanol (70%). Slides should then be polished using a lint-free cloth (or non-shredding tissue) and stored wrapped in aluminum foil at (–20°C) until use.
2. Phosphate-buffered saline (PBS): adjusted to pH 6.8 (Giemsa solution) or pH 7.2 (trypsin).
3. Trypsin stock solution (140X): dissolve 17.5 mg trypsin 1:250 (Difco) in PBS (pH 6.8). Store 500-μL aliquots at (–20°C) for up to 6 mo.
4. Giemsa stain (cat. no. 1.09204.0500 Merck). Dissolve 5 mL in 100 mL PBS (pH 7.2) and filter before use.
5. Routine microscope with phase-contrast (PC) illuminator and the following objectives: ×10 (phase contrast), ×40 (phase contrast), and ×50 (brightfield–dry) for slide evaluation and preliminary analysis.

2.3. G-Banding and FISH

1. Image analysis system for G-banding and FISH (*see* **Note 1**).
2. Laboratory oven for slide aging (G-banding) or slide drying (FISH).
3. Coplin jars, 100 mL (glass), for staining and washing.
4. 4X SSC: 35.1 g NaCl, 17.7 g Na citrate made up to 1 L. Adjust to pH 7.2.
5. 0.5X SSC, 2X SSC, and so forth: dilute from 4X SSC stock but monitor pH.

2.4. FISH Only

1. Ethanol: absolute, 90%, 70%. Can be used twice, then discarded.
2. Pepsin stock solution: dissolve 250 mg pepsin (Sigma cat. no. P7012) in 12.5 mL deionized H_2O. Freeze 500-μL aliquots (–20°C) and store for up to 6 mo.
3. Pepsin working solution: Dilute 500 μL stock solution in 100 mL deionized H_2O containing 1 mL of 1 *N* HCl; store at (–20°C) for up to 6 mo.
4. Formaldehyde solution: 1% formaldehyde in PBS (pH 7.2) containing 50 m*M* $MgCl_2$.
5. Acetone, for use in mild pretreatment.
6. Hybridization buffer: Hybrisol VII (Qbiogene). Store at room temperature (contains formamide).
7. Cold competitor DNA for prehybridization with probes containing repeat sequences: Cot-1 DNA, 1 μg/μL (Roche); store at –20°C.
8. Nail varnish (clear).
9. Rubber cement.
10. Hybridization chamber: sealed container with an internal shelf to separate slides (above) from humidifier (e.g., water-impregnated towels).
11. Hybridization bed: prewarmed freezer block kept in incubator at 37°C; use during application of probes to slides.
12. Wash solution: 4X SSC with 0.1% Tween-20, molecular biology grade (Sigma). Slides can be popped into wash solution between any steps to prevent drying out.
13. Plastic cover slips for probe detection (Qbiogene).
14. Mounting medium: Dissolve 50 ng/mL 4', 6-diamidino-2-phenylindole dihydrochloride (DAPI) in Vectashield antifade mounting medium (Alexis).
15. Cover slips: glass, grade 0, 22 × 60 mm.
16. Chromosome painting probes: store at (–20°C) unless otherwise stated (*see* **Note 2**).

17. Research microscope with the following brightfield objectives with as high numerical apertures as budgetary limitations permit: ×10 (oil), ×50 Epiplan (dry), ×63 Zeiss Plan-Neofluar (oil), ×63 Zeiss Planapochromat (oil), or equivalents from other manufacturers. Ideally, a cytogenetics research microscope should be equipped with an automatic filter wheel and configured to an appropriate FISH imaging system (*see* **Note 1**).

3. Methods

3.1. Harvesting and Slide Preparation

Mammalian cells in continuous culture typically divide every 1–3 d. The metaphase stage of mitosis, the only cell cycle stage when chromosomes are clearly visible, usually lasts less than 1 h, severely reducing the number of cells available for conventional cytogenetic analysis. Accordingly, the fraction of dividing cells must be enriched by exposure of growing cultures to colcemid or some other mitotic blocking agent for a few hours, or longer in the case of slow-growing cells. It is, therefore, important to ensure that cell cultures are in their logarithmic growth phase by feeding and, if necessary, diluting/seeding out. Neglect of this simple precaution is an all-too-common cause of failed harvests. It is difficult to overstate just how crucial initial harvesting and slide preparation is to subsequent success with both G-banding (*see* **Subheading 3.2.**) and FISH (*see* **Subheading 3.3.**). Harvesting is often the step least rewarded by success. For reasons that remain obscure, some cell lines resist successful harvesting. Furthermore, an hypotonic treatment that consistently yields good preparations with one cell line might be totally unsuitable for another of similar derivation. This inconvenient problem precludes use of standard harvesting protocols applicable to all cell lines, unlike DNA preparation, for example. It is therefore necessary to ascertain empirically which harvesting procedure is optimal for each cell line. This is achieved by harvesting, in parallel, cell aliquots that have been exposed to a range of hypotonic conditions (viz. with a variety of different buffers and incubation times and, if need be, incubation temperatures, etc.) (*see* **Table 1** for an example). Cytogenetic harvesting is exquisitely sensitive to the biological variability inherent in living systems and must often be repeated several times before satisfactory results are achieved (*see* **Note 3**).

In contrast to hypotonic treatment, fixation permits standardization. Although some deterioration occurs, fixed cells can be stored several years at (–20°C) until required. Immediately prior to slide-making, cell suspensions should be washed in fixative. Slide-making is performed by dropping suspension onto ice-cold, precleaned slides held at a slight angle atop a prefrozen (–20°C) freezer cold block. Two drops aimed at the slide region immediately under the frosted zone and at the lower middle, respectively, should result in

Table 1
Data Sheet (see Note 4) for OCI-Ly-19 (DSMZ ACC 528) Cell Line

Harvest	Hypotonic treatment						Results[a]			Use tube?	Quantities of slides and suspensions[b]			
Tube[c]	Col[d]	KCl[d]	NaCit[d]	Other[d]	Temp[e]	Time[e]	MI	Spr	Qual		GTG	Giemsa	FISH	Store[f]
Harvest #1[c]: 9/30/02														
a	3 h	100%	—	—	RT	7 min	A	A	AB	Yes	8	1	6	Rest material
b	3 h	100%	—	—	RT	1 min	A	C	C	No	—	—	—	—
c	3 h	80%	20%	—	RT	7 min	A	BA	B	Yes	—	—	—	—
d	3 h	80%	20%	—	RT	1 min	A	C	C	No	—	—	—	—
Action[g]	Harvest #1 tube a, satisfactory therefore discard culture.													

Note: Data are those of an actual experiment, the harvesting of cells from the OCI-Ly-19 cell line, processed to prepare the G-banding and FISH slides images shown in **Fig. 1.**

Abbreviations: Col, colcemid; MI, mitotic index; Qual, quality; RT, room temperature; Spr, spreading.

[a]To assess the efficacy of harvest conditions, it is necessary to compare their relative efficiencies in yielding metaphases (MI), which are well spread without excessive breakage (Spr) and in which chromosome morphology is satisfactory (Qual).

[b]Indicates what should be done with suspensions from each harvest tube (e.g., mixing for slide-making and/or storage) and how many slides are to be prepared for G-banding (GTG), solid staining (Giemsa) and FISH.

[c]To avoid subsequent confusion, it is essential to identify each harvest, which, in turn, is prepared by mixing labeled tubes yielding acceptable preparations.

[d]Both the times of exposure and the concentrations of colcemid (and hypotonic buffers) should be noted.

[e]Both the temperature and duration of hypotonic treatments are crucial and should be recorded.

[f]Here, it can be indicated whether any of the harvest tubes/mixtures are suitable for storage as suspensions at (−20°C).

[g]Decisions regarding the need for repeating the harvest and, if so, how may be written in this box. In the case of OCI-Ly-19, the first harvest (tube a) was deemed adequate and remaining tubes were discarded.

figure-of-eight spreading patterns that facilitate both G-banding and FISH. Once made, slides can be variously stored for a few years at (–80°C), for short intervals at room temperature for FISH or aged overnight at 60°C for G-banding.

1. Add colcemid (final concentration of 40 ng/mL) to growing cultures for 2–4 h.
2. As an alternative to colcemid treatment, incubate cells overnight with FUDR to improve chromosome morphology (*see* **Note 4**).
3. Suspension cell cultures: aliquot cells (e.g., four times in 10-mL tubes), centrifuge (5 min at 400*g*), and discard supernatant.
4. Adherent cell cultures: Shake vigorously to remove mitoses and retain supernatant in centrifuge tube (50 mL). Meanwhile, rinse remaining adherent cells with serum-free medium or PBS and discard wash. Add sufficient trypsin/EDTA to cover the cells and incubate briefly (5–15 min) with intermittent light agitation. When cells are ready (i.e., "rounded up"), shake vigorously and remove by rinsing with supernatant from the centrifuge tube. Then, centrifuge aliquots as with suspension cultures. (The serum present in the culture medium will act to inactivate residual trypsin activity.)
5. Resuspend cell pellets gently by manual agitation. Add 5–20 vol from various working hypotonic solutions (20:1, 1:1, etc.). Incubate paired aliquots at (initially) room temperature for 1 min and 7 min, respectively. (*See* **Table 1** for example.)
6. Centrifuge and discard supernatant. Resuspend cells gently and carefully add ice-cold fixative, at first dropwise, and then faster, until the tube is full.
7. Store refrigerated for 1–2 h.
8. Equilibrate to room temperature (RT) to minimize clumping, then centrifuge (5 min at 400*g*). Repeat.
9. Store fixed cells overnight at 4°C.
10. Next day, equilibrate to RT, then centrifuge (5 min at 400*g*). Repeat twice.
11. Resuspend cells in sufficient fixative to yield a lightly opaque suspension. Typical cell concentrations range from 2 million to 8 million cells per milliliter.
12. Remove four precleaned slides (one per harvest tube) from storage at (–20°C) and place on a plastic-covered freezer block held at a slight incline away from the operator by insertion of a pipet.
13. Locally humidify by breathing heavily on slides.
14. Holding the pipet approx 30 cm above the slides, place two drops of cell suspension onto each slide—the first immediately below the frosted zone and the second about two-thirds along the slide. Do not flood.
15. Lift slides in pairs for speed. Breathe on them again to maximize spreading.
16. (Optional) To improve spreading, gently ignite residual fixative by igniting fixative (with a camping stove or Bunsen burner). Do not allow slide to get hot, as this could spoil subsequent G-banding and FISH.
17. Label and air-dry. Stand slides vertically until dry.
18. Examine slides by phase-contrast microscopy and assess each hypotonic treatment individually (*see* **Note 3**).

19. Prepare slides from successful treatments, mixing cell suspensions if more than one is deemed adequate. Label.
20. Store unused cell suspensions at –20°C in sealed 2-mL microfuge tubes filled to the brim to exclude air. Under such conditions, suspensions remain stable for several years; we have performed FISH successfully using 5-yr-old suspensions. Suspensions cryopreserved in this way must be thoroughly washed in fresh fixative prior to slide preparation. After sampling, suspensions should be refilled to the brim, marking the original level to control dilution.

3.2. Trypsin G-Banding (see Note 5)

Although several banding methods are in use, the standard procedure involves G-banding by trypsin pretreatment *(20)*. G-Banding selectively depletes the chromatin of certain proteins to produce strong lateral bands after staining with Giemsa (*see* **Fig. 1A,B**). Analysis of chromosomes harvested using the above-described technique should typically reveal some 300 bands, although with stretched or submaximally condensed (prometaphase) chromosome preparations, over 1000 bands might be discerned.

1. Fresh slides are unsuitable for immediate G-banding. Slides must be first aged by baking overnight at 60°C. About six to eight slides containing an adequate supply of well-spread metaphases in which the chromosome morphology is deemed adequate should be prepared for each cell line.
2. First prepare three Coplin jars, one each for 500 μL trypsin in 70 mL PBS (pH 7.2), ice-cold PBS (pH 6.8) to stop enzymatic activity, and 5% Giemsa in PBS (pH 6.8).
3. The Coplin jar containing trypsin in PBS should be placed in a water bath at 37°C and equilibrated to 37°C before use.
4. To determine optimal trypsin incubation times, dip the first slide halfway into the trypsin for 10 s and the whole slide for the remaining 10 s to test, in this case, for 10-s and 20-s trypsinization times, respectively.
5. Immediately stop trypsin activity by immersion in cold PBS for a few seconds.
6. Stain in Giemsa solution for 15 min.
7. Rinse briefly in deionized H_2O and carefully blot-dry using paper towels (e.g., as used for Southern blotting).
8. Examine microscopically (*see* **Note 6**). Scan for likely metaphases at low power. Examine those selected at higher power using the Epiplan dry objective. From the chromosome banding quality, decide whether the suitable trypsin time lies within the 10- to 20-s range spanned by the test slide.

If satisfactory, repeat **steps 1–7**. If unsatisfactory, repeat **steps 1–8** using longer (e.g., 30–45 s) or shorter (e.g., 3–6 s) typsin test times, as appropriate until the optimal incubation time becomes apparent.

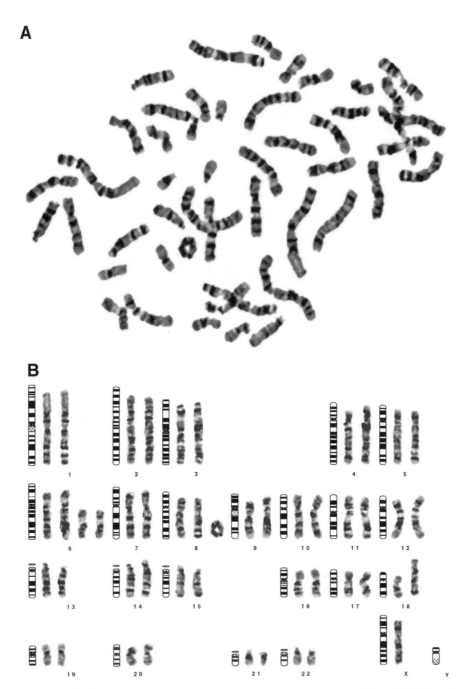

Fig. 1A,B. Cytogenetic characterization of a human lymphoma cell line (OCI-Ly-19).
The images depict a G-banded metaphase, karyogram, and FISH analysis of a cell line
OCI-Ly-19 established in 1987 from a 27-yr-old female patient with B-cell non-Hodgkin's

3.3. FISH (see Notes 7 and 8)

Chromosome painting describes FISH using long heterogeneous mixtures comprised of DNA sequences from multiple contiguous loci, none of which need be specified. Painting probes usually cover entire chromosomes or substantial parts thereof and can be used singly or in combinations—the latter maximizing the informational possibilities (e.g., by confirming a translocation suspected after G-banding). Hybridization with painting probes for chromosomes 8 and 14 is shown in **Fig. 1C**. Whichever probe combination is adopted, it is usually necessary to counterstain the chromosomes. The standard counterstain is 6-diamidino-2-phenylindole dihydrochloride (DAPI), which yields a deep blue color, more intense at the centromeric heterochromatin, in particular that of chromosomes 1, 9, and 16 and in the terminal long-arm region of the Y chromosome. In better preparations, DAPI generates negative G-bands that, with the aid of most image analysis programs, could be readily converted into G-bands, albeit rather faint ones. Painting probes can be produced by polymerase chain reaction (PCR) amplification of human chromosomal material retained by monochromosomal human/rodent hybrid cell lines. By exploiting human-specific repeat sequences (e.g., Alu) as primer targets, it is possible to

---

(Fig. 1 continued) lymphoma (B-NHL), diffuse large cell lymphoma (DLCL) at relapse *(26)*. The metaphase cell was analyzed and the chromosomes arranged to form the karyogram using a Quips image analysis system (Applied Imaging) configured to an Axioplan photomicroscope using a ×63 Planapochromat objective (Zeiss). G-Banding analysis. Image **(A)** depicts a G-banded metaphase preparation of the OCI-Ly-19 cell line. The ISCN karyotype *(27)* of the cell depicted in the karyogram **(B)** was 48<2n>X,−X,+6,+6,+8,t(4;8)(q32;q32),del(6)(q15)x2,r(8)(var),t(14;18) (q32;q21), add(18)(q23). (Note that the rearranged chromosomes are placed right of their normal homologs.) In this case, G-banding revealed the presence of an unambiguous primary change known to be recurrent in B-NHL/DLCL, a balanced, reciprocal translocation, t(14;18), whereby part of the long arm of chromosome 18 (breakpoint at band 18q21) is exchanged with the subterminal long-arm region of chromosome 14 (breakpoint at 14q32) generating a lengthened chromosome 14 homolog (generally referred to as "14q+"). Numerical changes included loss of one chromosome X and gains of chromosomes 6 (twice) and 8. The accompanying structural changes include multiple rearrangements of chromosome 8, including a ring chromosome that varies in size and a balanced translocation with chromosome 4, t(4;8), with breakpoints at 4q3 and 8q24. Both additional chromosome 6 carried identical deletions involving most of their long arm regions. Because the karyotype of OCI-Ly-19 has not been published, it cannot be used as positive evidence of authenticity. Thus, evidence of authenticity rests on the uniqueness of this karyotype—it is unlike any of those recorded in **ref.** *1* or in the DSMZ interactive website (www.dsmz.de)—and the appropriateness of any rearrangements within, to its supposed origin.

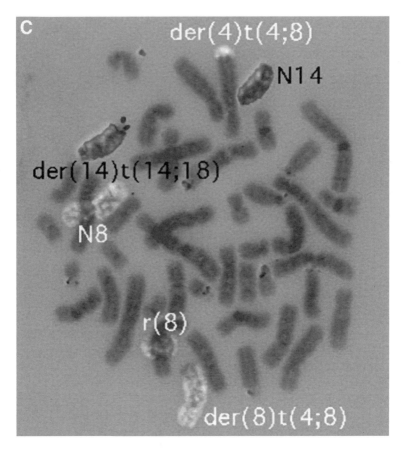

Fig. 1C. Four-color FISH. The FISH image (**C**) was captured using a cooled CCD camera (Cohu) configured to a Smart Capture imaging system (Applied Imaging). The four-color images were captured separately, merged, and the contrast-enhanced images rendered into gray tones suitable for printing. The FISH image shows the result of hybridizing metaphase chromosomes with painting probes for chromosome 8 (labeled with Spectrum Green, rendered white) and chromosome 14 (labeled with Cy3, rendered mid-gray), together with a single-locus probe prepared by labeling with Spectrum Red-d-UTP (Invitrogen), a BAC clone specific to the subtelomeric region of chromosome 18 (rendered black). The chromosomes were counterstained with DAPI and the resultant blue images rendered dark gray. Note the presence of the pair of terminal black signals on the der(14)t(14;18), absent from the N(ormal) 14, confirming the presence of chromosome 18 terminal long-arm material on the rearranged homolog. The t(14;18) juxtaposes the *BCL2* oncogene (at chromosome 18q21) with the immunoglobulin heavy chain (*IGH*) locus. Regulatory regions (enhancers) present at *IGH* that are actively transcribed in lymphatic cells switch on transcription of *BCL2*. Ectopic expression of *BCL2* is thought to promote neoplastic transformation in lymphatic cells. Additional rearrangements of chromosome 8 present in the t(4;8) and r(8)

amplify human DNA selectively. Such probes inevitably include significant amounts of human repeat DNA hybridizing indiscriminately across the genome, which must be suppressed. This is achieved by preincubating probe material together with unlabeled ("cold") human DNA enriched for repetitive sequences by a two-step denaturation–renaturation process. During renaturation, the most highly repetitive sequences (Cot-1 DNA) are the first to reanneal, allowing more complex, slower reannealing DNA to be digested away using single-strand-specific DNase-1. For this reason, most commercial painting probes include Cot-1 DNA.

Single-locus probes can be produced by labeling large-insert clones and are available commercially for a variety of neoplastic loci: FISH using a probe covering the subtelomeric region of chromosome 18 at band q23 is depicted in **Fig. 1C**. Such probes, which have become an important tool in chromosome analysis, hybridize to chromosome-arm-specific sequences present in the subtelomeric chromosome regions. These are favored sites of translocation and could be targets for instability—"jumping translocation" *(9)*. Unlike single-locus cDNA probes prepared by reverse transcription of specific mRNA that contain no repeat sequences, the BAC/PAC clones used to prepare such probes contain repeat sequences that require suppression by prehybridization with Cot-1 DNA.

The posthybridization stringency wash, which can be performed at either low temperatures including formamide, which lowers the stability of the DNA double helix, or at higher temperatures using low SSC concentrations alone, is critical to success. Stringency washing allows the operator to control the balance of probe signal intensity against background. The stability of DNA–DNA hybrids on FISH slides allows repeated cycles of stringency washing. For those starting with untested FISH probes, it is feasible to start off using a less stringent wash, which, if yielding unacceptable background levels, can be repeated at higher stringencies (i.e., at lower salt concentrations).

The FISH protocol described below is applicable to a wide variety of probes and, therefore, useful for those intending to combine probes from different sources. Indirectly labeled probes (e.g., with digoxygenin or biotin) require additional detection steps that can be plugged into the following protocol.

(Fig. 1C continued) ring chromosome serve to amplify the *MYC* oncogene, which is mapped to chromosome 8q24. *MYC* rearrangement leading to its overexpression is a common secondary change in DLCL with t(14;18) associated with tumor advancement. Similarly, loss of long-arm material effected by the 6q– deletions is also a recurrent secondary change in DLCL, although the putative tumor suppressor gene targeted by this deletion has yet to be unequivocally identified.

1. Use either fresh (1–7 d old) or archival slides stored at (–80°C).
2. Although not required, the background signal can be reduced by preincubation in pepsin solution for 2 min at 37°C (*see* **Note 7**).
3. Slide dehydration. Pass slides sequentially through an alcohol series for 2 min in 70% (two times), 90% (two times), and 100% ethanol in Coplin jars.
4. Dry slides overnight at 42°C.
5. Deproteinize in acetone for 10 min (to minimize background autofluorescence).
6. Slide denaturation. Place slides for 2 min at 72°C in 30 mL of 2X SSC plus 70 mL formamide. The temperature of this step is critical. Therefore, avoid denaturing too many slides simultaneously. If a high throughput is desired, slides should be prewarmed. Quench in prechilled (–20°C) 70% ethanol for 2 min.
7. Repeat **step 3** (the alcohol series).
8. Varnish slide label (to prevent subsequent eradication).
9. Place slide on prewarmed block at 37°C.
10. Remove probe from the freezer noting the concentration of labeled DNA. Add excess Cot-1 DNA (20–50X probe).
11. Probe denaturation: Place desired volume of probe into microfuge tube (sterile) and incubate in a "floater" for 5 min at 72°C in a water bath. (Important: If recommended by manufacturer, omit probe denaturation.)
12. Probe prehybridization. Collect probe by brief centrifugation, then incubate for 15–60 min at 37°C in a second water bath.
13. Probe application. Using shortened micropipet tips (sterile), carefully drop 8–12 µL of probe (making up the volume with Hybrisol, if necessary) onto each slide half. Thus, two hybridizations can be performed on each slide (separated by a drop of Hybrisol, to inhibit mixing). Cover slides carefully with glass cover slips, tapping out any bubbles, and seal with rubber cement.
14. Hybridization. Place slides carefully in moistened and sealed hybridization chamber. Leave overnight (or up to 72 h) in incubator (preferably humidified) at 37°C.
15. After hybridization, carefully remove rubber cement and cover slips in 2X SSC using tweezers.
16. Stringency washing. Wash slides for 5 min at 72°C in 0.5X SSC.
17. (Optional) For use with digoxigenin labeled probes; briefly prewash in wash solution at room temperature and shake to remove excess liquid. *Important*: Do not allow slides to dry out until dehydration (**step 18**). To each slide, apply 40 µL anti-digoxigenin antibody labeled with FITC (Qbiogene) and cover with plastic cover slip. Incubate for 15–30 min at 37°C in hybridization chamber. Wash for 5 min (three times) in wash solution at room temperature in subdued light.
18. Dehydration (alcohol series): Dehydrate slides as described in **step 3**, but performed in subdued light.
19. Mounting and sealing. Using abbreviated micropipet tips, to ensure even bubble-free coverage carefully place three 30-µL drops of DAPI/Vectashield mountant along the slide. Apply cover slip and tap out any large bubbles using the blunt end of a pencil or equivalent. Seal with nail varnish. Allow varnish to dry.

20. Visualization. Slides should be visualized at high power under oil immersion with a ×63 objective with a high numerical aperture. Although Zeiss supplies immersion oil specifically designed for fluorescent microscopy (518F), its propensity to flocculate spontaneously and at low temperatures renders it unsuitable for routine application to slides stored at 4°C.
21. Analysis and interpretation: *see* **Notes 8** and **9**.

4. Notes

1. Image analysis systems. The ability to reposition chromosomes at a mouse-click afforded by image analysis systems assists dissection of unresolved markers—benefiting both speed and accuracy. Karyograms can be subsequently printed with comparable expeditiousness, obviating the need for laborious cut-and-paste routines. FISH imaging systems are available from several manufacturers, based either on PC or Macintosh platforms. For further information, consult the website of Applied Imaging (www.aicorp.com/) or Metasystems (www.metasystems.de/), which supply a variety of such systems. Imaging systems confer significant benefits, including amplification of weak signals, merging of differently colored signals, contrast enhancement, background reduction, generation of G-bands from DAPI counterstain, and rapid documentation and printing.

2. FISH probes. Because it is seldom possible to resolve more complex rearrangements, chromosome painting should be used by those wishing to maximize detail and accuracy. Most painting and satellite DNA probes obtained from larger manufacturers yield satisfactory results. For those using untested probes, it is useful first to calibrate these using normal chromosomes. This effort is usually well invested. Some probes generate unnecessarily bright signals. Knowing this beforehand allows such probes to be "stretched" by dilution with Hybrisol. All too often, probes arrive that yield inadequate or inappropriate signals. Timely ascertainment of such problems not only facilitates refund or replacement but could also prevent the pursuit of false trails inspired by probes that hybridize to more than one region.

3. Slide-making. Slides for analysis should fulfill three criteria: sufficient metaphases, adequate chromosome spreading, and good morphology (i.e., large but undistended chromatids lying in parallel). To document progress in harvesting procedures and aid evidence-based searches for their improvement, we use a standard data sheet that records progress toward these ideals. An actual example is shown in **Table 1**, which presents harvesting data for the cell line OCI-Ly-19, the subsequent G-banding and FISH analysis of which are presented in **Figs. 1** and **2**. In this case, reasonable preparations were obtained at the first attempt using the standard protocol (**Subheading 3.1., step 3**). Although all four hypotonic combinations yielded adequate numbers of metaphases (A), only tubes -a and -c yielded satisfactory spreading, but only one tube (-a) yielded good chromosome morphology (AB) and was used for subsequent slide preparation. A total of 15 slides were prepared: 8 for G-banding, 1 for Giemsa staining alone (to check for the presence of small chromosomal elements that G-banding sometimes render

invisible), and 6 for FISH. In addition, the remaining cell suspension in fixative was stored (−20°C) for future use. Slides with sparse yields of metaphases are unsuitable for FISH where probe costs are often critical. For slowly dividing cell lines (doubling times > 48 h), colcemid times can be increased first to 6 h, then to 17 h (overnight), simultaneously reducing colcemid concentrations by half to minimize toxicity. However, paucity of metaphases is usually the result of depletion by overly harsh hypotonic treatments. Contrary to most published protocols, we find that reducing hypotonic exposures to 1 min and, if necessary, performing this step in microfuge tubes to facilitate speedy centrifugation to reduce total hypotonic times still further is often effective. Insufficient spreading results in tight metaphases with an excess of overlapping chromosomes; such cells might be amenable to FISH but are useless for G-banding. In such cases, spreading can sometimes be improved by harsher hypotonic treatment, whether by increasing the proportion of KCl to 100% or by increasing the hypotonic time up to 15 min, or by performing the latter at 37°C instead of RT. However, paradoxically, not a few cell lines yield their best spreading at 1 min, indicating how little we understand the underlying biological processes involved. Gentle flaming often assists spreading and, contrary to received wisdom, has little or no deleterious effect on G-banding or FISH. In our hands, "dropping from a height" effects scant improvement in spreading, although offensively heavy breathing, performed both immediately before and after dropping, is beneficial, by increasing local humidity levels. Excessive spreading, on the other hand, is often cured by reducing the proportion of KCl, or by reducing hypotonic treatment times, or by retaining more of the original medium from the first centrifugation (**Subheading 3.1.**, **steps 3** and **4**).

4. Harvesting with FUDR. As a general rule, the best morphologies are produced by hypotonics containing 50% or less Na citrate. Excessive amounts of the latter tend to yield fuzzy irregular morphologies that produce disappointing results with G-banding and FISH alike. Some types of cell, and derived cell lines alike, consistently yield short stubby chromosomes that appear refractory to all attempts at improvement. In such cases, it might be helpful to try FUDR pretreatment. Accordingly, treat cultures overnight with FUDR/uridine. The next morning, resuspend in fresh medium with added thymidine to reverse the blockade and harvest 7–9 h later.

5. G-Banding. As a general rule, good chromosomes yield good G-banding. Exceptions include chromosomes that are too "young" (puffed up or faint banding) or "over the top" (poor contrast or dark banding). Artificial aging by baking overnight at 60°C not only speeds up results but eliminates variations in optimal trypsin times because of climatic or seasonal variations in temperature or humidity. For those desperately requiring a same-day result, aging times could be shortened to 60–90 min by increasing the hot plate/oven temperature to 90°C. Trypsin G-banding is a robust technique and problems unconnected with poor chromosome morphology are rare. Those used to working with one species should note, however, that chromosomes of other species could exhibit

higher/lower sensitivities to trypsin. Losses in tryptic activity occur after about 6 mo among aliquots stored at (–80°C), which should then be discarded in favor of fresh stocks.

6. Karyotyping. G-Banding lies at the center of cytogenetic analysis. The ability to recognize each of the 24 normal human chromosome homologs necessarily precedes analysis of rearrangements. Because the majority of human cancer cell lines carry chromosome rearrangements, the choice of cell lines for learning purposes is critical. Learning should be performed using either primary cultures of normal unaffected individuals (e.g., lymphocyte cultures) or B-lymphoblastoid cell lines known to have retained their diploid character. Those intent on acquiring the ability to perform karyotyping are strongly advised to spend some time in a laboratory where such skills are practiced daily (e.g., a routine diagnostic laboratory).

7. FISH signals and noise. FISH experiments are sometimes plagued by high background signals, or "noise." BAC/PAC clones including repeat DNAs will deliver signals at other loci carrying similar sequences ("cross-hybridization"). Commercial probes are usually, but by no means always, relatively free of this problem. Increasing the wash stringency (**Subheading 3.3., step 16**) by reducing the SSC concentration to 0.1X might help. Alternately, adding Cot-1 DNA to the hybridization mix might help to reduce hybridization noise. Among noncommercial probes, excessive noise could often be cured by reducing the probe concentration. Normal DNA concentrations for single-locus probes should range from 2–6 ng/µL to 10–20 ng/µL for painting probes. Assuming that it is not the result of "dirty" slides, nonspecific noise could be caused by either autofluorescence or protein–protein binding after antibody staining, which might be reduced by additional slide pretreatment in pepsin solution (**Subheading 3.3., step 2**). Incubate slides for 2 min in acidified pepsin solution at 37°C. Rinse in PBS (pH 7.2) for 3 min at RT. Postfix slides, held flat, in 1% formaldehyde solution for 10 min at RT using plastic cover slips. Rinse in PBS (pH 7.2) for 3 min at RT. Continue with **step 3** of **Subheading 3.3.** Weak FISH signal intensity might arise because the probe itself is inherently weak, the wash too stringent, or the chromosomes insufficiently denatured. To test for these alternatives, repeat the stringency wash (**Subheading 3.3., step 16**) but with either 2X or 1X SSC in the wash buffer. In parallel, repeat the slide denaturation (**Subheading 3.3., step 6**) increasing the denaturation time to 4 min. When neither alteration brings any improvement and the probe is new and untested or old and infrequently used, it is likely that the probe is inherently weak. (Even large-insert clones sometimes deliver puzzlingly weak signals that are thus attributed to problems in the accessibility of their chromosomal targets.) For those equipped with advanced imaging systems incorporating a camera of high sensitivity, it is often possible to capture images from probe signals invisible to the naked eye. In the case of new commercial probes, the supplier should be contacted. Probes with larger targets often cross-hybridize to similar DNA sequences present on other chromosomes. It is important first to identify patterns of cross-hybridization by FISH onto normal chromosomes to

avoid misinterpreting the latter as rearrangements. Some resource centers, notably BAC/PAC Resources, helpfully list cross-hybridization patterns for some clones.

8. FISH analysis. The first aim of FISH is to characterize those rearrangements of interest present that resist analysis by G-banding. This inevitably requires both intuition and luck. Clearly, the need for the latter is reduced where G-banding is optimized. The most difficult rearrangements to resolve are unbalanced ones involving multiple chromosomes. Sometimes, however, originally reciprocal translocations appear unbalanced because of loss or additional rearrangement of one partner. In such cases, the identity of the "missing partner" might be often guessed at from among those chromosomes where one or more homologs appear to be missing. Having identified the chromosomal constituents of cryptic rearrangements, the next task is to reconcile FISH with G-banding data enabling breakpoint identification. In cases where chromosome segments are short or their banding patterns nondescript, this aim might be frustrated. The International System for Chromosome Nomenclature (ISCN) enables almost all rearrangements to be described with minimal ambiguity in most cases *(27)*.

9. Use of cytogenetic data. Having successfully completed cytogenetic analysis of a tumor cell line to the point of ISCN karyotyping, the question of what to do with the data arises. The first question to be addressed is identity: Has the cell line in question been karyotyped previously and, if so, does the observed karyotype correspond with that previously reported? In our experience, complete correspondence between cell line karyotypes is rare, even where their identity has been confirmed by DNA fingerprinting. First, among complex karyotypes, complete resolution might be unnecessary and is, indeed, rarely achieved. This leaves significant scope for uncertainty and differences in interpretation. ISCN karyotypes are inferior to karyogram images in this regard. Wherever possible, consult the original journal or reprint, as photocopies seldom permit reproduction of intermediate tones, which are the "devil in the detail" of G-banding. Second, a minority of cell lines might evolve karyotypically during culture in vitro. This instability could effect numerical or structural changes. Such a cell line is CCRF-CEM, derived from a patient with T-cell leukemia, which has spawned a multitude of subclones—all cytogenetically distinct *(12)*—and, sometimes following cross-contamination events, masquerading under aliases. Those wishing to compare their karyotypes with those derived at the DSMZ can consult either the DSMZ descriptive catalog *(28)* or website, which features an interactive database facilitating searches (www.dsmz.de/).

Acknowledgments

We wish to thank our colleagues, Maren Kaufmann—several of whose suggestions are silently incorporated in the foregoing protocols—for her expert technical work and Dr. Stefan Nagel for his critical reading of the manuscript.

References

1. Drexler, H. G. (2001) *The Leukemia–Lymphoma Cell Line FactsBook*, Academic, London, p. 2.
2. Heim, S. and Mitelman, F. (1995) *Cancer Cytogenetics*, 2nd ed., Wiley–Liss, New York, passim.
3. Reya, T., Morrison, S. J., Clarke, M. F., and Weissman, I. L. (2001) Stem cells, cancer, and cancer stem cells. *Nature* **414,** 105–111.
4. Drexler, H. G, MacLeod, R. A. F., and Uphoff, C. C. (1999) Leukemia cell lines: in vitro models for the study of Philadelphia chromosome-positive leukemia. *Leukemia Res.* **23,** 207–215.
5. Drexler, H. G., MacLeod, R. A. F., Borkhardt, A., and Janssen, J. W. G. (1995) Recurrent chromosomal translocations and fusion genes in leukemia–lymphoma cell lines. *Leukemia* **9,** 480–500.
6. Marini, P., MacLeod, R. A. F., Treuner, C., et al. (1999) SiMa, a new neuroblastoma cell line combining poor prognostic cytogenetic markers with high adrenergic differentiation. *Cancer Genet. Cytogenet.* **112,** 161–164.
7. Drexler, H. G., Matsuo, Y., and MacLeod, R. A. F. (2000) Continuous hematopoietic cell lines as model systems for leukemia–lymphoma research. *Leukemia Res.* **24,** 881–911.
8. Tosi, S., Giudici, G., Rambaldi, A., et al. (1999) Characterization of the human myeloid leukemia-derived cell line GF-D8 by multiplex fluorescence in situ hybridization, subtelomeric probes, and comparative genomic hybridization. *Genes Chromosomes Cancer* **24,** 213–221.
9. MacLeod, R. A. F., Spitzer, D., Bar-Am, I., et al. (2000) Karyotypic dissection of Hodgkin's disease cell lines reveals ectopic subtelomeres and ribosomal DNA at sites of multiple jumping translocations and genomic amplification. *Leukemia* **14,** 1803–1814.
10. Vanasse, G. J., Concannon, P., and Willerford, D. M. (1999) Regulated genomic instability and neoplasia in the lymphoid lineage. *Blood* **94,** 3997–4010.
11. Küppers, R. and Dalla-Favera, R. (2001) Mechanisms of chromosomal translocations in B cell lymphomas. *Oncogene* **20,** 5580–5594.
12. MacLeod, R. A. F., Dirks, W. G., Matsuo, Y., Kaufmann, M., Milch, H., and Drexler, H. G. (1999) Widespread intraspecies cross-contamination of human tumor cell lines arising at source. *Int. J. Cancer* **83,** 555–563.
13. Drexler, H. G., Dirks, W. G., and MacLeod, R. A. F. (1999) False human hematopoietic cell lines: cross-contaminations and misinterpretations. *Leukemia* **13,** 1601–1607.
14. Nelson-Rees, W. A., Daniels, D. W., and Flandermeyer, R. R. (1981) Cross-contamination of cells in culture. *Science* **212,** 446–452.
15. Markovic, O. and Markovic, N. (1998) Cell cross-contamination in cell cultures: the silent and neglected danger. *In Vitro Cell Dev. Biol. Anim.* **34,** 1–8.
16. MacLeod, R. A. F., Dirks, W. G., and Drexler, H. G. (2002) Persistent use of misidentified cell lines and its prevention. *Genes Chromosomes Cancer.* **33,** 103–105.

17. Stacey, G. N., Masters, J. R. W., Hay, R. J., Drexler, H. G., MacLeod, R. A. F., and Freshney, R. I. (2000) Cell contamination leads to inaccurate data: We must take action now. *Nature* **403,** 356.
18. Caspersson, T., Zech, L., and Johansson, C. (1970) Differential binding of alkylating fluorochromes in human chromosomes. *Exp. Cell Res.* **60,** 315–319.
19. Sumner, A. T., Evans, H. J., and Buckland, R. A. (1971) New technique for distinguishing between human chromosomes. *Nature New Biol.* **232,** 31–32.
20. Seabright, M. (1973) Improvement of trypsin method for banding chromosomes. *Lancet* **1,** 1249–1250.
21. Rowley, J. D. (1973) A new consistent chromosomal abnormality in chronic myelogenous leukaemia identified by quinacrine fluorescence and Giemsa staining. *Nature* **243,** 290–293.
22. Nowell, P. C. and Hungerford, D. A. (1960) A minute chromosome in human granulocytic leukemia. *Science* **132,** 1497.
23. Cremer, T., Lichter, P., Borden, J., Ward, D. C., and Manuelidis, L. (1988) Detection of chromosome aberrations in metaphase and interphase tumor cells by in situ hybridization using chromosome-specific library probes. *Hum. Genet.* **80,** 235–246.
24. Lichter, P., Cremer, T., Borden, J., Manuelidis, L., and Ward, D. C. (1988) Delineation of individual human chromosomes in metaphase and interphase cells by in situ suppression hybridization using recombinant DNA libraries. *Hum. Genet.* **80,** 224–234.
25. Lichter, P. (1997) Multicolor FISHing: what's the catch? *Trends Genet.* **12,** 475–479.
26. Chang, H., Blondal, J. A., Benchimol, S., Minden, M. D., and Messner, H. A. (1995) p53 mutations, c-myc and bcl-2 rearrangements in human non-Hodgkin's lymphoma cell lines. *Leukemia Lymphoma* **19,** 165–171.
27. ISCN (1985) An International System for Human Cytogenetic Nomenclature: report of the Standing Committee on Human Cytogenetic Nomenclature. *Birth Defects Orig. Artic. Ser.* **21,** 1–117.
28. Drexler, H. G., Dirks, W., MacLeod, R. A. F., Quentmeier, H., Steube, K. G., and Uphoff, C. C. (2001) *DSMZ Catalogue of Human and Animal Cell Lines* (8th ed.), DSMZ, Braunschweig, Germany.

6

Human and Mouse Hematopoietic Colony-Forming Cell Assays

Cindy L. Miller and Becky Lai

Summary

Hematopoietic stem cells present in small numbers in certain fetal organs during development and in adult bone marrow produce a heterogeneous pool of progenitors that can be detected in vitro using colony-forming cell (CFC) assays. Hematopoietic progenitor cells, when cultured in a semisolid methylcellulose-based medium that is supplemented with suitable growth factors, proliferate and differentiate to produce clonal clusters (colonies) of maturing cells. The CFCs are then classified and enumerated *in situ* by light microscopy. Protocols for the detection and enumeration of myeloid multipotential progenitors and committed progenitors of the erythroid, monocyte, and granulocyte lineages in samples from human peripheral blood, bone marrow, and cord blood as well as mouse fetal liver and bone marrow are described.

Key Words: Hematopoietic progenitors; bone marrow; peripheral blood; fetal liver; colony-forming cell assays; CFU-GEMM; BFU-E; CFU-E; CFU-GM; CFU pre-B.

1. Introduction

During fetal development and in the adult bone marrow, a small number of hematopoietic stem cells (HSCs) undergo self-renewal cell divisions and proliferate to produce a heterogeneous compartment of hematopoietic progenitors. Progressive proliferation and differentiation steps result in the production of large numbers of mature blood cells including T- and B-lymphoid cells, natural killer (NK) cells, dendritic cells, monocyte/macrophages, granulocytes, red blood cells, and platelets.

Numerous in vitro and in vivo assays have been developed to characterize and quantify hematopoietic cells at various stages of differentiation. The most definitive assays to detect HSCs with extensive potential for self-renewal,

From: *Methods in Molecular Biology, vol. 290: Basic Cell Culture Protocols, Third Edition*
Edited by: C. D. Helgason and C. L. Miller © Humana Press Inc., Totowa, NJ

proliferation, and multilineage differentiation involve the transplantation of test cells into host animals and detection of donor-derived hematopoietic cells weeks to months later. The limiting dilution competitive repopulating unit (CRU) assay is carried out using xenogeneic, immunocompromised recipients to detect human HSCs, and irradiated congenic strains to quantify mouse HSCs *(1–3)*. The in vitro long-term culture-initiating cell (LTC-IC) *(4,5)* and cobblestone area forming cell (CAFC) *(6,7)* assays quantify primitive cells capable of continuously producing myeloid cells for a minimum of 4–5 wk when cultured on a suitable feeder layer. Clonogenic assays have been developed to detect hematopoietic progenitors, termed colony-forming cells (CFCs), in vitro. Colony-forming unit-blast (CFU-blast) *(8,9)*, high proliferative potential-CFC (HPP-CFC) *(10)*, and CFU-granulocyte, erythroid, monocyte/macrophage, megakaryocyte (CFU-GEMM) are representative of progenitors with multilineage differentiation potential and limited self-renewal capacity. More mature hematopoietic CFCs that have no (or minimal) self-renewal capacity and are committed to mature into cells of one or two hematopoietic lineages include CFU-GM, CFU-M, CFU-G, burst-forming unit erythroid (BFU-E), CFU-E, CFU-megakaryocyte (CFU-Mk), and mouse CFU-pre-B-cell (CFU-pre-B). It is very important to emphasize that although colony assays can potentially detect cells that sustain short-term hematopoiesis in vivo, none are suitable for the quantification of HSCs.

Colony-forming cell assays are performed by placing hematopoietic cell suspensions into a semisolid matrix such as methylcellulose, collagen, agar, or fibrin clots supplemented with appropriate nutrients and growth factors. The semisolid medium allows individual progenitors to divide and differentiate to produce a discrete colony-containing mature progeny cells after a suitable culture period. The researcher then determines the CFC types and numbers based on the morphological features of the colony *in situ* using an inverted microscope. Alternatively, gelling agents such as collagen can be dehydrated, fixed, and treated with cytochemical or immunocytochemical stains *(11)*. For example, immunostaining with anti-CD41 antibody is used to definitively identify CFU-Mk *(12)*. The number of colonies obtained is linearly proportional to the CFC content in the input cell suspension provided that sufficiently low numbers of cells are plated, and the culture media and culture conditions are optimal.

Methylcellulose is a relatively inert polymer that forms a stable gel with good optical clarity at a final of concentration of 0.9 to 1.5%. Large batch-to-batch differences in the ability of various media components, including methylcellulose, fetal bovine serum (FBS), and bovine serum albumin (BSA), to support CFC growth make careful prescreening of multiple batches a requirement. Commercially available recombinant hematopoietic cytokines have largely replaced conditioned medium as the source of colony-stimulating factors (CSFs). Many of the cloned cytokines can stimulate hematopoietic cells in vitro at different stages of maturation. For example, G-CSF and interleukin (IL)-11 exert biological effects on primitive cells as well as lineage committed granulocyte and megakaryocyte progenitors, respectively. IL-3 and IL-6 support the proliferation of multipotential myeloid progenitors, whereas the action of cytokines including erythropoietin (Epo), M-CSF, and IL-5 is believed to be primarily on specific cell lineages. Stem cell factor (SCF) (also known as c-kit ligand and Steel factor) shows minimal activity when used as a single factor, but synergizes with other cytokines to promote proliferation and differentiation of most hematopoietic progenitor types. The reader is encouraged to consult the published literature for more information (i.e., **refs. *13–15***).

Colony-forming cell assays are useful tools for the study of the biological properties of hematopoietic progenitors, for the identification of stimulatory and inhibitory molecules that affect their growth, for preliminary assessment of hematopoiesis in transgenic or knockout mouse strains, and for drug-toxicity screening. These assays are also used to study and quantify progenitors in samples from patients with leukemia and myeloproliferative disorders. In cell processing laboratories, CFC analyses are routinely performed to evaluate the functional integrity of hematopoietic cells following cell manipulations such as CD34$^+$ cell enrichment, T-cell depletion, and cryopreservation.

This chapter describes methods for the preparation of red blood cell (RBC)-depleted and mononuclear cell suspensions from human samples and assays for detection of human CFU-E, BFU-E, CFU-GM, and CFU-GEMM in methylcellulose-based media. Methods for detection of mouse BFU-E, CFU-GM, CFU-GEMM, and CFU-pre-B in bone marrow (BM) and fetal liver (FL) samples are also presented.

2. Materials

2.1. Human CFC Assays

1. 2.6% Methylcellulose stock solution: 2.6% methylcellulose in Iscove's Modified Dulbecco's Medium (IMDM), 40 mL per bottle (MethoCult™; cat. no. 04100, StemCell Technologies Inc. [STI], Vancouver, Canada, www.stemcell.com). Store at –20°C.
2. Fetal bovine serum (FBS) for human CFC assays (cat. no. 06250, STI). Store in aliquots at –20°C.
3. 10% Bovine serum albumin (BSA) (cat. no. 09300, STI). Store in aliquots at –20°C.
4. 200 mM L-Glutamine stock solution: L-glutamine in phosphate-buffered saline (PBS) (cat. no. 07100, STI). Store in aliquots at –20°C.
5. 10^{-2} M 2-Mercaptoethanol (2-ME) stock solution: Prepare a 10^{-1} M solution by adding 0.1 mL 2-ME (cat. no. M7522, Sigma-Aldrich, www.sigmaaldrich.com) in a total of 14.3 mL PBS. Dilute 1/10 to prepare 10^{-2} M stock. Store in aliquots at –20°C for up to 6 mo.

6. Iscove's MDM (IMDM) (cat. no. 36150, STI) or suppliers, including Invitrogen Gibco, (www.invitrogen.com) and Sigma. Store at 2–8°C.

7. Recombinant human (rh) cytokine stock solutions (*see* **Note 1**): Cytokines are available from various suppliers (i.e., STI, R&D Systems [www.rndsystems. com], BioSource [www.biosource.com]). Reconstitute according to manufacturer's instructions. Prepare individual stock solutions at concentrations of 5 µg/mL stem cell factor (rhSCF), 1 µg/mL interleukin-3 (rhIL-3), 1 µg/mL granulocyte–macrophage colony stimulating factor (rhGM-CSF), and 300 U/mL erythropoietin (rhEpo) in IMDM with 0.1% BSA. Store in working aliquots at –20°C for up to 6 mo.

8. Complete human methylcellulose-based medium (human MC medium) for human CFU-E, BFU-E, CFU-GM, and CFU-GEMM (*see* **Notes 2** and **3**): To prepare 100 mL of human MC medium, first thaw a 40-mL bottle of 2.6% methylcellulose stock solution at room temperature or in the refrigerator and then add the following individual components directly to the bottle: 30 mL FBS (final 30%), 10 mL of 10% BSA (final 1%), 1 mL of 200 mM L-glutamine (final 2 mM), 1 mL of 10^{-2} M 2-ME (final 10^{-4} M), 1 mL each of the stock solutions of rhSCF (final 50 ng/mL), rhIL-3 (final 10 ng/mL), rhGM-CSF (final 10 ng/mL), rhEpo (final 3 U/mL), and 14 mL IMDM. The final MC concentration will be 1% in a final volume of 100 mL (*see* **Note 4**). Mix components thoroughly and aliquot into tubes for storage at –20°C (*see* **Note 5**). Stable for at least 1 yr.

9. 2% FBS in IMDM (IMDM/2% FBS). This solution is prepared by adding 2 mL FBS to 98 mL IMDM. Store in working aliquots at 2–8°C for up to 1 mo.

10. Ficoll-Paque™ Plus (Ficoll), density 1.077 g/mL (Amersham BioScience, [www.amersham.com] or cat. no. 07907/07957, STI). Store at room temperature.

11. Ammonium chloride solution (cat. no. 07800, STI). Store at –20°C and working aliquots for 1 wk at 2–8°C.

12. 0.4% Trypan blue dye (cat. no. 07050, STI) and 3% acetic acid (cat. no. 07060, STI) for viable and nucleated cell counts respectively.

13. Human BM and peripheral blood (PB) samples. Cells are collected using heparin as the anticoagulant following procedures and handling precautions approved by the institution.

2.2. Mouse CFC Assays

1. Complete mouse MC medium for BFU-E, CFU-GM, CFU-GEMM (mouse MC medium) (*see* **Note 3**) (cat. no. 03434, STI) containing 1% MC, 15% FBS, 1% BSA, 200 µg/mL transferrin, 10 µg/mL insulin, 2 mM L-glutamine, 10^{-4} M of 2-ME, 50 ng/mL recombinant mouse (rm)SCF, 10 ng/mL rmIL-3, 10 ng/mL rhIL-6, and 3 U/mL erythropoietin. Aliquot into tubes (*see* **Note 5**) and store at –20°C for up to 2 yr.

2. Complete mouse MC medium for CFU pre-B assays (mouse pre-B MC medium) (*see* **Note 3**) (cat. no. 03630, STI) containing 1% MC, 30% FBS, 2 mM L-glutamine, 10^{-4} M 2-ME, and 10 ng/mL rhIL-7 (*see* **Note 6**). Aliquot into tubes (*see* **Note 5**) and store at –20°C for up to 1 yr.

3. IMDM/2% FBS: *see* **Subheading 2.1.**, **item 9**.
4. Day 14.5 postcoitum (pc) pregnant mice. Set up cages containing female and male mice in the afternoon, 15 d prior to the desired date of use. The following morning, check for copulation plugs (considered d 0.5 pc) and remove the mated females to a separate cage.
5. C57Bl/6 mice: typically 6–12 wk old, and of either sex (*see* **Note 7**).

2.3. Equipment and Culture Supplies for Human and Mouse CFC Assays

1. Micropipettors and 20-μL, 200-μL, and 1000-μL sterile tips.
2. Culture supplies: 1-, 5-, and 10-mL sterile pipets; 6-, 15-, and 50-mL sterile tubes; 100-mm Petri dishes or square bacterial dishes; 3-mL Luer-lock syringes.
3. 35-mm Low-adherence Petri culture dishes (cat. no. 27100/27150, STI) (*see* **Note 8**).
4. 60-mm Gridded dishes (cat. no. 27120/27121, STI).
5. 16-Gage blunt-end needles (cat. no. 28110, STI)
6. Automated cell counter or Neubauer hemocytometer.
7. Biosafety cabinet approved for Level II handling of biological material.
8. Incubator set at 37°C with 5% CO_2 in air and >95% humidity (*see* **Note 9**).
9. Inverted microscope equipped with ×10 or ×12.5 eyepiece objectives; ×2, ×4, and ×10 planar objectives; a moveable stage holder for 60-mm dishes.
10. Sterile sets of fine, sharp scissors and forceps for animal dissection. Sterilize by autoclaving for 40 min at 121°C.

3. Methods

The following sections describe methods for (1) preparation of human and mouse hematopoietic cell samples, (2) setup of CFC assays, and (3) identification and enumeration of CFCs. Processing of the cell sample is often required to deplete RBCs that can obscure colonies and make colony counts inaccurate, to deplete accessory cells (i.e., macrophages) that produce endogenous factors in cultures that can potentially inhibit or promote CFC growth, and to yield sufficient colony numbers for accurate assessment in samples where the CFC frequency is expected to be very low. Cell separation technologies and fluorescent-activated cell sorting (FACS) methodologies routinely used to enrich hematopoietic progenitors (i.e., enrich human CD34+ cells or mouse Sca-1+ cells) are beyond the scope of this chapter and will not be presented. General considerations for performing CFC assays (*see* **Note 10**) and procedures to isolate individual colonies or cells from the entire culture for special applications are discussed (*see* **Note 11**). All cell culture procedures should be performed using sterile technique in a certified biosafety cabinet, and universal procedures for handling potentially biohazardous materials should be followed.

3.1. Human CFC Assays

For the varied reasons discussed earlier, most human samples require processing prior to plating in CFC assays. For example, Ficoll density separation of PB, cord blood (CB), and BM samples enriches mononuclear cells (and thus CFCs) by removing RBCs and other nonprogenitor cell types. If total nucleated cell suspensions are desired, RBCs are lysed using ammonium chloride treatment. Although this works well for BM samples, it does not work for CB because these samples contain large numbers of nucleated RBC precursors that are not lysed by this treatment.

3.1.1. Human Mononuclear Cell Isolation

1. Measure and record PB, BM, or CB start sample volume.
2. Dilute sample with an equal volume of IMDM/2%FBS and mix well by pipetting up and down four to five times or vortexing.
3. Add 15 mL of Ficoll per 50-mL conical tube or 3 mL per 14-mL tube for smaller sample volumes.
4. Slowly layer 30 mL of the cell suspension per 50-mL tube or 6 mL of the cell suspension per 14-mL tube onto the surface of the Ficoll by resting the tip of the pipet against the side of the tube. It is important to avoid mixing the cell suspension into the Ficoll density medium, as poor cell recoveries could result (*see* **Note 12**). Centrifuge at room temperature for 30 min at 400*g* with the brake "off."
5. Using a sterile pipet, carefully remove the cells from the interface between the plasma/medium layer and the Ficoll. Transfer cells to a 15-mL tube and dilute cell suspension with a minimum of 2 vol (>1:2 ratio) of IMDM/2% FBS. Mix well by pipetting up and down four to five times and then centrifuge for 10 min at 300*g* with the brake "on."
6. Carefully decant off supernatant, leaving approx 0.5 mL of medium on the cell pellet. Vortex to resuspend the cells, fill tube with IMDM/2% FBS, and mix well by vortexing or by pipetting up and down four to five times. Centrifuge tube(s) for 10 min at 300*g* with the brake "on."
7. Carefully decant off supernatant. Add 1–3 mL of IMDM/2% FBS to the cell pellet and make a single-cell suspension by vortexing or by pipetting up and down four to five times.
8. Measure and record processed sample volume. Perform a nucleated cell count using an automated cell counter or manually using 3% acetic acid and a Neubauer chamber. (For detailed instructions on performing manual cell counts, *see* Chapter 1). A dilution of 1:20 to 1:50 is usually suitable. Calculate and record the cell concentration. The following yields of mononuclear cells from the start samples can be expected: $(1–2) \times 10^6$ per mL PB, $(1–2.5) \times 10^6$ per mL CB, and $(0.5–1) \times 10^7$ per mL BM.

3.1.2. Ammonium Chloride Treatment of Human BM Samples

1. Measure and record volume of start BM sample to be processed. Perform a cell count if estimation of cell recovery is required. Add a 4:1 v/v ratio of ammonium chloride solution (i.e., 2 mL heparinized BM and 8 mL ammonium chloride solution).

2. Mix well by inverting tube three to four times or by vortexing gently. Place tube on ice for a total of 10 min with mixing as in step 1 after approx 5 min of incubation. The majority of RBCs should now be lysed. Fill tube with IMDM/2% FBS and centrifuge for 10 min at 300g with the brake "on."

3. Carefully decant off supernatant, leaving approx 0.5 mL of medium on the cell pellet. Vortex to resuspend the cell pellet, fill tube with IMDM/2% FBS, and mix well by vortexing or by pipetting up and down four to five times. Centrifuge tube for 10 min at 300g with the brake "on." Repeat this wash step once more.

4. Carefully decant off supernatant. Add 2 mL of IMDM/2% FBS and make a single-cell suspension by vortexing or by pipetting up and down four to five times. Measure and record processed sample volume. Perform a nucleated cell count and calculate the cell concentration and cell recovery. Percent cell recovery is calculated using the following formula: (Cell concentration × Volume of processed sample) divided by (Cell concentration × Volume of start sample) × 100. A recovery of 60–80% of the nucleated cells from the start sample of normal BM can be expected.

3.1.3. Setup of Human CFC Assays

1. To identify assays, label lids of 35-mm Petri dishes at the edge using a *permanent* fine felt marker.

2. Thaw aliquots of human MC medium (*see* **Note 3**) at room temperature or under refrigeration. Vortex tubes to ensure that all components are thoroughly mixed.

3. Dilute hematopoietic cells to 10 times the final concentration required in IMDM/2% FBS (*see* **Table 1** and **Note 13**). For example, to achieve a final concentration of 1×10^5 cells per 35-mm dish, dilute cells to 1×10^6 cells/mL in IMDM/2% FBS.

4. Add 0.3 mL of cells to 3 mL of complete human MC medium for duplicate cultures or 0.4 mL of cells to 4 mL of MC medium for triplicate cultures.

5. Vortex tubes and let stand for 2–5 min to allow bubbles to rise.

6. Using a 3-mL Luer-lock syringe and 16-G blunt-end needle, draw up approx 1 mL and expel completely to remove most of the air from syringe. Draw up approx 3 mL and carefully dispense 1.1 mL into each 35-mm dish. Distribute methylcellulose evenly by *gently* tilting and rotating each dish. Avoid getting MC on lids or up the sides and break any large bubbles using a dry sterile micropipettor tip.

7. Place the two labeled 35-mm Petri dishes into a 100-mm dish. Add a third 35-mm dish (without lid) containing 3–4 mL of sterile water to help maintain a high humidity over the culture period. Larger Petri dishes or square bacterial culture dishes can be used as outer dishes when three or more replicate cultures are setup.

8. Place cultures in incubator maintained at 37°C, 5% CO_2 in air and > 95% humidity (*see* **Note 9**) for 14–16 d. Evaluate cultures microscopically after 7–10 d of incubation to check for possible contamination, dehydration, and adequate colony numbers (*see* **Note 10**). If isolation of individual colonies or harvesting of total cells from the culture is desired for certain applications, then cultures should be incubated for shorter times (*see* **Note 11**). If assays cannot be counted at d 14–

Table 1
Recommended Input Cell Numbers for Human CFC Assays

Cell source	Recommended input cell concentration in 1.1 mL per 35-mm dish[a]
Bone marrow—ammonium chloride treated	5×10^4
	$(2 \times 10^4 – 1 \times 10^5)$
Bone marrow—mononuclear cells[b]	2×10^4
	$(1 \times 10^4 – 5 \times 10^4)$
Cord blood—mononuclear cells[b]	1×10^4
	$(5 \times 10^3 – 2 \times 10^4)$
Peripheral blood—mononuclear cells[b]	2×10^5
	$(1 \times 10^5 – 4 \times 10^5)$
CD34[+]-enriched cell suspensions (BM, CB, MPB)[c]	1000
	$(500 – 2 \times 10^3)$

[a]The recommended input cell concentration should yield 30–120 colonies per culture using normal samples from the various tissues. If the progenitor frequency cannot be estimated (i.e., samples from leukemic and drug-treated patients), two or three input cell doses within the range shown in parenthesis should be set up.
[b]Mononuclear cells isolated using Ficoll-Paque.
[c]*See* **Note 13**.

16, transfer cultures to an incubator maintained at 33°C, 5% CO_2 in air and >95% humidity and count within 3–4 d (*see* **Note 10**).

3.1.4. Identifying and Counting Human CFCs

1. To prepare a reusable gridded template, draw a centered "+" on the bottom of a 60-mm gridded dish and place a mark on these lines corresponding to the outer edges of a 35-mm dish using a fine permanent marker.
2. Keeping cultures as level as possible, center the 35-mm dish within the gridded 60-mm dish and place on the movable stage of an inverted microscope. Scan the entire dish on low power (×2 objective) by moving the stage vertically up and down and then laterally (helps minimize "motion nausea"). Note the relative distribution of the colonies.
3. Count CFU-E on the entire plate using high power (×4 objective). BFU-E, CFU-GM, and CFU-GEMM are then counted using a lower power (×2 objective). A higher power (×4 or ×10 objective) is used to identify cell types within a colony for the purpose of confirming colony classification. It is important to continuously refocus to identify colonies that are present in different planes and at the outer edges of the cultures. Observe that some CFU-GM colonies have two or more focal points. For descriptions of human CFC, refer to **Fig. 1** and **Table 2**.

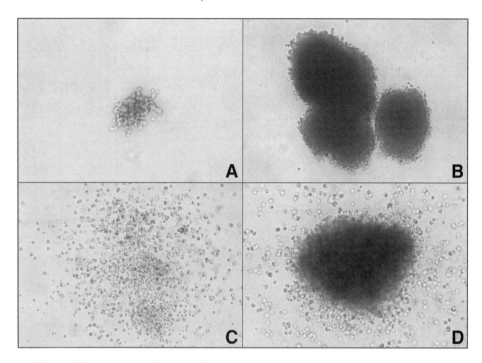

Fig. 1. Micrographs of human hematopoietic colonies taken after 14 d of culture in MethoCult™ H4434: **(A)** CFU-E (original at ×125 magnification) containing ≤200 small hemoglobinized erythroblasts that have a reddish color when viewed microscopically; **(B)** BFU-E (original at approx ×125 magnification) containing 3 clusters and ≥ 200 cells. Mature BFU-E with limited proliferation capacity contain three to eight erythroid clusters (or equivalent cell numbers), whereas the immature BFU-E form larger colonies **(C)**. This CFU-GM (original at ×125 magnification) contains three distinct focal clusters of cells. Note morphological similarity among cells within each cluster **(D)**. CFU-GEMM (original ×25 magnification) containing erythroid cells surrounded by 20 or more granulocyte and monocyte lineage cells.

3.2. Mouse CFC Assays

3.2.1. Isolation of Mouse BM Cells

1. Sacrifice mice according to protocols approved by institution. Position mouse on its back and wet fur thoroughly with 70% isopropyl alcohol to decrease the possibility of contaminating cell preparations.
2. Using nonsterile scissors, cut a slit in the fur just below the rib cage, being careful not to cut through the peritoneal membrane.
3. Firmly grasp skin and peel back to expose hind limbs.
4. Using sterile sharp dissecting scissors, cut the knee joint in the center. Trim away ligaments and excess tissue from both the femur and tibia of each leg.

Table 2
Description of Human and Mouse CFCs and Cytokine Combinations Used for CFC Assays

CFC class	Cytokine(s)[a] (optional)[b]	CFC description	
		Human	Mouse
CFU-E	Epo	Produces one or two clusters of erythroblasts, ≤ 200 cells	Produces one or two clusters of erythroblasts with minimum of 8 (8–32) cells; detectable in 2- to 3-d cultures
BFU-E	IL-3 + Epo + SCF	Produces three or more clusters of erythroblasts and >200 total cells.	Produces 30 or more erythroid cells
CFU-GM[c]	IL-3 + GM-CSF + SCF (G-CSF, M-CSF, IL-6, IL-5)	Produces 20 or more granulocyte (CFU-G), monocyte (CFU-M) or granulocyte and monocyte (CFU-GM) cells.	Produces 30 or more granulocyte (CFU-G), monocyte (CFU-M), or granulocyte and monocyte (CFU-GM) cells
CFU-GEMM	Cytokines to support each lineage. IL-3 + GM-CSF + SCF + Epo (G-CSF, M-CSF, IL-6, Tpo)	Produces a minimum of 20 cells (usually larger) and contains erythroid cells as well as granulocytic, monocytic, and megakaryocytic lineages cells.	Produces a minimum of 30 cells (usually larger) and erythroid, granulocytic, monocytic, and megakaryocytic lineages cells
CFU-Mk	Tpo + IL-3 +IL-6 (SCF, IL-11)	Produces three or more megakaryocytes.	Produces three or more megakaryocytes.
CFU-pre-B	IL-7 (SCF)	Not detected in methylcellulose-based medium	Produces 30 or more B-lymphocyte lineage cells

[a]Minimal cytokine combination used for detection of these progenitors in vitro.
[b]Cytokines indicated in parenthesis can be added (or substituted in some applications) as desired.
[c]Total CFU-GM= CFU-GM + CFU-G + CFU-M.

5. Grasp the femur with forceps and cut near the hip joint. Similarly, grasp the tibia with forceps and cut it near the ankle joint. Use of sharp scissors will help prevent splitting of the bone.
6. Transfer the bones to a sterile Petri dish or to a tube containing IMDM/2% FBS and place on ice if cells cannot be isolated within approx 1 h.
7. Using sharp sterile scissors, trim the ends of the long bones to expose the interior marrow shaft, ensuring that there is no tissue blocking the openings.
8. Using a syringe and 21-G needle (smaller 22-G or 23-G needles can be used for tibia) containing 1–3 mL of medium, insert the bevel of needle into the marrow shaft and flush BM into a tube. The bone will appear white if all marrow is removed. Repeat flush step if required. The same medium can be used to isolate BM from the bones (femora and tibias) of one to three animals.
9. To make a single-cell suspension, disrupt cell aggregates by drawing suspension up and down three to four times using a syringe and 21-G needle. Perform a nucleated cell count using 3% acetic acid (*see* **Subheading 3.1.1.**). If the cell concentration is approx 10^7 per milliliter or greater, it is not necessary to wash or concentrate the cells before use. The expected cell yields are approx 0.8×10^6 per tibia and $(1.2–1.5) \times 10^6$ per femur ($[3–5] \times 10^7$ cells per four bones). Keep the tube of cells with medium on ice until use and setup CFC assays as soon as possible (*see* **Note 10**).

3.2.2. Isolation of Day 14.5 pc Fetal Liver Cells

1. Sacrifice timed-pregnant female mice according to protocols approved by institution. Wet fur with 70% isopropyl alcohol and make an approx 1-cm abdominal incision using nonsterile scissors, being careful not to cut peritoneal membrane. Peel back the pelt.
2. Using sterile scissors, make an incision through the abdominal membrane. Remove the intact uteri containing the fetuses by cutting at both the top and base of each uterus. Place them into a 50-mL tube containing sterile medium or PBS to remove any maternal blood.
3. In a biosafety cabinet, place the uteri into a sterile 100-mm dish (no medium). Dissect the fetuses away from the yolk sac, amnion, and placenta. Transfer the embryos to a new dish containing medium and rinse thoroughly to remove any blood.
4. Transfer individual fetus to another sterile dish and position so dark red FL is facing upward. Use fine forceps to tease the FL free from the surrounding tissue being careful to avoid the heart.
5. Combine FLs (approximately three to six) in 2 mL of IMDM/2% FBS in a 14-mL tube. Disrupt FLs using 5 mL pipet or a syringe/16-G blunt-end needle. Add an additional 1–2 mL of medium and disrupt small cell aggregates using a syringe/ 21-G needle (*see* **Note 14**).
6. Let tube stand for 3 min to allow tissue fragments to settle and then transfer supernatant cell suspension to another tube.

7. To wash cells, fill tube with IMDM/2% FBS, mix by pipetting up and down four to five times and centrifuge for 7 min at 400*g* at room temperature.

8. Carefully decant supernatant, leaving approx 0.5 mL of medium on the cell pellet. Resuspend cells by vortexing, add 2–3 mL medium, and pipet up and down four to five times to achieve a single-cell suspension; then record the volume. Perform a nucleated cell count using 3% acetic acid (*see* **Subheading 3.1.1.**). An average yield of 1×10^7 nucleated cells per FL can be expected.

3.2.3. Setup of Mouse CFC Assays

The methods for setting up mouse BFU-E, CFU-GM, and CFU-GEMM, and mouse CFU-pre-B assays are similar to those outlined for human CFC (*see* **Subheading 3.1.3.**). The differences are outlined below.

3.2.3.1. Mouse BFU-E, CFU-GM, and CFU-GEMM Assay

1. To identify mouse assays, label lids of 35-mm Petri dishes at the edge using a *permanent* fine felt marker.

2. Thaw aliquots of complete mouse MC medium (*see* **Note 3**) at room temperature or under refrigeration. Vortex tubes to ensure that all components are thoroughly mixed.

3. Dilute mouse BM or FL cells to 10 times the final concentration required in IMDM/2% FBS (*see* **Table 3**). For example, to achieve a final concentration of 2×10^4 cells per 35-mm dish, dilute cells to 2×10^5 cells/mL in IMDM/2% FBS.

4. Add 0.3 mL of cells to 3 mL of mouse MC medium for duplicate cultures or 0.4 mL of cells to 4 mL of MC medium for triplicate cultures

5. Setup cultures as described in **Subheading 3.1.3.**, **steps 5–7** and incubate for 12 d. Evaluate cultures microscopically after 6–7 d of incubation to check for possible contamination, dehydration, and adequate colony numbers (*see* **Note 10**). If isolation of individual colonies or harvesting of total cells from the culture is desired for certain applications, then cultures should be incubated for shorter times (*see* **Note 11**). If assays cannot be counted at d 12, transfer cultures to an incubator maintained at 33°C, 5% CO_2 in air and > 95% humidity and count within 2–3 d (*see* **Note 10**).

3.2.3.2. Mouse CFU-Pre-B Assay

1. To identify mouse CFU-pre-B assays, label lids of 35-mm Petri dishes at the edge using a *permanent* fine felt marker.

2. Thaw aliquots of mouse pre-B MC medium (*see* **Note 3**) at room temperature or under refrigeration. Vortex tubes to ensure that all components are thoroughly mixed.

3. Dilute mouse BM or FL cells to 10 times the final concentration required in IMDM/2% FBS (*see* **Table 3**). For example, to achieve a final concentration of 5×10^4 cells per 35-mm dish, dilute cells to 5×10^5 cells/mL in IMDM/2% FBS.

4. Add 0.3 mL of cells to 3 mL of mouse pre-B MC medium for duplicate cultures or 0.4 mL of cells to 4 mL of MC medium for triplicate cultures

Table 3
Recommended Input Cell Numbers for Mouse CFC Assays and CFC Frequencies in Mouse BM and D 14.5 FL

Tissue	Input cell doses per culture[a] (1.1 mL per 35-mm dish)		CFC per 10^5 cells[b]			
	Myeloid CFC	CFU-pre-B	BFU-E	CFU-GM	CFU-GEMM	CFU-pre-B
Bone marrow	2×10^4 ([1–5] $\times 10^5$)	5×10^4 ([0.5–2] $\times 10^5$)	40 ± 15	320 ± 80	15 ± 5	230 ± 30
D 14.5 pc fetal liver	2×10^4 ([1–5] $\times 10^4$)	1×10^5 ([1–2] $\times 10^5$)	45 ± 15	275 ± 50	15 ± 10	5 ± 2

[a]If progenitor frequency cannot be estimated, two or three input cell doses within the suggested range (in parentheses) should be set up (see **Notes 7** and **10**).

[b]Values represent the mean ± 1 standard deviation for CFC numbers in BM or FL samples from C57Bl/6 mouse strains.

83

5. Setup cultures as described in **Subheading 3.1.3.**, **steps 5–7** and incubate for 7 d. Evaluate cultures microscopically after 5 d of incubation to check for possible contamination, dehydration, and adequate colony numbers (*see* **Note 10**). If isolation of individual colonies or harvesting of total cells from the culture is desired for certain applications, then cultures should be incubated for shorter times (*see* **Note 11**). If assays cannot be counted at d 7, transfer cultures to an incubator maintained at 33°C, 5% CO_2 in air and > 95% humidity and count within 2–3 d (*see* **Note 10**).

3.2.4. Identifying and Counting Mouse CFCs

3.2.4.1. IDENTIFYING AND COUNTING MOUSE BFU-E, CFU-GM, AND CFU-GEMM

1. Perform **steps 1** and **2** outlined in **Subheading 3.1.4.** (*see* **Subheading 3.1.4.**, **step 3** for general counting information, as well as **Table 2** and **Fig. 2**).
2. Count total colonies (colonies containing 30 cells) in the entire dish on low power (×2 objective). Use a higher magnification as required to confirm colony size. Then, count BFU-E and CFU-GEMM in the entire dish using the ×4 objective. CFU-GM numbers are determined as follows: Total CFU-GM = Total colonies minus (BFU-E+CFU-GEMM). Alternatively, CFCs can be identified and counted in the entire dish using the ×4 objective.

3.2.4.2. IDENTIFYING AND COUNTING MOUSE CFU-PRE-B

1. Perform **steps 1** and **2** outlined in **Subheading 3.1.4.** (*see* **Subheading 3.1.4.**, **step 3** for general counting information, as well as **Table 2** and **Fig. 2**).
2. Count total CFU-pre-B (colonies containing 30 cells) in the entire dish using the ×4 objective. It is important to continually refocus to identify small CFU-pre-B colonies.

4. Notes

1. Some of the human cytokines such as IL-6, G-CSF, IL-7, and Tpo show species crossreactivity and can be used for mouse CFC assays. Other cytokines (i.e., IL-3, GM-CSF, and SCF) are more species-specific and, therefore, cytokines of the same species as the cells being analyzed should be used.
2. If the researcher chooses to prepare methylcellulose-based medium using components from suppliers other than those listed, several factors must be taken into consideration. There is large variability among raw materials (i.e., methylcellulose powder, FBS, and BSA) from different suppliers and from one batch to another for their ability to support the growth of CFCs. As such, samples from several batches of each component should be obtained and compared for their ability to support growth of the maximal number of colonies. The selected components should then be combined and retested. Once components have been selected, sufficient amounts to last several years should be purchased because component screening is very time-consuming and labor-intensive.

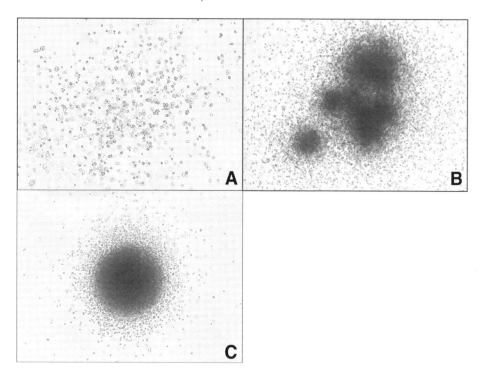

Fig. 2. Micrographs of mouse hematopoietic colonies taken after 12 d in MethoCult™ M3434 (**A,B**) and 7 d in MethoCult M3630 (**C**). (**A**) BFU-E (original at ×125 magnification), diffuse colony containing multiple clusters of small erythroid cells, no reddish color is seen because this MC medium does not support visible hemoglobinization. (**B**) Large CFU-GM colony (original at ×50 magnification) containing four large cell clusters. Monocyte cells tend to be larger than granulocyte cells. CFU-GEMM (not shown) are typically large colonies and erythroid, granulocyte, and monocyte cells should be clearly identifiable. Clusters of large irregular-shaped megakaryocyte cells (2–10 cells) are often present. (**C**) CFU-pre-B colonies (colony shown, original at ×125 magnification) vary in size and contain from 30 to several thousand cells per colony. Most colonies contain very small uniformly shaped cells.

3. Special handling procedures are required when working with methylcellulose (MC)-based media. The freezing process causes focal areas where the MC becomes more concentrated and "lumps" can form if the media is thawed rapidly (i.e., at 37°C). Because of the unique properties of the MC solution, lower temperatures are required to dissolve these lumps. Therefore, MC-based media should always be thawed at room temperature (this requires approx 4 h) or at 2–8°C (i.e., overnight in the refrigerator). If accidentally thawed at 37°C, place on ice or in a refrigerator for 1–2 h to dissolve the lumps.

4. The advantage of using base MC medium and adding individual components is that, if desired, the formulation can be modified (substitute or add other cytokines, add drugs, etc.) by adding the desired components and then adjusting the required volume of IMDM to achieve a final volume of 100 mL. If no changes to the basic formulation are desired, prepared complete human MC medium can be purchased from STI (MethoCult, GF H4434).

5. The MC solutions are very viscous and syringes and large bore needles (i.e., 16-gage) should be used for accurate aliquoting and dispensing (use of blunt-end needles also prevents needle prick injuries). To aliquot bottles of MC, mix well by shaking vigorously for 30–60 s and let stand for 2–5 min to allow bubbles to rise. Immerse the needle end just below the surface of the MC and slowly draw up the medium. Expel the medium back into the bottle to remove the air bubbles present in the syringe. Repeat twice more before drawing up the final volume to be dispensed plus an extra volume (*example:* if dispensing 3 mL, then draw up to the 4-mL mark and dispense from the 4-mL mark to the 1-mL mark for an accurate 3-mL volume). Dispense the MC medium, cap the tubes tightly, and store at –20°C.

6. Addition of rmSCF at 5–20 ng/mL to rhIL-7 containing MC medium could increase the numbers of CFU-pre-B detected. However, the addition of rmSCF also promotes myeloid growth within the cultures and this phenomenon is increased when higher input cell numbers are used because of endogenous cytokine production. Preliminary experiments using two to three different cell densities (i.e., 0.5×10^5, 1×10^5, and 2×10^5 per culture) with 10 ng/mL rhIL-7 in the presence and absence of rmSCF can be used to establish optimal conditions.

7. Assay conditions were established using 6- to 12-wk-old C57Bl/6 mice. Other laboratory mouse strains (Balb/c, DBA, and C3H) have similar progenitor frequencies and the indicated cell plating concentrations (*see* **Table 3**) are appropriate. If CFC frequency cannot be estimated (i.e., other mouse strains, possible effects of age on CFC frequency, transgenic and knockout mice where hematopoiesis might be perturbed, drug or cytokine treated mice), assays should be setup using two to three cell doses that vary by twofold.

8. It is important to use Petri culture dishes that have been screened for low adherence because excessive cell adherence can inhibit colony growth and make it difficult to distinguish individual colonies.

9. It is important to routinely monitor the temperature, CO_2, and humidity levels, as well as to regularly clean the incubator. Small chamber incubators (i.e., approx 12 ft^3) with a water pan placed in the bottom of the chamber give more uniform temperature and humidity than the large-chamber water-jacketed incubators. A small amount of copper sulfate added into the water pan inhibits bacterial and fungal growth. The temperature and CO_2 levels should be monitored independently from incubator gages using in-chamber thermometers and gas monitors (i.e., Fyrite CO_2 device), respectively.

10. General considerations for attaining accurate and reproducible results when performing hematopoietic CFC assays.

a. Cell preparations. It is advisable to set up assays using freshly isolated cells. When this is not feasible, it is important to perform preliminary experiments to establish optimal assay conditions, document all cell processing information, and include appropriate controls in each experiment. For example, human CFC assays can be done using cryopreserved cells and samples stored for 24 h. Mouse BFU-E, CFU-GM, and CFU-GEMM determinations where intact bones or cell suspensions are stored overnight in IMDM/10% FBS are also possible. Mouse CFU-pre-B analysis should be done as soon as possible, as we have noted decreased numbers in samples that were stored for 6–8 h prior to setting up the assay.

b. Input cell concentrations and colony numbers. Sufficient cells should be plated to yield approx 30–120 colonies per 35-mm dish (1.1 mL culture). Too many colonies (overplating) causes inaccuracies by inhibition of progenitor proliferation resulting from depletion of essential nutrients and accumulation of toxic cellular metabolic products and counting errors because of difficulty in identifying individual colonies. Too few colonies might not yield statistically accurate data. The accuracy can be increased by setting up more replicates or by enriching progenitor numbers in the input cell sample.

c. Culture conditions. It is important to maintain correct incubation conditions (*see* **Note 9**). Scanning of dishes midway during the incubation period is important to identify cultures that might need to be discarded and set up again because of contamination (visible fungal or bacterial growth giving medium a cloudy appearance), dehydration (decreased volume and irregular appearance), or inadequate colony numbers (overplating or too few colonies). If cultures cannot be counted at the end of the appropriate time, incubation at a lower temperature (i.e., 33°C) and high humidity will maintain colony morphology. However, colonies should be counted as soon as possible.

d. Colony enumeration. Practice is required to gain competence in CFC identification and enumeration. Recounting the same dishes on consecutive days and comparative counting with co-workers (same dishes) and with researchers at other institutions (CFC assays set up with the same cryopreserved cell suspension or mouse strain and culture conditions) is recommended. The *Atlas of Human Hematopoietic Colonies* is available from STI to assist in human CFC identification.

11. For certain applications such as cytogenetic analysis, DNA, RNA, and protein analyses, and replating experiments to detect progenitor self-renewal capacity, individual colonies or cells from the entire culture can be isolated. The cultures are usually incubated for shorter time periods to ensure high viability of cells within the colonies (approx 9–12 d for human cells, approx 5–8 d for mouse cells). A well-isolated colony is identified using an inverted microscope (can be placed within a biosafety cabinet if desired). Individual colonies are "plucked" in the smallest possible volume using a micropipettor and 200-μL pipet tips. The colony is placed in sterile 0.5- or 1.5-mL microtubes containing the appropriate wash medium. Entire MC cultures are harvested by adding 1–2 mL of medium (i.e., PBS or IMDM/2% FBS) to the dish and then gently mixing using a

pipettor and 1000-μL tips. The rinsing step is repeated several times and all washes are combined into a 15-mL conical tube for a single dish or into a 50-mL conical tube when two to three dishes are harvested. Sufficient wash medium is added to microtubes or tubes to dilute MC by 5- to 10-fold. Centrifuge tubes for 10 min at 300g and microtubes for 3–5 min at 300g. Cells are washed at least once more before use.

12. To provide a "cushion" on the Ficoll surface, place 3–4 mL of IMDM/2% FBS into a tube, pipet vigorously, and lift pipet tip above the liquid surface when expelling medium to create a "froth." Layer approx 0.5 cm of this "froth" onto the surface of the Ficoll prior to slowly adding the cell suspension as described in **Subheading 3.1.1., step 4**.

13. Appropriate input cell numbers for human CFC assays can be estimated if the CD34$^+$ cell content of the sample is known (i.e., by anti-CD34 antibody immunostaining and FACS analysis). Approximately 10 to 20% of the CD34$^+$ cell population are BFU-E, CFU-GM, and CFU-GEMM. For example, for a BM sample containing 1% CD34$^+$ cells, a setup of 5×10^4 cells per 35-mm dish should yield 50–100 colonies.

14. Alternatively, FL cells can be isolated by placing the livers into a 70-μm nylon mesh cell strainer (placed inside a sterile 35-mm Petri dish containing 2 mL IMDM/2% FBS). Mince the tissue with scissors, being careful not to cut through the membrane and then gently press the tissue through the membrane using a syringe plunger. Rinse the nylon membrane thoroughly with medium to ensure maximal cell recovery. Transfer cells to an appropriate size tube and continue with the remainder of the steps described.

References

1. Szilvassy, S. J., Humphries, R. K., Lansdorp, P. M., Eaves, A. C., and Eaves, C. J. (1990) Quantitative assay for totipotent reconstituting hematopoietic stem cells by a competitive repopulation strategy. *Proc. Natl. Acad. Sci. USA* **87,** 8736–8740.
2. Conneally, E., Cashman, J., Petzer, A., and Eaves, C. (1997) Expansion in vitro of transplantable human cord blood stem cells demonstrated using a quantitative assay of their lympho-myeloid repopulating activity in nonobese diabetic-scid/scid mice. *Proc. Natl. Acad. Sci. USA* **94,** 9836–9841.
3. Szilvassy, S. J., Nicolini, F. E., Eaves, C. J., and Miller, C. L. (2002) Quantitation of murine and human hematopoietic stem cells by limiting dilution analysis in competitively repopulated hosts, in *Hematopoietic Stem Cell Protocols* (Klug, C. A. and Jordan, C. T., eds.), Humana, Totowa, NJ, pp. 167–187.
4. Sutherland, H. J., Lansdorp, P. M., Henkelman, D. H., Eaves, A. C., and Eaves, C. J. (1990) Functional characterization of individual human hematopoietic stem cells cultured at limiting dilution on supportive marrow stromal layers. *Proc. Natl. Acad. Sci. USA* **87,** 3584–3588.
5. Miller, C. L. and Eaves, C. J. (2002) Long-term culture-initiating cell assays for human and murine cells, in *Hematopoietic Stem Cell Protocols* (Klug, C. A. and Jordan, C. T., eds.), Humana, Totowa. NJ, pp. 123–141.

6. Ploemacher, R. E., van der Sluijs, J. P., van Beurden, C. A., Baert, M. R., and Chan, P. L. (1991) Use of a limiting-dilution type long-term cultures in frequency analysis of marrow-repopulating and spleen colony-forming hematopoietic stem cells in the mouse. *Blood* **78,** 2527–2533.

7. de Haan, G. and Ploemacher, R. (2002) The cobble-area-forming cell assay, in *Hematopoietic Stem Cell Protocols* (Klug, C. A. and Jordan, C. T., eds.), Humana, Totowa, NJ, pp. 143–151.

8. Nakahata, T. and Ogawa, M. (1982) Identification in culture of a class of hemopoietic colony-forming units with extensive capability to self-renew and generate multipotential hemopoietic colonies. *Proc. Natl. Acad. Sci. USA* **79,** 3843–3847.

9. Brandt, J. E., Baird, N., Lu, L., Srour, E., and Hoffman, R. (1988) Characterization of a human hematopoietic progenitor cell capable of forming blast cell containing colonies in vitro. *J. Clin. Invest.* **82,** 1017–1027.

10. McNiece, L. K., Robinson, B. E., and Quesenberry, P. J. (1988) Stimulation of murine colony-forming cells with high proliferative potential by the combination of GM-CSF and CSF-1. *Blood* **72,** 191–195.

11. Dobo, I., Allegraud, A., Navenot, J. M., Boasson, M., Bidet, J. M., and Praloran, V. J. (1995) Collagen matrix: an attractive alternative to agar and methylcellulose for the culture of hematopoietic progenitors in autologous transplantation products. *J. Hematother.* **4,** 281–287.

12. Hogge, D., Fanning, S., Bockhold, K., et al. (1997) Quantitation and characterization of human megakaryocyte colony-forming cells using a standardized serum-free agarose assay. *Br. J. Haematol.* **96,** 790–800.

13. Ogawa, M. (1993) Differentiation and proliferation of hematopoietic stem cells. *Blood* **81,** 2844–2853.

14. Krystal, G., Alai, M., Cutler, R. L., Dickeson, H., Mui, A. L., and Wognum, A. W. (1991) Hematopoietic growth factor receptors. *Hematol. Pathol.* **5,** 141–162.

15. Kaushansky, K. and Drachman, J. G. (2002) The molecular and cellular biology of thrombopoietin: the primary regulator of platelet production. *Oncogene* **21,** 3359–3367.

7

Isolation and Culture of Murine Macrophages

John Q. Davies and Siamon Gordon

Summary

The two most convenient sources of primary murine macrophages are the bone marrow and the peritoneal cavity. Resident peritoneal macrophages can readily be harvested from mice and purified by adherence to tissue culture plastic. The injection of Bio-Gel polyacrylamide beads or thioglycollate broth into the peritoneal cavity produces an inflammatory response allowing the purification of large numbers of elicited macrophages. The production of an activated macrophage population can be achieved by using Bacillus–Calmette–Guerin as the inflammatory stimulus. Resident bone marrow macrophages can be isolated following enzymatic separation of cells from bone marrow plugs and enrichment on 30% fetal calf serum containing medium or Ficoll-Hypaque gradients. Bone marrow-derived macrophages can be produced by differentiating nonadherent macrophage precursors with medium containing macrophage colony-stimulating factor.

Key Words: Macrophage; murine; culture; peritoneum; bone marrow; resident; Bio-Gel; thioglycollate; BCG.

1. Introduction

This chapter describes established methods for the isolation and in vitro propagation of primary murine macrophages from various sites. Macrophages (Mφ) are central players in both the innate and adaptive immune systems and are attractive cells to study in culture because of their wide range of cellular functions. They are crucial phagocytes involved in important cytotoxic activities and in the destruction of micro-organisms (1). Mφ present antigen to primed T-lymphocytes and secrete a number of important cytokines (2) and chemokines (3), which regulate a wide range of immune responses. In view of their ability to adapt to local environments, Mφ display a striking diversity of phenotype and activation states depending on the site of origin and the method of isolation. This heterogeneity must be considered when Mφ are isolated for use

From: *Methods in Molecular Biology, vol. 290: Basic Cell Culture Protocols, Third Edition*
Edited by: C. D. Helgason and C. L. Miller © Humana Press Inc., Totowa, NJ

Table 1
Selected Macrophage Antigen Markers/Corresponding Antibodies

Marker[a]	Species[b]	Clone	Supplier/reference
CD68 (macrosialin)	Mouse	FA-11	Serotec/**ref. 5**
CD11b (CR-3)	Mouse	5C6	Serotec
CD18	Mouse	C71/16	Serotec
CD14	Mouse	rmC5-3	PharMingen
MHC II	Mouse	TIB120	ATCC
CD32	Mouse	2.4G2	PharMingen
F4/80	Mouse	F4/80	Serotec
SR-A (scavenger receptor)	Mouse	2F8	Serotec
Sialoadhesin	Mouse	3D6.113	Serotec
Mannose receptor	Mouse	MR5D3	Serotec/**ref. 6**

[a]Several of these markers can be expressed on dendritic cells, neutrophils, or other cells
[b]All antibody (Ab) reagents listed for mouse antigens are rat.
Source: Data from **refs. 4** and **4a**.

in any experimental system, at which point, the population of cells concerned can easily be characterized by the use of a number of monoclonal antibodies (mAbs) to important antigen markers (*see* **Table 1**).

The first subsection deals with common methods of isolating and culturing various murine Mϕ from the peritoneal cavity. The second section deals with the isolation of murine Mϕ from bone marrow.

1.1. Isolation and Culture of Primary Peritoneal Macrophages

This is perhaps the most convenient source of primary mouse Mϕ. Resident peritoneal Mϕ (RPMϕ) are free-living phagocytes within the peritoneal cavity. A few million resident Mϕ can be harvested from one mouse *(7)*. Should a larger number of cells be required, elicited Mϕ can be produced by injecting sterile inflammatory agents such as thioglycollate or Bio-Gel polyacrylamide beads into the peritoneal cavity. Elicited Mϕ are thought to be more immature than resident Mϕ, having been recruited as monocytes from the blood *(8)*. For experiments requiring activated Mϕ a convenient agent to use as an inflammatory stimulant is *Mycobacterium bovis* Bacillus–Calmette–Guerin (BCG), Pasteur strain. The purification of Mϕ from fluid following peritoneal lavage is easily performed utilizing the Mϕ ability to adhere firmly to tissue culture plastic. Mϕ adhesion requires the engagement of a number of important receptors, including the scavenger receptor (SR) *(9)* and the type 3 complement receptor (CR-3) *(10)*.

1.2. Isolation and Culture of Murine Bone Marrow Macrophages

The bone marrow is a source of both mature Mφ and Mφ precursors. Large numbers of Mφ can be derived from bone marrow precursors, by harvesting immature cells from the femurs of mice and culturing them with specific growth factors *(11,12)*. This results in a fairly homogeneous population of cells, which is often more desirable than the more difficult process of isolating specialized resident bone marrow cells (RBMMφ) by mechanical and/or enzymatic means.

2. Materials

2.1. Isolation and Culture of Primary Peritoneal Macrophages

2.1.1. Harvesting of Resident Peritoneal Cells

1. Pathogen-free mice (*see* **Note 1**).
2. 70% Ethanol in water.
3. Sterile scissors and forceps.
4. Dissecting board.
5. Sterile phosphate-buffered saline—Ca^{2+}-Mg^{2+} free (PBS) (Gibco, Invitrogen Ltd).
6. 10-mL Syringe, 19- and 25-gage needle (one each per mouse).
7. 50-mL Polypropylene tubes (Falcon®, Becton Dickinson Labware).

2.1.2. Purification of Resident Peritoneal Macrophages by Adhesion

1. RPMI 1640 (Gibco, Invitrogen Ltd).
2. Heat-inactivated (HI) fetal calf serum. To heat inactivate fetal calf serum (FCS), place in a 56°C water bath for 30 min. Filter using a 37-mm Serum Acrodisc syringe filter (Gelman Laboratory, Pall Corp.).
3. Media supplements: penicillin, streptomycin, glutamine (Gibco, Invitrogen Ltd).
4. Macrophage culture medium: RPMI 1640 is supplemented with 10% HI FCS (R^{10}) containing 50 IU of penicillin, 50 µg streptomycin, and 2 mM glutamine (PSG) per milliliter (Gibco, Invitrogen Ltd).
5. 24-Well tissue culture plates (Falcon, Becton Dickinson Labware).
6. Sterile PBS (37°C).

2.1.3. Bio-Gel Polyacrylamide Bead Elicited Peritoneal Cells (BgPMφ)

1. Bio-Gel P-100 polyacrylamide beads (fine, hydrated size 45–90 µm; Bio-Rad, Richmond, CA).
2. 70-µm Cell strainer (Falcon, Becton Dickinson Labware).
3. 1-mL Syringe and 26-gage needle.

2.1.4. Thioglycollate Broth Elicited Peritoneal Macrophages (TPMφ)

Brewer's complete thioglycollate broth (Difco Laboratories, West Molesy, UK). To prepare this, suspend 15 g of dehydrated thioglycollate medium in 500 mL of distilled water in a 1-L Erlenmeyer flask. Heat over a flame until dissolved and remove from flame immediately after boiling. The solution should change from brown to red. Aliquot into 25- or 50-mL bottles and auto-clave for 20 min at 121°C. Age the solution for 1–2 mo in a dark room at room temperature before use, as this will augment the yield of inflammatory cells. This is thought to be the result of an increase in glycation products *(13)*. Prior to use, ensure that the broth is clear; any cloudiness indicates contamination and renders it unusable.

2.1.5. BCG Recruited Macrophages (BCGMφ)

Mycobacterium bovis BCG, Pasteur strain (kindly provided by Dr. Genevieve Milon, Pasteur Institute, Paris, France). Store 10^7 colony-forming units (CFUs) in 0.5 mL PBS at –80°C (*see* **Note 2**).

2.2. Isolation and Culture of Murine Bone Marrow Macrophages

2.2.1. Isolation and Culture of Resident Murine Bone Marrow Macrophages (RBMMφ)

1. Macrophage culture medium (*see* **Subheading 2.1.2.4., item 4**).
2. RPMI 1640 (Gibco, Invitrogen Ltd).
3. Endotoxin-free, HI FCS (Gibco, Invitrogen Ltd).
4. RPMI 1640 containing 30% HI FCS (R^{30}).
5. Collagenase D type I (Roche Diagnostics).
6. DNase type I (Roche Diagnostics).
7. RPMI 1640 containing 0.05% Collagenase D and 0.001% DNase I. This solution should be prepared fresh by adding 5 mg Collagenase D and 100 µg DNAse I to 10 mL RPMI 1640. Filter-sterilize using a 32-mm/0.2-µm Acrodisc syringe filter (Gelman Laboratory, Pall Corp.) and 10-mL syringe. Stock solutions should be kept at –20°C.
8. 25-Gage needles and 5-mL syringes.
9. Sterile forceps and strong scissors.
10. Tube rotator/rocker.
11. 100 × 15-mm Petri dishes.
12. Sterile 11-mm sterile circular glass cover slips. To sterilize, immerse slides in ethanol, and in the tissue culture hood, individually pull out and flame. Place individually into each 24-well dish.
13. 24-Well tissue culture plates (Falcon, Becton Dickinson Labware).
14. 50-mL Polypropylene tubes (Falcon, Becton Dickinson Labware).

2.2.2. Production of Bone Marrow-Derived Macrophages (BMDMφ)

1. Cell strainers, 70 µm (Falcon, Becton Dickinson Labware).
2. HEPES (*N*-[2-hydroxyethyl]piperazine-N^1-[2-ethanesulfonic acid]) (Gibco, Invitrogen Ltd).

Table 2
Phenotype of Resident, Elicited, and Activated Murine Peritoneal Macrophages

	Resident	Thioglycollate	Bio-Gel	BCG
Total peritoneal cell yield per mouse ($\times 10^6$)	7	15–20	10–15	10
Macrophages (% of total)	40	80	60	60
Adherence on TCP at 24 h in SCM	+	+	+/–	+
F4/80 expression (F4/80 Ab)	++	+	+	+
Mannose receptor	++	++	++	+++
Macrosialin/murine CD68 (FA11 Ab)	+	++	+	+
MHC class II (TIB 120 Ab)	–	+	+	+++
Induced respiratory burst (superoxide generation)	–	+	+	+
Constitutive NO production	–	–	–	+

Note: –, not detectable; +/–, might or might not be detectable; +, detectable at low levels; ++, detectable at moderate levels; +++, detectable at high levels.
   Abbreviations: TCP = tissue culture plastic; SCM = serum-containing medium; NO = nitric oxide.
   Source: Adapted from **ref. 7**.

3. Sterile PBS containing 10 mM EDTA and 4 mg/mL Lidocaine-HCL (Sigma).
4. 15-cm Bacterial plastic (BP) dishes (Falcon, Becton Dickinson Labware).
5. L-Cell conditioned medium. To make this, grow mouse L929 fibroblasts (ATCC no. CCL-1) in modified Eagle's medium (MEM) containing 5% FCS in T175 flasks (Falcon, Becton Dickinson Labware). Harvest the cells using 5 mM EDTA containing 0.25% trypsin in PBS every 7 d and replate at one-tenth density. Collect conditioned medium, spin at 1500g for 10 min, filter, and store frozen in aliquots at –80°C until needed *(14)*. Alternatively, recombinant mouse macrophage colony-stimulating factor (M-CSF) (R&D Systems) can be used at a concentration of 0.5–1.5 ng/mL.
6. RPMI 1640 supplemented with 10% HI FCS (*see* **Subheading 2.1.2.**, **item 4**), 10 mM HEPES, 15% (v/v) L-cell conditioned medium (or recombinant M-CSF).

3. Methods
3.1. Isolation and Culture of Primary Peritoneal Macrophages

The methods described in this section should allow the reader to (1) harvest resident peritoneal cells from mice, (2) purify RPMφ by adhesion, (3) produce elicited Mφ using Bio-Gel polyacrylamide beads or (4) thioglycollate broth, and (5) produce activated Mφ using BCG. **Table 2** provides a summary of the Mφ phenotypes produced using these methods. For the general conditions and cell culture materials required for Mφ cell culture, *see* **Note 3**.

3.1.1. Harvesting of Resident Peritoneal Cells

1. Sacrifice the mice by CO_2 asphyxiation or cervical dislocation. Pin onto dissection board with abdomens up and sterilize with 70% ethanol.
2. Make a small off-center skin incision over the caudal half of the abdomen with scissors and expose the underlying abdominal wall by retraction.
3. Resterilize with 70% ethanol and lifting the abdominal wall with sterile forceps, inject 10 mL sterile PBS into the caudal half of the peritoneal cavity using a 25-gage needle (beveled side up) (*see* **Note 4**).
4. Remove the pins and gently shake the entire body for 10 s.
5. Slowly withdraw saline containing resident peritoneal cells by inserting a 19-gage needle, beveled side down, into the cranial half of the peritoneal cavity (*see* **Note 5**).
6. Store the cell suspension on ice until required (*see* **Note 6**).

3.1.2. Purification of Resident Peritoneal Macrophages by Adhesion

1. Plate the resident peritoneal cells in Mϕ culture medium at 3×10^5 cells per well in a 24-well tissue culture plate and incubate for 60 min at 37°C.
2. Remove the nonadherent cells by washing five times in 500 µL warm PBS, using a gentle swirling action.

The adherent cells should consist of a population of cells, more than 90% of which should be Mϕ *(7)* (*see* **Note 7**).

3.1.3. Bio-Gel Polyacrylamide Bead Elicited Peritoneal Cells (BgPMϕ)

Fauve et al. first reported this method of recruiting inflammatory cells *(15)*. Bio-Gel beads cannot be phagocytosed or digested by Mϕ yielding cells free of intracellular debris *(16)*. This makes these Mϕ especially suitable for studies involving phagocytosis. In this laboratory, up to 1×10^7 cells have been successfully elicited per mouse *(17)*. Bio-Gel-elicited Mϕ have a number of distinct phenotypic characteristics and appear to adhere less well to tissue culture plastic, following overnight incubation in serum containing medium, than RPMϕ or thioglycollate broth elicited Mϕ (Stein, unpublished observations).

1. Wash 2 g of Bio-Gel beads twice in 20 mL endotoxin-free water or PBS. Pellet by centrifugation for 5 min at 400 *g* and resuspend in 100 mL PBS to give a 2% (v/v) solution. Autoclave at 15 lb/m^2 for 20 min before use.
2. Inject mice intraperitoneally (ip) with 1 mL of the above solution (*see* **Note 8**).
3. Harvest the peritoneal cells on d 4 or 5 and purify by adhesion as outlined for RPMϕ (*see* **Subheading 3.1.2.**) (*see* **Note 9**).

3.1.4. Thioglycollate Broth Elicited Peritoneal Macrophages (TPMϕ)

This procedure is one of the easiest and least expensive methods of obtaining murine Mϕ. Each mouse should yield approx 2×10^7 cells, of which up to

80% can be Mφ *(7)*. TPMφ ingest large amounts of the inflammatory agent (agar) but retain active endocytic and phagocytic function on isolation. In addition, thioglycollate broth often contains low levels of lipopolysaccharide (LPS), which could affect Mφ behavior in subsequent experiments *(7)*.

1. Inject 1 mL of Brewer's complete thioglycollate broth into the peritoneal cavity of the mouse (*see* **Note 8**).
2. Harvest thioglycollate elicited peritoneal cells from the peritoneal cavity after 4–5 d and purify TPMφ by adhesion to tissue culture plastic as outlined for RPMφ (*see* **Subheading 3.1.2.**).

3.1.5. BCG Recruited Macrophages (BCGMφ)

The recruitment of immunologically activated Mφ can be achieved by injecting BCG (Pasteur strain) into the peritoneal cavity of mice *(18)*. T-Cell products, such as interferon-γ (IFN-γ) activate BCGMφ. They express high levels of Major Histocompatibility Complex Class II molecules and produce endogenous levels of nitric oxide in serum-containing medium *(19)*. These cells are useful for the investigation of activated Mφ responses to various stimuli such as LPS or lipoteichoic acid.

1. Thaw stocks, sonicate, and resuspend so that approx 10^7 CFUs are present in 0.5 mL sterile PBS that is injected ip (*see* **Note 2**).
2. Harvest the BCG elicited peritoneal cells, 4–6 d following injection and purify BCGMφ by adhesion as described for RPMφ (*see* **Subheading 3.1.2.**).

3.2. Isolation and Culture of Murine Bone Marrow Macrophages

The methods described in the following subsections should allow the reader to (1) isolate resident bone marrow cells from mice and (2) produce BMDMφ using L-cell conditioned medium as a readily available source of M-CSF.

3.2.1. Isolation and Culture of Resident Murine Bone Marrow Macrophages (RBMMφ)

Approximately 1% of bone marrow cells are mature RBMMφ *(20)*. These cells are found within erythroid cell clusters, express specific adhesion molecules such as sialoadhesin, and characteristically have long, fragile plasma membrane processes, which ramify through the marrow stoma. It is, therefore, important to handle these cells carefully if one is to maintain their viability *(21)*.

1. To isolate bone marrow cells from murine femurs, first sacrifice the animal by cervical dislocation or CO_2 asphyxiation.
2. Sterilize the abdomen and hind legs with 70% ethanol. Expose the muscle over each hind leg by incising and reflecting the skin over the leg and abdomen and using scissors remove the muscles attaching the hind limb to the pelvis and tibia.

3. Remove the femur from the mouse by cutting through the tibia below the knee joint as well as the pelvic bone close to the hip joint (*see* **Note 10**).
4. Store femurs in RPMI 1640 on ice before use. Place bones in 70% ethanol for 1 min to ensure sterility, wash twice in sterile PBS, and then remove both epiphyses using strong scissors and forceps (*see* **Note 11**).
5. While holding the bone over a Petri dish, flush out the bone marrow cells by forcing 5 mL RPMI 1640 containing 0.05% Collagenase D and 0.001% DNase through the central bone marrow canal using a syringe and 25-gage needle, until the bones are white.
6. Collect the bone marrow plugs from two femurs, which are now in 10 mL of the RPMI 1640/enzyme solution and transfer to a 50-mL polypropylene tube. Place at 37°C with gentle rotation or shaking.
7. After 1 h, stop enzyme digestion by adding 100 µL FCS to the 10-mL cell suspension (1% final FCS concentration). At this stage, there should be a homogeneous population of cells.
8. Layer 5 mL of the above cell suspension over 10 mL R^{30} in a 50-mL tube and leave to stand for 1 h at room temperature (*see* **Note 12**).
9. Aspirate the uppermost 14 mL medium. The remaining 1 mL contains >90% clusters containing RBMMφ.
10. Pool clusters from three mice in RPMI 1640 and centrifuge at 100g for 10 min.
11. Resuspend the clusters in 1.2 mL macrophage culture medium and add 100 µL to sterile 11-mm circular glass cover slips in 24-well plates.
12. Incubate for 3 h at 37°C in 5% CO_2, during which time, >90% of the clusters become firmly adherent.
13. Remove nonadherent cells by washing five times with 1 mL PBS using a Pasteur pipet. At this stage, there is a population of extensively spread cells attached to variable numbers of small refractile hemopoietic cells.
14. Further separation from the underlying adherent RBMMφ can be achieved by incubating in PBS for an additional 30 min at room temperature, followed by *gentle*, direct flushing with a Pasteur pipet.

The RBMMφ have delicate plasma membrane processes and phagocytic inclusions, distinguishing them from adherent contaminating neutrophils and monocytes, which can make up as much as 50% of the cell numbers *(20)*. Approximately 1×10^5 cells per two femurs can be expected using this protocol.

3.2.2. Production of Bone Marrow-Derived Macrophages (BMDMφ)

Bone marrow-derived macrophages are derived from nonadherent Mφ precursors. Maintenance, differentiation, and growth of these precursors requires serum-supplemented medium containing recombinant M-CSF. An alternative cost-saving measure is to use L-cell conditioned medium (LCM) as a source of M-CSF *(14)*.

1. Flush femurs with RPMI 1640 (with no enzymes added) as described for RBMMφ (*see* **Note 13**).

2. Collect the bone marrow plugs from two femurs, which are in 10 mL of RPMI 1640, into a 50 mL polypropylene tube. Mechanically disrupt the marrow plugs by passing through a 19-G needle twice, filter the resulting cell suspension using a 70-μM cell strainer (optional) and centrifuge at $400g$ for 5 min.

3. Resuspend the cells in 50 mL RPMI 1640 containing 10 mM HEPES, 10% FCS, and 15% (v/v) LCM. Plate into two 15-cm BP dishes (25 mL per dish).

4. Replace medium on d 3 and 6 with RPMI 1640 containing 10 mM HEPES, 10% FCS, and 15% (v/v) LCM.

5. On d 7, harvest the BMDMϕ by incubating with 10 mL PBS containing 10 mM EDTA and 4 mg/mL Lidocaine-HCL for 10 min, followed by vigorous pipetting.

6. Collect the resulting cell suspension in a 50-mL polypropylene tube and quench with an equal volume of RPMI 1640 containing 10 mM HEPES, 10% FCS, and 15% (v/v) LCM. Centrifuge the cells at $400g$ for 5 min and plate at 1×10^6 Mϕ per well of a six-well BP dish.

The cells are now mature, but proliferating BMDMϕ that can be used in a wide range of assays (*see* **Note 14**). Typically, $(2–6) \times 10^7$ Mϕ per two femurs can be expected using this protocol *(22)*. The BMDMϕ can be kept for at least up to 3 wk in culture, provided that fresh medium containing LCM is added every 3–4 d. During this period, the BMDMϕ might require subculturing to prevent detachment from the underlying BP (*see* **Note 15**).

4. Notes

1. For most applications, we routinely use C57BL/6 or BALB/c mice at 8–12 wk of age, with good results. Variability in cell numbers isolated between wild-type strains is usually minimal. Significant alterations in cell numbers can occur in certain knockout animals, which should be taken into account during initial experiment planning.

2. BCG can also be obtained from the American Type Culture Collection (Pasteur strain, ATCC no. 35734). Grow to mid-log phase in endotoxin-free Middlebrook 7H9 medium (Middlebrook; Difco Laboratories Inc.) at 37°C and store frozen at –80°C until use. CFUs can be determined by serial dilution of BCG cultures on Middlebrook 7H11 agar plates. Frozen stocks of BCG should be thawed, sonicated briefly in a water bath (two times bursts of 10 s at 15% power; Sonicator®, Heat Systems), and diluted into sterile PBS so that 10^7 CFUs are present in 0.5 mL.

3. Culture Mϕ in a humidified incubator at 37°C containing 5% humidified CO_2. Mϕ adhere firmly to substrata when cultured on tissue culture plastic (TCP) or bacteriological plastic (BP) in the presence of serum. Mϕ cultured on BP differentiate more rapidly than on TCP. Mϕ generally require PBS containing 5 mM EDTA and 4 mg/mL Lidocaine-HCL treatment for effective detachment on TCP (trypsin ineffective), whereas 5–10 mM EDTA in PBS alone is usually effective for detachment from BCP. Should the need arise, culture Mϕ in Teflon-coated flasks or bags for suspension cells.

4. During insertion of needles into the peritoneal cavity, it is important to prevent contamination of the resident population of cells by blood caused by penetrating vascular structures. To avoid this complication, always try to lift the abdominal wall away from the underlying organs using forceps. Carefully penetrate the caudal half of the body wall with a 25-gage needle, *beveled side up* while pushing on the plunger at all times so that fluid is injected as soon as the peritoneal cavity is entered.

5. As the needle is removed, there is usually some loss of fluid, which is usually blocked fairly quickly by omental fat. During collection of peritoneal cells, the needle should be inserted *beveled side down* into the cranial half of the abdominal cavity. This should avoid omental fat blocking the needle.

6. If a large number of mice are to be used, it is wise not to pool all of the specimens. First, examine small aliquots under a phase-contrast microscope to ensure that there are not large numbers of contaminating red blood cells. Red blood cells are smaller and flatter (saucer shaped) than leukocytes.

7. Should there be a need for greater purity, the cells can be lifted using PBS containing 10 mM EDTA and the adhesion step can be repeated. This is generally only required when isolating Mϕ RNA or detecting protein where the result can be significantly altered by small numbers of contaminating red blood cells/other leukocytes. Expect to lose up to 20% of the original cell number.

8. Successful peritoneal injection requires some expertise. The technique can be practiced on sacrificed mice. Prepare the syringe containing Bio-Gel in suspension attached to a 26-gage needle. Restrain the mouse firmly and inject into the lower left or right quadrant of the abdomen, as there are no vital organs in this area. The midline and a line perpendicular to it passing through the umbilicus demarcate this quadrant. The needle should be angled at 45° to the skin and no resistance should be encountered to the passage of the needle.

9. The washing steps following adhesion remove the beads and the large number of associated neutrophils also elicited during this procedure. As an additional measure to aid separation, pass the newly harvested Bio-Gel elicited peritoneal cells through a 70-μm cell strainer (Falcon, Becton Dickinson Labware) to remove beads, before the adhesion step. Failure to remove the beads could lead to cell loss during the washing procedures, as significant numbers of Mϕ could adhere firmly (but not internalize) to the Bio-Gel beads during the adhesion process.

10. It is important to ensure that the femurs are sealed by the joints at both ends and that no cracks have been introduced during the isolation procedure to ensure that cell recovery, viability, and sterility are not compromised by leakage of the marrow or exposure to the 70% ethanol during the subsequent washing phase.

11. To ensure a clean horizontal cut above the epiphysis, grasp the diaphysis of the femur firmly just above the epiphysis with the strong forceps and cut the bone just below it, using the forceps as a guide. This will prevent the bone fracturing longitudinally.

12. Clusters of RBMMϕ can also be isolated using a Ficoll-Hypaque cushion (Pharmacia, Uppsala, Sweden) *(20)*. Overlaying cell suspensions on FCS or

Ficoll-Hypaque is more easily done by running cell suspensions down the side of the polypropylene tube held at a 45° angle. In order to isolate satisfactory numbers of viable RBMMφ, it is important to avoid vigorous pipetting and repeated centrifugation, as these cells are extremely fragile.

13. Sterile PBS can be used to flush the marrow cavity instead of RPMI 1640 without affecting cell yield or viability.

14. BMDMφ isolated using this protocol are excellent cells to perform various adhesion, phagocytosis, and endocytosis assays. The measurement of secreted products that are indicative of the degree of Mφ activation, such as tumor necrosis factor-α and nitric oxide, can also be performed. In our laboratory's experience, BMDMφ do not release superoxides, even after PMA (phorbol myristate acetate) stimulation.

15. Subculture BMDMφ no more than a 1:1 dilution. Should the BMDMφ detach from the BP because of overgrowth, aspirate the medium with cells, and replate. Do not discard the detached cells, as these are still viable BMDMφ, which will adhere once on a new substratum.

Acknowledgments

We thank Dr. Hsi-Hsien Lin for his critical reading of the manuscript as well as helpful discussions with Dr. Leanne Peiser and Subhankar Mukherjee. Work in the laboratory of Siamon Gordon is supported in part by the Medical Research Council and the Welcome Trust. John Davies is funded by an Oxford Nuffield Medical Fellowship.

References

1. Leijh, P. C. J., Van Furth, R., and Van Swet, T. L. (1986) In vitro determination of phagocytosis and intracellular killing by polymorphonuclear and mononuclear phagocytes, in *Handbook of Experimental Immunology* (Weir, D. M., ed.), Blackwell Scientific, Oxford, pp. 46.1–46.21.

2. Stein, M. and Gordon, S. (1991) Regulation of tumor necrosis factor (TNF) release by murine peritoneal macrophages: role of cell stimulation and specific phagocytic plasma membrane receptors. *Eur. J. Immunol.* **21,** 431–437.

3. Adams, D. H. and Lloyd, A. R. (1997) Chemokines: leucocyte recruitment and activation cytokines. *Lancet* **349,** 490–495.

4. Peiser, L., Gough P. J., Darley, E., and Gordon, S. (2000) Characterisation of macrophage antigens and receptors by immunohistochemistry and fluorescent analysis: expression, endocytosis and phagocytosis, in *Macrophages* (Paulnock, D. M., ed.), Oxford University Press, Oxford, pp. 61–91.

4a. Martinez-Pomares, L., Platt, N., McKnight, A. J., da Silva, R. P., and Gordon, S. (1996) Macrophage membrane molecules: markers of tissue differentiation and heterogeneity. *Immunobiology* **195,** 407–416.

5. Smith, M. J. and Koch, G. L. (1987) Differential expression of murine macrophage surface glycoprotein antigens in intracellular membranes. *J. Cell Sci.* **87(Pt. 1),** 113–119.

6. Martinez-Pomares, L., Reid, D. M., Brown, G. D., et al. (2003) Analysis of mannose receptor regulation by IL-4, IL-10, and proteolytic processing using novel monoclonal antibodies. *J. Leukocyte Biol.* **73,** 604–613.

7. Haworth, R. and Gordon, S. (1998) Isolation and measuring the function of professional phagocytes: murine macrophages, in *Methods in Microbiology* (Kaufmann, S. and Kabelitz, D., eds.), Academic, London, Vol. 25, pp. 287–311.

8. Fortier, A. H. (1994) Isolation of Murine Macrophages. In *Current Protocols in Immunology* (Coligan, J. E., Kruisbeek, A. M., Margulies, D. H., Shevach, E. M., and Strober, W., eds.), Green Publishing/Wiley, New York, unit 14.1.1.

9. Fraser, I., Hughes, D., and Gordon, S. (1993) Divalent cation-independent macrophage adhesion inhibited by monoclonal antibody to murine scavenger receptor. *Nature* **364,** 343–346.

10. Rosen, H. and Gordon, S. (1987) Monoclonal antibody to the murine type 3 complement receptor inhibits adhesion of myelomonocytic cells in vitro and inflammatory cell recruitment in vivo. *J. Exp. Med.* **166,** 1685–1701.

11. Brunt, L. M., Portnoy, D. A., and Unanue, E. R. (1990) Presentation of Listeria monocytogenes to CD8+ T cells requires secretion of hemolysin and intracellular bacterial growth. *J. Immunol.* **145,** 3540–3546.

12. Stanley, E. R., Cifone, M., Heard, P. M., and Defendi, V. (1976) Factors regulating macrophage production and growth: identity of colony-stimulating factor and macrophage growth factor. *J. Exp. Med.* **143,** 631–647.

13. Li, Y. M., Baviello, G., Vlassara, H., and Mitsuhashi, T. (1997) Glycation products in aged thioglycollate medium enhance the elicitation of peritoneal macrophages. *J. Immunol. Methods* **201,** 183–188.

14. Hume, D. A. and Gordon, S. (1983) Optimal conditions for proliferation of bone marrow-derived mouse macrophages in culture: the roles of CSF-1, serum, Ca^{2+}, and adherence. *J. Cell Physiol.* **117,** 189–194.

15. Fauve, R. M., Jusforgues, H., and Hevin, B. (1983) Maintenance of granuloma macrophages in serum-free medium. *J. Immunol. Methods* **64,** 345–351.

16. Gordon, S. (1995) The macrophage. *Bioessays* **17,** 977–986.

17. Mahoney, J. A., Haworth, R., and Gordon, S. (2000) Monocytes and macrophages, in *Haematopoietic and Lymphoid Cell Culture* (Dallman, M. J. and Lamb, J. R., eds.), Cambridge University Press, Cambridge, pp. 121–146.

18. Ezekowitz, R. A., Austyn, J., Stahl, P. D., and Gordon, S. (1981) Surface properties of bacillus Calmette–Guerin-activated mouse macrophages. Reduced expression of mannose-specific endocytosis, Fc receptors, and antigen F4/80 accompanies induction of Ia. *J. Exp. Med.* **154,** 60–76.

19. Haworth, R., Platt, N., Keshav, S., et al. (1997) The macrophage scavenger receptor type A is expressed by activated macrophages and protects the host against lethal endotoxic shock. *J. Exp. Med.* **186,** 1431–1439.

20. Crocker, P. R. and Gordon, S. (1985) Isolation and characterization of resident stromal macrophages and hematopoietic cell clusters from mouse bone marrow. *J. Exp. Med.* **162,** 993–1014.

21. Handel-Fernandez, M. E. and Lopez D. M. (2000) Macrophages in tissues, fluids and immune response sites, in *Macrophages* (Paulnock, D. M., ed.), Oxford University Press, Oxford, pp. 1–30.
22. Peiser, L., Gough, P. J., Kodama, T., and Gordon, S. (2000) Macrophage class A scavenger receptor-mediated phagocytosis of Escherichia coli: role of cell heterogeneity, microbial strain, and culture conditions in vitro. *Infect. Immun.* **68,** 1953–1963.

8

Isolation and Culture of Human Macrophages

John Q. Davies and Siamon Gordon

Summary

Methods to isolate and culture human monocyte-derived macrophages and alveolar macrophages are described. Monocytes are obtained from buffy-coat preparations by Ficoll density gradient centrifugation, followed by adhesion-mediated purification on tissue culture or gelatin-coated plastic. The monocytes differentiate into macrophages in vitro by culturing in medium containing autologous human fibrin-depleted plasma. Alveolar macrophages can be purified from bronchoalveolar fluid samples by adhesion to tissue culture plastic. If resected lung tissue is available, alveolar macrophages can be obtained by mechanically disrupting the lung parenchyma, followed by adhesion-mediated purification.

Key Words: Macrophage; monocyte; human; culture; alveolar macrophage; bronchoalveolar fluid.

1. Introduction

The methodologies used to isolate and culture macrophages from human tissue, whether normal or diseased, are fairly well established. However, the study of these tissue-specific mature macrophages in humans is only available to those laboratories having appropriate ethical approval and close ties to a surgical department. Luckily, large numbers of human macrophages (Mφ) are fairly easily obtained from circulating blood monocytes. Although this might limit the study of certain tissue-specific questions in many instances, monocyte-derived Mφ (MDMφ) are frequently the most useful tool for studying Mφ function in humans, as appropriate human Mφ cell lines are not available.

Of the specialized tissue-specific mature Mφ, alveolar Mφ (AMφ) are, on the whole, relatively easy to obtain from bronchoalveolar lavage (BAL) fluids or lung tissue and a method for their isolation is therefore described in this chapter. Apart from their obvious phagocytic properties, AMφ appear to have

From: *Methods in Molecular Biology, vol. 290: Basic Cell Culture Protocols, Third Edition*
Edited by: C. D. Helgason and C. L. Miller © Humana Press Inc., Totowa, NJ

an interesting but important suppressive regulatory role on the immune system in the lung, thought to be vital in preventing chronic immune responses to airborne antigens *(1,2)*. Resident Mφ in the gut also demonstrate a similar phenotype *(3)*.

The first subsection of this chapter deals with common methods of isolating and culturing human MDMφ from whole-blood or buffy-coat preparations. The second subsection deals with the isolation of AMφ from bronchoalveolar fluid or resected lung tissue.

1.1. Isolation and Culture of Human Monocyte-Derived Macrophages

Human Mφ are most frequently obtained by purification of circulating blood monocytes, followed by in vitro differentiation into mature Mφ. The protocol discussed in this chapter obtains circulating white blood cells by density gradient centifugation from buffy-coat preparations. Monocytes are then separated from lymphocytes by adherence to tissue culture plastic (routine protocol) or gelatin-coated plastic (alternative protocol), which should achieve a cell purity of 90–95% *(4)*. Before attempting such methods, the investigator should be aware that a variable degree of Mφ activation could occur during this adhesion process *(5,6)*. Thus, to enable the generation of very pure monocyte populations of a low-activation state, alternative, more costly methods of monocyte isolation might be needed. Examples of these isolation methods, which will not be discussed further, include centrifugal counterflow elutriation *(7,8)*, which requires specialized equipment, and magnetic negative immunoselection *(9)*, for which freshly isolated peripheral blood monocytes are incubated with magnetic beads conjugated with antibodies to remove unwanted nonmonocytic cells.

1.2. Isolation and Culture of Human Alveolar Macrophages

The lung contains numerous macrophages in a number of distinct anatomical sites, which serve to regulate immune responses to continuous antigenic challenge. These include alveolar, interstitial, and airway Mφ. The existence of the so-called pulmonary interstitial Mφ in human lungs has been described *(10)*, but its true existence appears to be controversial *(11)*. Alveolar Mφ (AMφ), which are found at the interface of air and tissue in the alveoli and alveolar ducts, are probably best isolated from bronchoalveolar lavage fluid. In normal subjects, there should generally be no need to purify the isolated cell population, which should be ≥95% AMφ *(12)*. AMφ from bronchoalveolar lavage fluid contaminated with other cell types under pathological conditions will require simple methods of purification, which will be discussed.

2. Materials

2.1. Isolation and Culture of Primary Human Monocytes

2.1.1. Isolation of White Blood Cells From Buffy Coats

1. Human blood, buffy-coat fraction (*see* **Note 1**), screened for human immunodeficiency virus (HIV) and hepatitis B viruses (e.g., National Blood Service, Bristol, UK).
2. Heat-inactivated (HI) autologous human fibrin-depleted plasma. Stored at –20°C; use immediately (*see* **Note 2**).
3. 50-mL Polypropylene tubes (Falcon®, Becton Dickinson Labware).
4. PBS: Ca^{2+}- and Mg^{2+}-free phosphate-buffered saline (Gibco, Invitrogen). One 500-mL bottle at room temperature and two bottles on ice.
5. Ficoll-Hypaque (Pharmacia, Uppsala, Sweden).
6. Hemocytometer.
7. Media supplements: penicillin, streptomycin, glutamine (Gibco, Invitrogen Ltd).
8. RPMI 1640 containing 50 IU penicillin, 50 µg streptomycin, and 2 mM glutamine (PSG) per milliliter (Gibco, Invitrogen Ltd). Store at 4°C; use within 1 mo.
9. 37-mm Serum Acrodisc syringe filter (Gelman Laboratory, Pall Corp.)
10. X-Vivo 10 serum-free medium (Bio-Whittaker, Walkersville, MD). Store at 4°C; use within 1 mo.
11. Monocyte adhesion medium (MoAM): RPMI 1640 + 7.5% HI autologous human fibrin-depleted plasma (*see* **Note 3**), 2 mM glutamine, 50 U/mL penicillin, 50 µg/mL streptomycin. Store at 4°C, use within 1 mo.

2.1.2. Purification of Primary Human Monocytes by Adhesion to Tissue Culture Plastic (Routine Isolation Method)

1. 15-cm Tissue-culture-treated dishes (Falcon, Becton Dickinson Labware).
2. RPMI 1640 (Gibco, Invitrogen) heated in water bath to 37°C.
3. Ca^{2+}- and Mg^{2+}-free PBS (Gibco, Invitrogen).
4. Monocyte adhesion medium: *see* **Subheading 2.1.1., item 11**.
5. Mφ culture medium: X-Vivo 10 (BioWhittaker, Cambrex, USA) + 1% HI autologous human fibrin-depleted plasma (*see* **Note 3**), 2 mM glutamine, 50 U/mL penicillin, 50 µg/mL streptomycin. Store at 4°C; use within 1 mo.

2.1.3. Purification of Primary Human Monocytes by Adhesion to Gelatin-Coated Surfaces (Alternative Isolation Method)

1. 2% Gelatin Solution Type B, from bovine skin, cell culture tested (Sigma–Aldrich, Inc.).
2. 150-mm Bacterial plastic culture dishes (Corning [Falcon], Becton Dickinson Labware).
3. Ca^{2+}- and Mg^{2+}-free PBS (Gibco, Invitrogen).
4. Ca^{2+}- and Mg^{2+}-free PBS containing 5 mM EDTA.
5. Monocyte adhesion medium: *see* **Subheading 2.1.1., item 11**.
6. Mφ culture medium: *see* **Subheading 2.1.2., item 5**.

2.2. Isolation and Culture of Human Alveolar Macrophages

2.2.1. Isolation of Human Alveolar Macrophages From Bronchoalveolar Fluid Samples

1. Bronchoalveolar fluid, on ice (usually 100–200 mL per donor).
2. Hanks' balanced salt solution/(HBSS) (Gibco, Invitrogen Ltd) on ice.
3. HI fetal calf serum (FCS) (Gibco, Invitrogen Ltd) (*see* **Note 4**).
4. Alveolar Mϕ culture medium: RPMI 1640 supplemented with 50 IU/mL penicillin, 100 µg/mL streptomycin, 0.5 µg/mL amphotericin B (Fungizone) (Gibco, Invitrogen Ltd) and 10% HI FCS. Store at 4°C; use within 1 mo.
5. 50-mL Polypropylene tubes (Falcon, Becton Dickinson Labware).
6. Sterile gauze.
7. Hemocytometer.
8. Trypan blue and Wright–Giemsa stains (Sigma–Aldrich, Inc.).
9. Six-well flat-bottom tissue culture dishes (Falcon, Becton Dickinson Labware).

2.2.2. Isolation of Human Alveolar Macrophages From Resected Lung Tissue

1. Freshly resected lung tissue, nonpathological area (confirm by frozen-section histology; this might not, of course, be feasible, in which case, select the most "normal" appearing area).
2. Polycarbonate dessicator/vacuum jar (Nalgene™) and pump
3. 100 × 15-mm Sterile Petri dish.
4. Sterile scalpel blade with holder.
5. RPMI 1640/PSG/EDTA: 5 mM EDTA in RPMI 1640 (Gibco, Invitrogen Ltd) containing 50 IU of penicillin, 50 µg streptomycin, and 2 mM glutamine (PSG) per milliliter, buffered with 5 M sodium bicarbonate to a pH of 7.4. Filter-sterilize before use.
6. Alveolar Mϕ culture medium: RPMI 1640 supplemented with 50 IU/mL penicillin, 100 µg/mL streptomycin (Gibco, Invitrogen Ltd), and 10% HI FCS.
7. 70-µm Cell strainer (Falcon, Becton Dickinson Labware).
8. 50-mL Polypropylene tubes (Falcon, Becton Dickinson Labware)
9. Six-well flat-bottom tissue culture dishes (Falcon, Becton Dickinson Labware).

2.2.3. In Vitro Culture of Alveolar Macrophages

1. Alveolar Mϕ culture medium: RPMI 1640 supplemented with 50 IU/mL penicillin, 100 µg/mL streptomycin (Gibco, Invitrogen Ltd), and 10% HI FCS.
2. 96-Well flat-bottom tissue culture dishes (Falcon, Becton Dickinson Labware).

3. Methods

3.1. Isolation and Culture of Primary Human Monocytes

The methods described in this subsection should allow the reader to (1) isolate primary human monocytes from buffy coats or whole blood, (2) purify primary human monocytes by adhesion to tissue culture plastic (routine protocol) or gelatin-coated surfaces (alternative protocol), and (3) produce and maintain mature Mφ in culture. All procedures should be performed in a tissue culture hood to ensure sterility.

3.1.1. Isolation of White Blood Cells From Buffy Coats

1. Add 25 mL of room temperature PBS to four 50-mL polypropylene tubes. Decant the buffy coat from one donor equally (approx 25 mL) into each tube, giving a 1:1 dilution. Mix by gentle inversion.
2. Place 15 mL Ficoll-Hypaque at room temperature into six 50-mL polypropylene tubes. Gently overlay with approx 33 mL of the diluted buffy coat.
3. Centrifuge at 900g for 30 min, at room temperature, *no brake*.
4. Remove all except approx 5 mL of supernatant into fresh tubes. Keep the supernatant to be used as donor-specific human fibrin-depleted plasma after heat inactivation (*see* **Note 2**).
5. Remove the last 5 mL of supernatant and discard using a sterile 5-mL pipet.
6. Carefully aspirate the interface cells with a 5-mL pipet (*see* **Note 5**).
7. Place the cells into fresh tubes, two interfaces per 50-mL tube. Dissociate cells by pipetting up and down.
8. Fill tubes with ice-cold PBS and spin for 7 min at 250g, 4°C with half brake. Aspirate supernatant carefully, avoiding disturbing the cell pellet.
9. Resuspend each pellet in 5 mL ice-cold PBS and pipet up and down to dissociate the cells. Pool all the cells into one 50-mL tube.
10. Fill the 50-mL tube with ice-cold PBS and spin at 250g, 4°C, with full brake. Repeat this wash step three or four times until the solution is clear, indicating the absence of platelets (*see* **Note 6**).
11. Resuspend in 40 mL ice-cold MoAM and count. This protocol should yield (3–8) × 10^8 peripheral blood monocytic cells per buffy coat, or (5–15) × 10^5 cells/mL of whole blood *(4)*.

3.1.2. Purification of Primary Human Monocytes by Adhesion to TCP

1. Plate approx 1 × 10^8 primary blood monocytic cells in 20 mL MoAM in a 15-cm tissue-culture-treated dish (*see* **Note 7**).
2. Incubate for 45–90 min at 37°C in a humidified incubator.
3. Remove MoAM and nonadherent cells. Wash the residual partly adherent lymphocytes from the adherent cell layer by adding 15 mL RPMI 1640 (37°C) and swirling the dish gently (*see* **Note 8**).
4. Gently aspirate the MoAM, repeating the process five to eight times until only strongly adherent cells remain (*see* **Note 9**).
5. Add 20 mL MoAM per 15-cm dish and leave overnight in the 5% CO_2, 37°C incubator.
6. Detach the monocytes on d 1 by removing MoAM and incubating with 10 mL PBS containing 5 mM EDTA for 10–20 min at room temperature. Remove PBS/EDTA, which contains cells, and transfer to a 50-mL polypropylene tube.

Repeatedly pipet the PBS/EDTA directly onto the dish surface to dislodge remaining cells.

7. Transfer the cell suspension into a 50-mL polypropylene tube already containing 10 mL MoAM, rinse the dish with 10 mL PBS to remove residual cells, and transfer to the same 50-mL polypropylene tube. Collect the cells by centrifugation for 7 min at 250g at room temperature.

8. Count cells with a hemocytometer and adjust density to $(2-10) \times 10^5$ cells/ mL in Mϕ culture medium (not MoAM). There should be approx $(3-8) \times 10^7$ monocytes per buffy coat at this stage *(4)*, with "normal" considerable donor variability.

3.1.3. Purification of Primary Human Monocytes by Adhesion to Gelatin-Coated Surfaces (Alternative Protocol)

Monocytes have high affinity for fibronectin immobilized on a gelatin-coated surface *(13)*. This method for isolating monocytes could prove to be useful if the cell viability/yield is significantly lower than expected when using the "routine" protocol.

1. Coat the 150-mm bacterial plastic culture dishes with the 2% gelatin solution in a tissue culture hood (*see* **Note 10**).
2. Add 10 mL Mϕ culture medium to 10 mL of the prepared cell suspension (*see* **Subheading 3.1.1., step 11**).
3. Place in a 37°C, 5% CO_2 tissue culture incubator for 45–90 min.
4. Check for cell adhesion and then remove the medium and non-adherent cells after gentle swirling/shaking. Add 15 mL warm RPMI 1640, gently swirl, and remove. Repeat three to four times (*see* **Note 8**).
5. Add 20 mL Mϕ culture medium and incubate at 37°C, 5% CO_2 for 24 h during which time most of the cells (monocytes) should become detached. Wash the cells off using PBS and transfer to a 50-mL polypropylene tube (*see* **Note 11**).
6. Centrifuge the cells at 250g for 5 min, remove supernatant, and resuspend in RPMI 1640 to wash. Repeat, resuspending cells in 10–20 mL Mϕ culture medium.
7. Count cells and adjust to appropriate concentration ($[2-10] \times 10^5$ cells/mL).
8. Plate on tissue culture plastic. Use 10 mL of the above cell suspension in a 10 cm tissue culture dish.
9. Incubate for 7–12 d, replenishing the Mϕ culture medium every 3–4 d.

3.1.4. Production and Maintenance of Mature Human Mϕ in Culture

The above-isolated monocytes differentiate into Mϕ within the first few days of culture. Transient proliferation might be noticed. No change of medium is usually necessary for the first week, after which the medium is changed every 3 d. Monocyte-derived Mϕ can be maintained in culture using Mϕ culture medium for at least 2 wk, during which time occasional multinucleated cells might be identified. This morphology is greatly enhanced in the presence of IL-4 or IL-13. The use of serum-free medium usually results in monocyte death

Table 1
Selected Human Macrophage Antigen Markers
and Corresponding Antibodies

Marker[a]	Clone[b]	Supplier
CD68	EBM-11	Dako
CD11b	ICRF44	Serotec
CD18	YFC118.33	Serotec
CD14	UCHM1	Serotec
MHC II	CR3/43	Dako
CD32	AT10	Serotec
CD64	MCA756	Serotec

[a]Several of these markers may be expressed on dendritic cells, neutrophils, or other cells.
[b]All Ab reagents listed are mouse.
Source: Data from **refs. *14*** and ***15***.

by apoptosis unless supplemented by Mϕ CSF-1 and/or IL-4 *(4)*. We usually assess the Mϕ phenotype by fluorescent-activated cell sorting (FACS) analysis after 7–12 d in culture. When compared to freshly isolated monocytes, monocyte-derived Mϕ will upregulate expression of MHC class II and the pan-monocytic marker CD68 is expressed at high levels in both (*see* **Table 1**). Should there be poor viability and/or yield of human monocyte preparations, repeat the isolation procedure, as donor variability is said to be the most common cause. Consistently poor results indicate that a change in strategy should be implemented. Try changing the adhesion substratum (use the alternative protocol) or adhesion times during the isolation process, try additional washes to remove platelets, change the serum concentrations of MoAM or Mϕ culture medium, and check that all media components are endotoxin-free.

3.2. Isolation and Culture of Primary Human Alveolar Macrophages

The methods described in this subsection should allow the reader to isolate (1) primary human AMϕ from bronchoalveolar lavage fluid or (2) pneumonectomy specimens.

3.2.1. Alveolar Macrophage Isolation From Bronchoalveolar Lavage Fluid

There are numerous clinical protocols for collecting bronchoalveolar lavage fluids (BAL) in adults and children, the discussion of which is beyond the scope of this chapter. Generally, sterile warmed saline (which can be buffered) is introduced into the lung using a fiber-optic scope and then removed by suction. The saline removed contains secretions, cells, and protein from the lower respiratory tract (*see* **Note 12**). The use of divalent cation-free buffers and the addition of chelators of calcium such as EDTA might improve AMϕ yields *(16)*.

1. Filter the BAL fluid using a single layer of sterile gauze to remove mucus clumps and collect the flowthrough into 50-mL polypropylene tubes.
2. Collect the cells by centrifugation at 250*g* for 10 min. Pool cell pellets by resuspending in a total of 10 mL cold HBSS.
3. Remove a 100-μL aliquot. Perform cytospin. Stain with Wright–Giemsa stain and do a cell differential count *(17,18)*. Record results for each donor (*see* **Note 13**).
4. Wash the filtrate twice using the same volume of cold HBSS.
5. Adjust the cell count to approx 1×10^6/mL in alveolar Mϕ culture medium. Add 0.5 mL of the cell suspension to each well of a 6-well tissue culture plate already containing 0.5 mL of alveolar Mϕ culture medium (1 mL total volume).
6. Incubate for 40–90 min at 37°C, 5% CO_2.
7. Remove the medium and nonadherent cells after gentle swirling/shaking. Add 1 mL warm RPMI 1640, gently swirl, and remove. Repeat washing three to four times (*see* **Note 9**).
8. Lift the adherent alveolar macrophages by incubating with 1 mL PBS containing 5 m*M* EDTA for 5 min, followed by direct pipetting. Quench by adding 1 mL alveolar Mϕ culture medium. Assess cell viability by counting using 0.1% trypan blue.

In a normal donor, on average a 100-mL BAL yields 7.3×10^6 alveolar macrophages *(12)*. In smokers, the yield is significantly increased *(19)*, although Mϕ viability can often be significantly reduced.

3.2.2. Alveolar Macrophage Isolation From Pneumonectomy Specimens

Surgically removed human lung tissue is usually obtained from smokers during the removal of a tumor. The macroscopically normal areas of such lung resections, from which the AMϕ are obtained, might actually have significant abnormalities (inflammation or tumor infiltration) and it is recommended that this tissue be examined histologically before AMϕ isolation. Histologically "normal" areas should yield a similar AMϕ population as seen in smoking BAL samples *(20)*.

1. To isolate alveolar macrophages from lung alveoli, mince 5–20 g of lung tissue into small 2-mm pieces using the scalpel blade and sterile Petri dish.
2. Place the Petri dish into the sterile vacuum jar. Degas three times (in the tissue culture hood) for approx 30 s to remove air from the lung fragments. This facilitates further handling of the tissue.
3. Resuspend the lung fragments in 20 mL of 4°C RPMI 1640/PSG/EDTA and agitate vigorously three times for 10 s.
4. Remove lung fragments and debris by centrifugation at 100*g* for 1 min.
5. Pass cells in suspension through the 70-μm cell strainer and into a 50-mL polypropylene tube.

6. Collect cells by centrifugation at 250*g* for 10 min and resuspend in 5 mL of alveolar Mφ culture medium.
7. Count cells and adjust to approx 1×10^6 cells/mL. Purify alveolar macrophages by adhesion as in **Subheading 3.2.1., steps 5–8**.

This protocol using mechanical dissociation in the presence of EDTA should yield $(2–6) \times 10^6$ alveolar macrophages per gram of lung tissue *(21)*.

3.2.3. In Vitro Culture and Maintenance of Alveolar Macrophages

1. Adjust cells collected after purification to $(1–2) \times 10^6$ AMφ/mL in alveolar culture medium.
2. Add 100-μL aliquots of the resultant cell suspension into sterile 96-well tissue culture plates (i.e., for antigen assays).

The AMφ isolated in the above protocols initially appear rounded, with active membrane ruffling and prominent smokers' particles in secondary lysosomes. They will remain stable for several days in medium supplemented with FCS and will become more spread and elongated, but will not proliferate. If there is significant contamination by fibroblasts during the isolation of macrophages from resected lung tissue, these will overgrow the culture with time. This can be avoided by culturing AMφ on bacterial plastic, to which fibroblasts do not adhere well. Alveolar macrophages also grow well in suspension culture *(22)*, where they usually adhere to each other, forming spheres. This system of culture negates the effects adhesion could have on macrophage activation and could, in addition, simulate the alveolar environment more closely than a culture system on an adherent surface.

4. Notes

1. Buffy coats from blood banks are obtained by centrifugation of blood bank bags (400-mL volume) followed by removal of the upper layer, resulting in a cell-rich fraction (20–40 mL remaining). Adequate blood anticoagulation is *critical* at this stage (until Ficoll separation has occurred), otherwise monocytes bind to small blood clots. This will affect ultimate cell yield/activation state.
2. For preparation of the donor-specific (autologous) fibrin-depleted plasma, collect the pooled supernatants after Ficoll centrifugation. Heat inactivate (HI) by placing in a 56°C water bath for 30 min. Chill under running water and centrifuge for 10 min at 3000*g*. Decant supernatant from the large white fibrin pellet and filter using a 37-mm Serum Acrodisc syringe filter (Gelman Laboratory, Pall Corp.). More than one filter might be required. Aliquot into 50-mL polypropylene tubes and store at –20°C until use. We do not recommend more than one freeze–thaw cycle after heat inactivation. As clotting did not occur during the preparation of this material, fibrin-depleted plasma is a better term than "serum," as it lacks platelet products released during the coagulation process.

3. Remember that the autologous fibrin-depleted plasma is already 50% diluted in PBS when making supplemented medium.

4. To heat inactivate FCS, thaw the serum slowly to 37°C and mix the contents of the bottle thoroughly. Place the thawed bottle of serum into a 56°C water bath for 30 min. Swirl the serum every 5–10 min to ensure uniform heating and to prevent protein coagulation at the bottom of the bottle. Cool the serum immediately under running water and leave overnight at +4°C before filtering using a 37-mm Serum Acrodisc syringe filter (Gelman Laboratory, Pall Corp.) and 50-mL syringe to remove any precipitate. Aliquot into 50-mL polypropylene tubes and store at –20°C. It is recommended that HI serum may be refrozen once, thawed, and then used immediately because precipitate will increase as the serum stands in the thawed state, even if left in a refrigerator.

5. Many cells stick circumferentially to the tube. Carefully dislodge these cells using the 5-mL pipet before sucking off all the cells from the interface in as small a volume as possible. This is to ensure that there is minimal Ficoll contamination. Remember that monocytes and platelets collect on top of the Ficoll layer (lower density), whereas red blood cells and granulocytes collect at the bottom (higher density).

6. It is important to minimize platelet contamination, as platelets release substances that could activate Mφ. Do not contaminate any washing steps with EDTA, as this will cause platelets to adhere strongly to the monocytes, with resultant cytokine release.

7. We generally use two 15-cm culture dishes per buffy coat, in which case there is no need to count the cells at this stage.

8. It is very important to use warm RPMI 1640 (37°C). Cold wash medium will cause the monocytes to detach from the plate.

9. Check by phase-contrast microscopy after every second wash to ensure that adherent cells are not detaching while washing.

10. To coat bacterial plastic dishes with gelatin, add just enough to cover the dish and then remove excess using a sterile pipet. Allow dishes to dry in a biological safety cabinet before use. Ensure sterility at all times.

11. Should significant numbers of monocytes still remain attached to the gelatin, incubate for 5 min with PBS containing 10 mM EDTA, followed by vigorous direct pipetting. Five millimolars of EDTA containing 4 mg/mL Lidocaine-HCL (Sigma) is an alternative solution that will aid removal of monocytes and Mφ from culture surfaces.

12. Factors to consider in the project design prior to BAL collection to ensure experimental uniformity include the anatomical segment of the lung sampled, the volume of BAL fluid used, the processing of the sample for additional diagnostic purposes, and the adequacy of sampling (volume and cell type recovered). This should be discussed at length with the clinician performing the procedure.

13. For each donor, it is important to accurately record the volume and differential cell count for correlation with subsequent experimental results (enlist the help of an experienced cytologist!). This is particularly important when dealing with "normal" BAL samples that are not destined for macrophage purification by adhesion and where subsequent assays are performed on "whole cell" preparations.

Acknowledgments

We thank Dr. Hsi-Hsien Lin for his critical reading of the manuscript. Work in the laboratory of Siamon Gordon is supported in part by the Medical Research Council. John Davies is funded by an Oxford Nuffield Medical Fellowship.

References

1. Bilyk, N. and Holt, P. G. (1995) Cytokine modulation of the immunosuppressive phenotype of pulmonary alveolar macrophage populations. *Immunology* **86,** 231–237.
2. Upham, J. W., Strickland, D. H., Bilyk, N., Robinson, B. W., and Holt, P. G. (1995) Alveolar macrophages from humans and rodents selectively inhibit T-cell proliferation but permit T-cell activation and cytokine secretion. *Immunology* **84,** 142–147.
3. Smith, P. D., Smythies, L. E., Mosteller-Barnum, M., et al. (2001) Intestinal macrophages lack CD14 and CD89 and consequently are down-regulated for LPS- and IgA-mediated activities. *J. Immunol.* **167,** 2651–2656.
4. Mahoney, J. A., Haworth, R., and Gordon, S. (2000) Monocytes and macrophages, in *Haematopoietic and Lymphoid Cell Culture* (Dallman, M. J. and Lamb, J. R., eds.), Cambridge University Press, Cambridge, pp. 121–146.
5. Rosen, H. and Gordon, S. (1987) Monoclonal antibody to the murine type 3 complement receptor inhibits adhesion of myelomonocytic cells in vitro and inflammatory cell recruitment in vivo. *J. Exp. Med.* **166,** 1685–1701.
6. Fraser, I., Hughes, D., and Gordon, S. (1993) Divalent cation-independent macrophage adhesion inhibited by monoclonal antibody to murine scavenger receptor. *Nature* **364,** 343–346.
7. Wahl, L. M., Katona, I. M., Wilder, R. L., et al. (1984) Isolation of human mononuclear cell subsets by counterflow centrifugal elutriation (CCE). I. Characterization of B-lymphocyte-, T-lymphocyte-, and monocyte-enriched fractions by flow cytometric analysis. *Cell Immunol.* **85,** 373–383.
8. Faradji, A., Bohbot, A., Schmitt-Goguel, M., et al. (1994) Large scale isolation of human blood monocytes by continuous flow centrifugation leukapheresis and counterflow centrifugation elutriation for adoptive cellular immunotherapy in cancer patients. *J. Immunol. Methods* **174,** 297–309.
9. Flo, R. W., Naess, A., Lund-Johansen, F., et al. (1991) Negative selection of human monocytes using magnetic particles covered by anti-lymphocyte antibodies. *J. Immunol. Methods* **137,** 89–94.
10. Dehring, D. J. and Wismar, B. L. (1989) Intravascular macrophages in pulmonary capillaries of humans. *Am. Rev. Respir. Dis.* **139,** 1027–1029.
11. Brain, J. R. M. and Warner, A. (1997) Pulmonary intravascular macrophages, in *Lung Macrophages and Dendritic Cells in Health and Disease* (Lipscomb, M. R. S., ed.), Marcel Dekker, New York, Vol. 102, pp. 131–149.
12. Ettensohn, D. B., Jankowski, M. J., Duncan, P. G., and Lalor, P. A. (1988) Bronchoalveolar lavage in the normal volunteer subject. I. Technical aspects and intersubject variability. *Chest* **94,** 275–280.

13. Freundlich, B. and Avdalovic, N. (1983) Use of gelatin/plasma coated flasks for isolating human peripheral blood monocytes. *J. Immunol. Methods* **62,** 31–37.

14. Martinez-Pomares, L., Platt, N., McKnight, A. J., da Silva, R. P., and Gordon, S. (1996) Macrophage membrane molecules: markers of tissue differentiation and heterogeneity. *Immunobiology* **195,** 407–416.

15. Peiser, L., Gough P. J., Darley, E., and Gordon, S. (2000) Characterisation of macrophage antigens and receptors by immunohistochemistry and fluorescent analysis: expression, endocytosis and phagocytosis, in *Macrophages* (Paulnock, D. M., ed.), Oxford University Press, Oxford, pp. 61–91.

16. Brain, J. D. and Frank, R. (1973) Alveolar macrophage adhesion: wash electrolyte composition and free cell yield. *J. Appl. Physiol.* **34,** 75–80.

17. De Brauwer, E. I., Jacobs, J. A., Nieman, F., Bruggeman, C. A., and Drent, M. (2002) Bronchoalveolar lavage fluid differential cell count. How many cells should be counted? *Anal. Quant. Cytol. Histol.* **24,** 337–341.

18. Kini, S. (2002) *Color Atlas of Pulmonary Cytopathology*, Springer-Verlag, New York.

19. The BAL Cooperative Group Steering Committee. (1990) Bronchoalveolar lavage constituents in healthy individuals, idiopathic pulmonary fibrosis, and selected comparison groups. *Am. Rev. Respir. Dis.* **141,** S169–S202.

20. Kobzik, L. (1997) Methods to study lung macrophages, in *Lung Macrophages and Dendritic Cells in Health and Disease* (Lipscomb, M. R. S., ed.), Marcel Dekker, New York, Vol. 102, pp. 131–149.

21. Mason, R., Austyn, J., Brodsky, F., and Gordon, S. (1982) Monoclonal anti-macrophage antibodies: human pulmonary macrophages express HLA-DR (Ia-like) antigens in culture. *Am. Rev. Respir. Dis.* **125,** 586–593.

22. Helinski, E. H., Bielat, K. L., Ovak, G. M., and Pauly, J. L. (1988) Long-term cultivation of functional human macrophages in Teflon dishes with serum-free media. *J. Leukocyte Biol.* **44,** 111–121.

9

Development of T-Lymphocytes in Mouse Fetal Thymus Organ Culture

Tomoo Ueno, Cunlan Liu, Takeshi Nitta, and Yousuke Takahama

Summary

Fetal thymus organ culture (FTOC) is a unique and powerful culture system that allows intrathymic T-lymphocyte development in vitro. T-cell development in FTOC well represents fetal thymocyte development in vivo. Here, we describe the basic method for FTOC as well as several related techniques, including the reconstitution of thymus lobes with T-lymphoid progenitor cells, high-oxygen submersion culture, time-lapse visualization of thymic emigration, reaggregation culture, and retrovirus-mediated gene transfer to developing thymocytes in FTOC.

Key Words: T-lymphocytes; thymus; organ culture; FTOC; development; retrovirus; visualization; flow cytometry.

1. Introduction

Among the various lineages of hematopoietic cells, T-lymphocytes are the only cells whose development requires the environment of the thymus in addition to bone marrow or fetal liver. Recent studies have identified several molecules that take part in specifying the thymic environment. These molecules include interleukin (IL)-7, Delta-1, and class I/class II major histocompatibility complex (MHC) molecules. Despite the identification of these factors, it is still unclear whether any combination of the known molecules is sufficient for replacing the thymus environment that supports T-lymphocyte development. Thus, use of the thymic environment provides the most reliable and reproducible condition that supports the development of T-lymphocytes from the precursor cells.

The analysis of T-lymphocyte development in organ culture of mouse fetal thymus was first established by Owen (1,2) and Mandel (3,4) and later refined

From: *Methods in Molecular Biology, vol. 290: Basic Cell Culture Protocols, Third Edition*
Edited by: C. D. Helgason and C. L. Miller © Humana Press Inc., Totowa, NJ

mostly by Owen's group *(5,6)*. The fetal thymus organ culture (FTOC) technique offers a unique in vitro cell culture system in that functional T-cells are differentiated from immature progenitor cells. As such, T-cell development in FTOC closely reflects T-cell development during fetal ontogeny, even with respect to the time-course of differentiation *(7,8)*. FTOC allows the addition of various reagents, such as chemicals, antibodies, and viruses, for examining their effects on T-cell development.

This chapter describes a basic method for FTOC (**Subheadings 3.1.– 3.4.**) and several related techniques, including the reconstitution of thymus lobes with progenitor cells (**Subheading 3.5.**), high-oxygen submersion culture (**Subheading 3.6.**), time-lapse visualization of thymic emigration (**Subheading 3.7.**), reaggregation thymus organ culture (**Subheading 3.8.**), and retrovirus-mediated gene transfer to developing thymocytes in FTOC (**Subheading 3.9.**).

2. Materials

2.1. Isolation of Fetuses From Pregnant Mice

1. Timed pregnant C57BL/6 mice. Mice should be mated in an animal facility according to institutional guidelines. We usually place two female and one male mice in a cage in the evening (7–8 PM) and separate them in the morning (8–9 AM). Gestational days are tentatively designated by assigning the day at which mice are separated as d 0.5 and are confirmed on the day of experiment according to the size and many developmental features of fetuses (*see* **Note 1** and **refs. *9–11***).
2. Regular dissecting forceps and scissors. At least one set for non-sterile use to dissect skins, and two to three autoclaved sets for sterile use.

2.2. Preparation of Culture Wells

1. Sterile collagen sponges (Collagen sponge INTEGRAN Sheet type; Nippon Zoki Pharmaceutical Co., Ltd, Japan). Cut into small pieces (e.g., 1-cm square) and store dry at room temperature.
2. Polycarbonate (PC) filter membranes (Whatman, Nucleopore Corp.). PC membrane, cat. no. 110409, 13 mm in diameter. Autoclave to sterilize and store dry at room temperature.
3. 24-Well plates (16 mm in diameter, sterile).
4. Culture medium: RPMI 1640 supplemented with 10% fetal calf serum (FCS), 50 μM 2-mercaptoethanol, 10 mM HEPES, 2 mM L-glutamine, 1X nonessential amino acids, 1 mM sodium pyruvate, 100 U/mL penicillin, and 100 μg/mL streptomycin. All medium components except 2-mercaptoethanol were purchased from Gibco–BRL (Gaithersburg, MD). 2-Mercaptoethanol was purchased from Sigma Chemicals. FCS was pretreated for 30 min at 56°C and stored frozen in 50-mL aliquots. Screening of FCS is essential (*see* **Note 2**).

2.3. Isolation and Organ Culture of Fetal Thymus Lobes

1. Fetuses from timed pregnant mice (refer to **Subheading 2.1.**).
2. Type 7 forceps, biology grade (e.g., Dumont, Switzerland); stored sterile in 70% ethanol.
3. Dissecting microscope with zoom (e.g., ×7 to ×42 magnification), preferably equipped with fiber lights. The microscope should be placed in a clean hood.
4. Gauze sponges (e.g., Johnson and Johnson, 2 × 2-in. square, six to eight ply, sterile).
5. 100-mm Sterile plastic dishes.

2.4. Isolation of Single-Cell Suspensions From Fetal Thymus Organ Culture

1. Suspension buffer: PBS, pH 7.2, supplemented with 0.2% bovine serum albumin (BSA) and 0.1% NaN_3.
2. 1-mL Syringes.
3. 26-Gage needles.
4. 30-mm Plastic dishes.
5. Nylon mesh (approx 300 meshes/in.^{2}). Cut into small pieces of approx 5 mm square.

2.5. Optional Technique: Hanging-Drop Reconstitution of Deoxyguanosine-Treated Thymus Lobes With T-Precursor Cells

1. 2-Deoxyguanosine (D7145; Sigma, St. Louis, MO). Aliquots of a stock solution at 13.5 mM in PBS are stored frozen at –20°C and can be thawed at 37°C.
2. Terasaki 60-well plates (sterile).

2.6. Optional Technique: High-Oxygen Submersion Culture of Fetal Thymus Lobes

1. 96-Well round-bottom plates (sterile).
2. Plastic 3- to 5-L air bags and a heat-sealer.
3. Gas consisting of 70% O_2, 25% N_2, and 5% CO_2.

2.7. Optional Technique: Time-Lapse Visualization of Thymic Emigration Using Transparent Fetal Thymus Organ Culture

1. Cell culture devise at the stage under the microscope equipped with a digital charge-coupled device (CCD) camera. We use Axiovert S-100 microscope (Carl Zeiss, Jena, Germany) equipped with a C4742-95 digital CCD camera (Hamamatsu Photonics, Hamamatsu, Japan) and Openlab software (Improvision Inc., Lexington, MA).
2. CCL19 (R&D Systems, Minneapolis, MN). Aliquots of a stock solution at 10 μM in 0.1% BSA-containing PBS are stored frozen at –20°C.
3. Collagen acidic solution (3 mg/mL, pH = 3.0, Cellmatrix Type I-A; Nitta Gelatin, Osaka, Japan) is stored at 4°C. To make 10 mL collagen-based culture medium, 3.6 mL of Cellmatrix stock solution (final concentration = 1.08 mg/mL), 1 mL of

5X RPMI 1640 medium, 0.4 mL of alkaline solution containing 0.05 M NaOH, 0.2 M HEPES, 2.2% $NaHCO_3$, and 5 mL FCS-containing culture medium are mixed on ice immediately before use.

2.8. Reaggregate Thymus Organ Culture (RTOC)

1. Trypsin (0.5%)/5.3 mM EDTA solution (Gibco–BRL).
2. Ca^{2+}-free and Mg^{2+}-free PBS.

2.9. Optional Technique: Retroviral Gene Transfer Into Developing Thymocytes for the Fetal Thymus Organ Culture

1. 10-mL Syringes (sterile).
2. Syringe-driven filter (0.22-µm pore size, 16 mm in diameter, sterile).
3. Parafilm.
4. Plat-E cells *(12)* and a retrovirus vector pMRX-IRES-EGFP *(13)*. Culture medium for Plat-E cells is Dulbecco's modified Eagle's medium (DMEM) supplemented with 10% FCS, 100 U/mL penicillin G, 100 µg/mL streptomycin, 1 µg/mL puromycin, and 10 µg/mL blasticidin S. For transfection experiments, use the medium without puromycin and blasticidin S.
5. Polybrene (hexadimethrrine bromide) (Sigma).

3. Methods

3.1. Isolation of Fetuses From Pregnant Mice

1. All of the procedures should be performed under sterile conditions in a cell culture hood.
2. Prepare 100-mm sterile dishes, each containing 20–30 mL of culture medium (three dishes minimum).
3. Kill timed pregnant mice (usually used at d 14.5 or 15.5 of gestation) by CO_2 asphyxiation.
4. Wipe the abdomens of the mice with 70% ethanol and open them using the nonsterile set of scissors and forceps.
5. Take out fetus-filled uteri with a sterile set of scissors and forceps.
6. Transfer uteri to an empty 100-mm plastic dish.
7. Using a sterile set of sharp scissors and forceps, take out fetuses from uteri and transfer fetuses to a new dish containing culture medium.
8. Ascertain the gestational age of fetuses (*see* **Note 3**).
9. Wash out blood by transferring fetuses to new a dish containing fresh medium.
10. Repeat washing two to three times to remove blood. Gentle swirling of the dishes helps in removing the blood and other debris.
11. Count the number of fetuses and plan the experiment. For flow cytometry analysis, four to six fetal thymuses are usually used for one group of experiments. Fetuses can be temporarily stored in a refrigerator or on ice while preparing culture wells as in **Subheading 3.2.**

3.2. Preparation of Culture Wells

1. Cut collagen sponge into approx 1-cm^2 pieces using a clean set of sterile scissors and forceps.
2. Place one piece of the sponge in a culture well of a 24-well plate.
3. Fill the culture well with 1 mL culture medium.
4. Flip the sponge with forceps, so that the smooth side of the sponge faces up.
5. Place a piece of sterile PC membrane on each sponge. Flip the membrane with forceps, so that both sides of the membrane are completely wet with culture medium.
6. Gently remove 0.5 mL of the medium from each well using a 1-mL pipet. The final volume of the culture medium is 0.5 mL per well.

3.3. Isolation and Organ Culture of Fetal Thymus Lobes

1. Place a dissecting microscope in the culture hood.
2. Prepare a surgery dish by wetting a 2×2-in.^{2} gauge sponge in a 100-mm dish with approx 5 mL medium.
3. Wash two sterile no. 7 forceps with culture medium to remove all traces of ethanol, because fetal thymocytes tend to die following exposure to ethanol.
4. The following procedures (**step 5–9**) are done using no. 7 forceps under the microscope.
5. Place a fetus in the surgery dish under the microscope and turn the abdomen up (*see* **Fig. 1A,B**).
6. Raise the head (*see* **Fig. 1C**).
7. Gently open the chest and locate the two lobes of the thymus (*see* **Fig. 1D,E**).
8. The thymus lobes are removed from the body by raising them with forceps so that the whole lobe is lifted. The isolated lobes are placed on gauze, prewetted with culture medium, to remove blood (*see* **Fig. 1F** and **Note 4**).
9. Place thymus lobes onto the filter membrane in a culture well. Usually, four to six lobes are placed on each membrane (*see* **Fig. 1G**). Try to randomize the way the lobes are placed. For example, two lobes from one fetus should be divided into different groups when multiple experimental groups are set up.
10. Ascertain that the lobes are placed at the interface between the membrane and air. The lobes should not be sunk in culture medium (*see* **Note 5** for alternate method describing the addition of reagents to the cultures).
11. Add 1–2 mL of fresh culture medium to each empty well of the 24-well plate to minimize evaporation from the culture wells.
12. Place the culture plate in a 37°C, 5% CO_2 incubator.

3.4. Isolation of Single-Cell Suspensions From Fetal Thymus Organ Culture

1. Make a drop of 100 µL of the suspension buffer at the center of the reverse side of the lid of a 30-mm dish.
2. Transfer thymus lobes into the drop using no. 7 forceps. Count the number of lobes.
3. Place a small (approx 5 mm^2) piece of nylon mesh on the drop.

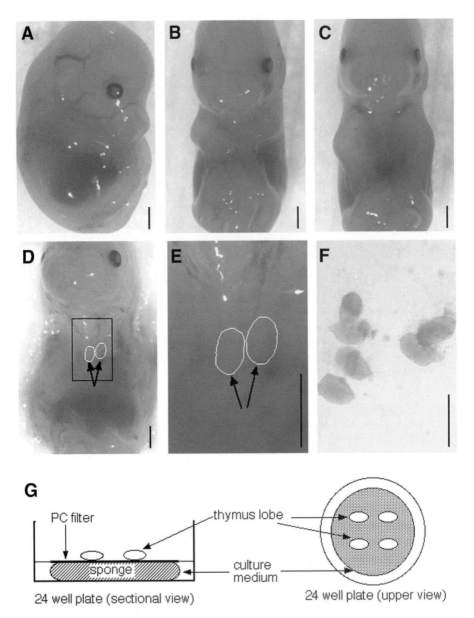

Fig. 1. Isolation of thymus lobes from fetal mice: (A) A fetus at gestational age d 14.5 from a C57BL/6 mouse is placed under dissecting microscope; (B) the fetus is turned so that the abdomen faces up; (C) the neck is raised up to expose the chest; (D) the chest is opened to expose two thymus lobes as shown by arrows; (E) high magnification of (D); arrows indicate two thymus lobes in the chest; (F) isolated thymus lobes; (G) diagram of culture well for FTOC. Scale bar = 1 mm.

4. Attach 26-gage needles to 1-mL syringes. Bend the tip (top 5 mm, 90° angle) of the needles, using forceps. Two needle/syringe sets are needed per group.

5. Gently tease the lobes by softly pressing the lobes with needles under a small piece of nylon mesh (approx 5 mm square) to release thymocytes. If needed, use a dissecting microscope.

6. Transfer the cell suspension to a plastic tube and count cell numbers. Use the cell suspensions for further examination of T-cell development (e.g., immunofluorescence and flow cytometry analysis) (*see* **Fig. 2** and **Notes 6–9**).

3.5. Optional Technique: Hanging-Drop Reconstitution of Deoxyguanosine-Treated Thymus Lobes With T-Precursor Cells

The hanging-drop-mediated reconstitution technique is useful for testing the developmental potential of T-precursor cells in fetal thymus lobes. T-Precursor cells from a given genetic background and/or with a given gene modification can be used for the reconstitution.

1. Thymus lobes from fetal mice at d 14.5 or d 15.5 of gestation are cultured as in **Subheading 2.3.** in the presence of 1.35 mM of 2-deoxyguanosine (dGuo) for 5–7 d (*see* **Note 5**). In a typical experiment, 10–20 thymus lobes are treated with dGuo (*see* **Note 10**).

2. Fill a 30-mm sterile dish with 3–4 mL of culture medium. Detach individual thymus lobes from the filter membrane into the medium using sterile forceps and a micropipet. Swirl the thymus lobes in the culture medium.

3. Transfer the lobes to fresh culture medium using a micropipet.

4. Diffuse away dGuo in a 37°C, 5% CO_2 incubator for about 1 h with two additional transfers into fresh medium.

5. Transfer 15 μL of culture medium containing one dGuo-treated thymus lobe per well of a Terasaki plate.

6. Add 20 μL culture medium containing T-precursor cells (e.g., 100–1000 fetal thymocytes or 1000–10,000 fetal liver cells).

7. Place the lid on the plate and gently invert.

8. Ascertain that thymus lobes are located at the bottom of the drop. If not, gently pipet the well.

9. Culture in a 37°C, 5% CO_2 incubator for 1 d.

10. Transfer the thymus lobes to a freshly prepared filter/sponge for regular thymus organ culture conditions (*see* **Subheading 2.3.**). Thymus lobes can be rinsed with fresh culture medium, in order to remove the cells that merely attach to the surface but do not enter the thymus organ.

11. Culture in a 37°C, 5% CO_2 incubator. Cultures can be evaluated in various ways, including cell number counting and flow cytometric examination of T-cell development. Typical results of T-cell development in this culture method can be found in **refs.** *6* and *14*.

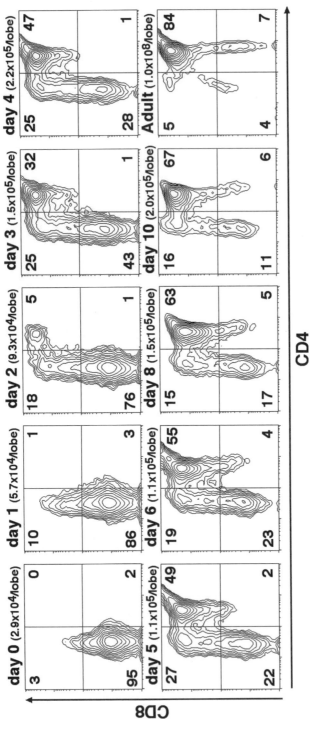

Fig. 2. T-Lymphocyte differentiation in FTOC. Contour histograms indicate CD4/CD8 two-color immunofluorescence profiles of thymocytes generated in FTOC. Day 14.5 fetal thymus lobes from C57BL/6 mice were organ-cultured for the indicated number of days. Numbers within the box indicate frequency of the cells in that box. Cell numbers recovered per thymus lobe are indicated in parentheses. The profile of cells isolated from an adult thymus is also shown.

124

3.6. Optional Technique: High-Oxygen Submersion Culture of Fetal Thymus Lobes

T-cell development in fetal thymus lobes may occur in a submersion culture under a high oxygen pressure. The method of a high-oxygen culture is useful for reconstitution of the thymus lobes using a limited number of T-precursor cells (*see* **Note 11**).

1. Fetal thymus lobes are placed in round-bottom wells of a 96-well plate (1 lobe/well). For the reconstitution of deoxyguanosine-treated thymus lobes, cells for the reconstitution are also included in the culture (*see* **Note 11**).
2. Spin the plate at $150g$ for 30 s to settle the thymus lobes at the very bottom of the well.
3. Place the culture wells in a plastic bag (3–5 L), fill the bag with a gas consisting of 70% O_2, 25% N_2, and 5% CO_2, and heat-seal the bag.
4. Place the bag in a 37°C, 5% CO_2 incubator. Cultures can be evaluated in various ways, including cell number counting and flow cytometric examination of T-cell development *(15)*.

3.7. Optional Technique: Time-Lapse Visualization of Thymic Emigration Using Transparent Fetal Thymus Organ Culture

To directly examine the mechanisms that mediate the emigration of newly generated T-cells out of the thymus, a time-lapse FTOC visualization system has been devised in which cell movement from the FTOC is directly monitored under a microscope and recorded using a digital CCD camera (*see* **Fig. 3A**) *(16)*. As shown in **Fig. 3B**, many cells are attracted out of the FTOC toward the spot of CCL19 within 1 d in culture. Most thymus emigrants are indeed mature T-cells *(16)*. The time-lapse visualization of FTOC is useful for the analysis of cellular movement during T-cell development.

1. Thymus lobes from d 15.5 C57BL/6 fetal mice are cultured for 5 d in standard FTOC conditions.
2. Thymus lobes are washed once, placed in a 30-mm dish, and submerged in 2 mL of ice-cold culture medium containing 1.08 mg/mL collagen.
3. Place the dish in a 37°C, CO_2 incubator for 5 min, to solidify the collagen.
4. An aliquot of CCL19 (10 μM, 5 μL) is spotted into the gel at approx 10 mm distant from the thymus lobe.
5. The dish is cultured at 37°C in 70% O_2 and 5% CO_2 atmosphere on the stage of microscope.
6. The culture is time-lapse monitored using a CCD camera.

A

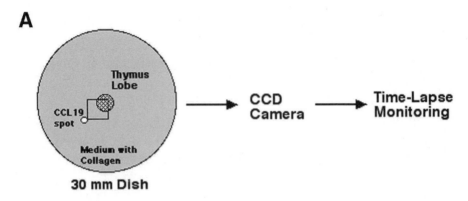

B

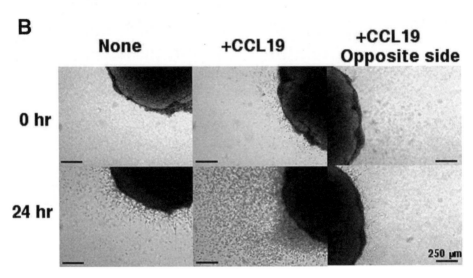

Fig. 3. Time-lapse visualization of thymic emigration using transparent FTOC: (**A**) a diagram of the culture; (**B**) edges of FTOC-cultured thymus lobes visualized at indicated time-points of the culture. The culture containing a CCL19 spot showed orientation-specific thymocyte emigration *(16)*.

3.8. Reaggregate Thymus Organ Culture

Reaggregate thymus organ culture (RTOC) provides a model in which the cellular interactions required for T-lymphocyte development can be studied under controlled in vitro conditions *(17)*. In this model, thymus lobes are depleted of endogenous T-cell progenitors by treatment with dGuo (*see* **Subheading 3.5.**). Surviving stromal cells are then enzymatically dissociated to generate single-cell suspensions. The cell slurry generated by centrifugation of

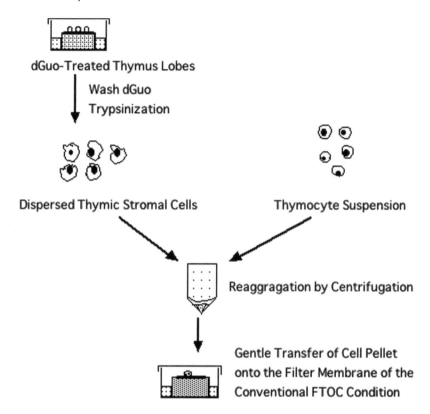

Fig. 4. Schematic diagram of reaggregate thymus organ culture.

a mixture of thymocytes and stromal cells reforms a structure resembling a thymus lobe-like structure (*see* **Fig. 4**).

3.8.1. Preparation of Thymic Stromal Cells

1. Culture d 15.5 fetal thymus lobes in the presence of 1.35 m*M* dGuo for 5–7 d to deplete them of lymphoid elements (*see* **Subheading 3.5.** and **Note 12**).
2. Fill a 30-mm sterile dish with 5 mL of culture medium. Transfer the dGuo-treated thymus lobes from the filter membrane to the culture medium using sterile forceps and a micropipet.
3. Transfer the lobes to Ca^{2+}-free and Mg^{2+}-free PBS with a micropipet.
4. Diffuse away dGuo at 37°C for 20 min.
5. Repeat step 3 and step 4 three times to wash out any residual dGuo.
6. Harvest the thymus lobes to a sterile 1.5-mL Eppendorf tube or a 24-well plastic well, and remove the supernatant.
7. Dissociate the thymus lobes by adding 1 mL of 0.125% trypsin–EDTA solution in Ca^{2+}-free and Mg^{2+}-free PBS for 30 min at 37°C.

8. Stop trypsinization by the addition of 1 mL of FCS-containing culture medium.
9. Disperse the stromal cells by vigorous pipetting.
10. Pass the dispersed stromal cell suspensions through 100-μm nylon mesh to remove the clumps.
11. Spin down and discard the supernatant.
12. Resuspend the cells in 200 μL FCS-containing culture medium and determine the cell number (*see* **Note 12**). If needed, cells can be stained with fluorescence-labeled antibodies and sorted by flow cytometry (*see* **Note 13**).

3.8.2. Formation of Reaggregates

1. Mix thymocyte populations of interest (*see* **Note 14**) with dispersed stromal cells at a ratio of 1:1 to 3:1 in a sterile 1.5-mL Eppendorf tube. Typically, $(3–5) \times 10^5$ thymocytes mixed with an equal number of thymic stromal cells are used.
2. Spin down the cells into a pellet at 1800 rpm (300g) for 5 min.
3. Gently remove the supernatant.
4. Disperse the cell pellet into a slurry by careful mixing with a micropipet and draw the slurry into a tip (or mix with a vortex mixer and draw into a fine, mouth-controlled glass capillary pipet).
5. Transfer and expel the slurry as a discrete standing drop on the surface of a PC filter prepared for conventional FTOC condition (*see* **Subheading 3.2.**). The cell "slurry" reaggregates will reform a thymus lobe-like structure within 12 h. Maintain the RTOC in a 37°C, 5% CO_2 incubator (*see* **Note 15**).

3.9. Optional Technique: Retroviral Gene Transfer Into Developing Thymocytes in Fetal Thymus Organ Culture

Retroviral gene transfer into developing thymocytes in FTOC provides a quick and economical method (versus germline transgenesis) to explore gene functions during T-cell development. Immature thymocytes can be efficiently and rapidly infected with a retrovirus using the spin-fection method. Gene-transferred cells can be readily detected and sorted using flow cytometry, by the coexpression of marker proteins such as green fluorescent protein (GFP). Retrovirus vectors expressing GFP along with a gene of interest using the internal ribosomal entry site (IRES) sequence have been widely used. A high-titer retrovirus can be produced by a transient transfection of the packaging cells with a retroviral plasmid. Plat-E packaging cells (*12*), combined with the pMRX-IRES-EGFP plasmid vector (*13*), are excellent for producing high-titer retroviruses. Other packaging cells and virus constructs can also be used.

3.9.1. Preparation of the Retroviral Supernatant

1. Set up the Plat-E cell culture. For a 10-cm dish, 2.5×10^6 cells are seeded in 10 mL of culture medium without puromycin and blasticidin S. Cells are cultured in a 37°C, 5% CO_2 incubator for 18–24 h.

2. Transfect Plat-E cells with retroviral plasmid DNA. For a 10-cm dish of Plat-E cells, 30 μg of DNA is introduced by the conventional calcium phosphate precipitation method (*see* **Note 16**). Twelve hours after the transfection, remove the supernatant containing precipitates, gently wash the cells with PBS, and add 10 mL of fresh medium.
3. Thirty-six hours after the transfection, collect culture supernatants containing retroviruses. The supernatant should be filtered through 0.2-μm syringe filters, and can be stored at –80°C or used immediately. After collecting the supernatant, cells can be used for further retroviral production. To do so, gently add 10 mL of fresh culture medium to the plate and continue culture in a 37°C, 5% CO_2 incubator. Retroviral supernatants can be collected every 12 h between 36 and 72 h after transfection (*see* **Note 17**).

3.9.2. Retroviral Infection of the Thymocytes

1. For gene transfer into CD4⁻CD8⁻ thymocytes, prepare a single-cell suspension of d 14 or 15 mouse fetal thymocytes (*see* **Subheading 3.4.**). For CD4⁺CD8⁺ thymocytes, prepare total thymocytes from neonatal mice (d 0 to 14). Add 500 μL retroviral supernatant (*see* **Note 18**) and 1.2 μL of 10 mg/mL polybrene (final concentration, 20 μg/mL) into each well of a 24-well plate containing the thymocyte suspension ($[1–10] \times 10^5$ cells/100 μL) in culture medium (*see* **Subheading 2.2.**).
2. Seal the plate with parafilm and spin at 1000*g* for 1 h at 30°C.
3. Transfer cells into a sterile 1.5-mL microtube, spin at 400*g* for 5 min, remove supernatant, and resuspend the cells in an appropriate volume (e.g., 100 μL) of fresh culture medium.
4. The developmental fate of retrovirus-infected thymocytes is assessed by transfer to FTOC (*see* **Note 19**).
5. Alternatively, infected cells can be cultured in a 37°C, 5% CO_2 incubator (*see* **Note 20**).

4. Notes

1. Timed pregnant mice may be purchased from various mouse suppliers. Generally, eight fetuses are expected from a pregnant C57BL/6 mouse. Because the numbers of fetuses can differ, it is necessary to check the number of fetuses in each mouse strain. If FTOC is an unfamiliar technique, preliminary organ cultures of d 15.5 fetal thymus lobes for 4–5 d are recommended. The fetuses and fetal thymuses are easiest to handle at d 15.5 of gestation.
2. It is important to screen the FCS for FTOC. We usually prescreen 10–20 independent lots of FCS by overnight suspension culture of adult thymocytes followed by determination of cell numbers recovered the following morning. The five or six best FCS lots that allow cell recovery close to 100% are selected for further screening in an actual test of T-cell development in FTOC. Progression along the CD4/CD8 developmental pathway yielding profiles and cell numbers

as shown in **Fig. 2** would be a good indication of expected T-cell development in culture and thus an acceptable FCS lot.

3. Fetuses with deviated developmental features as judged by size and other developmental signs such as the formation of hair follicles and crests in the limbs (*see* **refs.** *9–11*) should be eliminated. The deviation in developmental stage of the fetuses will dramatically affect the stages of T-cell development in the thymus (*see* **Fig. 2**).

4. This technique could be difficult for beginners. Adept handling of the forceps under the microscope needs practice.

5. When reagents are added, first remove 50 µL of culture medium. Then, add 50 µL (1:10 volume) of 10X concentrated reagents slowly and directly onto the lobes.

6. In order to examine T-cell development in FTOC, we generally use flow cytometry *(16)*. The two-color profiles of CD4/CD8 and CD25/CD44 are commonly used.

7. The advantages of FTOC for analyzing T-cell development include reproducibility and the convenience of in vitro cultures. Disadvantages include the limitation of cell numbers and necrotic cell death in the middle of the thymus lobe, which is not observed in the physiological thymus in vivo (*see* **Fig. 2**) *(7)*.

8. If FTOC is an unfamiliar technique, preliminary organ cultures of d 15.5 fetal thymus lobes for 4–5 d are recommended. The fetuses and fetal thymuses are easiest to handle at d 15.5 of gestation.

9. Neonatal thymus organ culture (NTOC) has been used for the analysis of positive selection signals inducing the generation of mature "single-positive" thymocytes *(18,19)*. NTOC of d 0 newborn thymus lobes is useful for in vitro stimulation of in vivo generated CD4+CD8+ thymocytes. However, it should be noted that, unlike FTOC, total cell numbers decrease during 4- to 5-d cultures in the NTOC condition *(7)*, which could complicate the interpretation of obtained results.

10. For the dGuo treatment *(20)*, fetal thymus lobes should be cultured with dGuo for at least 5 d. Otherwise, residual T-cell precursors retain their developmental potential and undergo T-cell development. Thymus lobes cultured for 7–8 d with dGuo are still capable of supporting T-cell development of reconstituted precursor cells.

11. High-oxygen submersion cultures of FTOC *(15)* are useful for reconstitution using limited numbers of progenitor cells, because the thymus lobes can be continuously cultured at the bottom of round or V-shaped culture wells and the entry of progenitor cells can occur efficiently during the culture with the help of gravity. However, it should be noted that T-cell development in this high-oxygen condition seems to occur more rapidly than T-cell development in vivo or in regular FTOC conditions.

12. To prepare the thymic stromal cells for RTOC, dGuo-treated d 14.5 to d 15.5 fetal thymus lobes can be used. Then, $(5–6) \times 10^4$ thymic stromal cells can be isolated from one dGuo-treated d 15.5 thymus lobe. Cell numbers obtained from

one dGuo d 15.5 thymus lobe are about 1.5-fold to 2-fold higher than the numbers from one dGuo d 14.5 thymus lobe.

13. The thymic stroma is made up of a number of different stromal cell types. To study the interactions between thymocytes and a defined thymic stromal cell population, such as MHC class II$^+$ thymic epithelial cells or MHC class II$^-$ mesenchymal cells, thymic stromal cells isolated from dGuo-treated fetal thymus lobes can be stained using anti-MHC II and anti-CD45 antibodies and purified by flow cytometry or magnetic cell sorting (MACS). Anti-CD45 antibody staining is used to deplete CD45$^+$ thymocytes and dendritic cells that survive even after the dGuo treatment.

14. Thymocytes for RTOC can be CD4$^-$CD8$^-$ double-negative (DN) thymocytes, CD4$^+$CD8$^+$ double-positive (DP) thymocytes, or even semimature CD4$^+$CD8$^-$/CD4$^-$CD8$^+$ single positive (SP) thymocytes, depending on the purpose of the experiment. Thymocyte populations can be prepared from adult thymuses, newborn thymuses, or fetal thymuses. Cells from different species can also be used. Cell sorting or MACS can be employed to purify thymocyte populations.

15. To form a reaggregate lobe on the filter membrane *(21)*, it is important to keep the surface of filter membrane dry and to keep the volume of the transferred cell slurry low, usually at 2–4 µL.

16. Mix 60 µL of 2M CaCl$_2$, 30 µL of DNA solution (1 µg/µL), and 360 µL of distilled water in a sterile 1.5-mL microtube. Add this solution quickly into 450 µL of 2X HBS (HEPES-buffered saline; 140 mM NaCl, 1.5 mM Na$_2$HPO$_4$, 50 mM HEPES [pH 7.05]) in a 1.5-mL microtube and mix by pipetting. Gently add this solution containing calcium phosphate–DNA coprecipitates onto precultured Plat-E cells. Thirty minutes later, check the formation of precipitates under the microscope. FuGene (Roche Applied Science), instead of the calcium phosphate coprecipitation, can also work for the transfection of Plat-E cells.

17. The efficiency of the transfection should be monitored after the collection of retroviruses. Transfected Plat-E cells can be trypsinized and analyzed for GFP expression by flow cytometer. In general, transfection efficiency ranges from 50% to 90%.

18. Frozen retroviral suspensions should be quickly thawed in a 37°C water bath immediately before use.

19. CD4$^-$CD8$^-$ thymocytes can be transferred to dGuo-treated fetal thymus lobes by the hanging-drop method (*see* **Subheading 3.5.** and **Fig. 5A**). CD4$^+$CD8$^+$ thymocytes should be reaggregated with dGuo-treated thymic stromal cells (*see* **Subheading 3.8.** and **Fig. 5B**). Retrovirus-infected cells present after FTOC can be detected by GFP expression using flow cytometry (*see* **Note 6**).

20. After 18–24 h of culture, retroviral infection can be evaluated by GFP expression (*see* **Fig. 5C**). It should be noted that GFP expression is not detectable immediately after the spin-fection and is generally detected 18–24 h after transfection. To maintain the developmental potential and survival of immature thymocytes, IL-7 (Sigma; final concentration, 1–5 ng/mL) can be added to the culture. GFP$^+$ cells can be purified by cell sorting and then be transferred to FTOC.

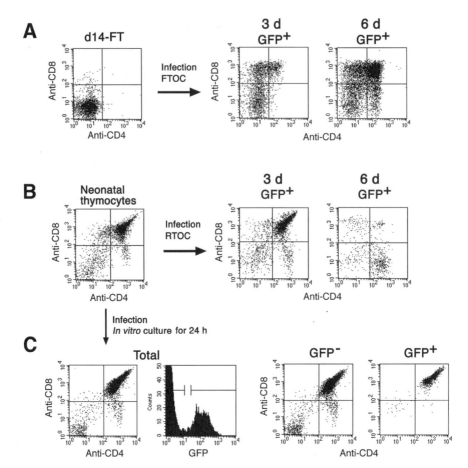

Fig. 5. In vitro reconstitution of the thymus by retrovirus-infected thymocytes: (**A**) Day 14.5 fetal thymocytes were infected with the pMRX-IRES-EGFP retrovirus and were cultured in a deoxyguanosine-treated fetal thymus for indicated number of days. Dot plots indicate CD4/CD8 immunofluorescence profiles. (**B**) Total thymocytes from neonatal mice were infected with the pMRX-IRES-EGFP retrovirus and reaggregated with thymic stromal cells. RTOC was cultured for indicated number of days. (**C**) Neonatal thymocytes in panel **B** were cultured in vitro for 24 h after infection. A histogram indicates GFP expression. The CD4/CD8 expression profiles of the GFP⁻ and GFP⁺ fractions are also shown.

References

1. Owen, J. J. T. and Ritter, M. A. (1969) Tissue interaction in the development of thymus lymphocytes. *J. Exp. Med.* **129,** 431–442.
2. Owen, J. J. T. (1974) Ontogeny of the immune system. *Prog. Immunol.* **2,** 163–173.
3. Mandel, T. and Russel, P. J. (1971) Differentiation of foetal mouse thymus. ultrastructure of organ cultures and of subcapsular grafts. *Immunology* **21,** 659–674.

4. Mandel, T. E. and Kennedy, M. M. (1978) The differentiation of murine thymocytes in vivo and in vitro. *Immunology* **35,** 317–331.
5. Jenkinson, E. J., van Ewijk, W., and Owen, J. J. T. (1981) Major histocompatibility complex antigen expression on the epithelium of the developing thymus in normal and nude mice. *J. Exp. Med.* **153,** 280–292.
6. Kingston, R., Jenkinson, E. J., and Owen, J. J. T. (1985) A single stem cell can recolonize an embryonic thymus, producing phenotypically distinct T-cell populations. *Nature* **317,** 811–813.
7. Takahama, Y. (2000) Differentiation of mouse thymocytes in fetal thymus organ culture, in *T Cell Protocols. Development and Activation* (Kearse, K. P., ed.), Humana, Totowa, NJ, pp. 37–46.
8. Takahama, Y., Hasegawa, T., Itohara, S., Ball, E. L., Sheard, M. A., and Hashimoto, Y. (1994) Entry of CD4⁻CD8⁻ immature thymocytes into the CD4/CD8 developmental pathway is controlled by tyrosine kinase signals that can be provided through T cell receptor components. *Int. Immunol.* **6,** 1505–1514.
9. Theiler, K. (1989) *The House Mouse.* Springer-Verlag, New York.
10. Kaufman, M. H. (1992) *The Atlas of Mouse Development,* Academic, San Diego, CA.
11. Butler, H. and Juurlink, B. H. (1987) *An Atlas for Staging Mammalian and Chick Embryos.* CRC, Boca Raton, FL.
12. Morita, S., Kojima, T., and Kitamura, T. (2000) Plat-E: an efficient and stable system for transient packaging of retroviruses. *Gene Therapy* **7,** 1063–1066.
13. Saitoh, T., Nakano, H., Yamamoto, N., and Yamaoka, S. (2002) Lymphotoxin-β receptor mediates NEMO-independent NF-κB activation. *FEBS Lett.* **532,** 45–51.
14. Tsuda, S., Rieke, S., Hashimoto, Y., Nakauchi, H., and Takahama, Y. (1996) IL-7 supports D-J but not V-DJ rearrangement of TCR-β gene in fetal liver progenitor cells. *J. Immunol.* **156,** 3233–3242.
15. Watanabe, Y. and Katsura, Y. (1993) Development of T cell receptor βα -bearing T cells in the submersion organ culture of murine fetal thymus at high oxygen concentration. *Eur. J. Immunol.* **23,** 200–205.
16. Ueno, T., Hara, K., Swope Willis, M., et al. (2002) Role for CCR7 ligands in the emigration of newly generated T lymphocytes from the neonatal thymus. *Immunity* **16,** 205–218.
17. Jenkinson, E. J., Anderson, G., and Owen J. J. T. (1992) Studies on T cell maturation on defined thymic stromal cell populations in vitro. *J. Exp. Med.* **176,** 845–853.
18. Takahama, Y., Suzuki, H., Katz, K. S., Grusby, M. J., and Singer, A. (1994) Positive selection of CD4⁺ T cells by TCR ligation without aggregation even in the absence of MHC. *Nature* **371,** 67–70.
19. Takahama, Y. and Nakauchi, H. (1996) Phorbol ester and calcium ionophore can replace TCR signals that induce positive selection of CD4 T cells. *J. Immunol.* **157,** 1508–1513.
20. Jenkinson, E. J., Franchi, L. L., Kingston, R., and Owen, J. J. T. (1982) Effect of deoxyguanosine on lymphopoiesis in the developing thymus rudiment in vitro: application in the production of chimeric thymus rudiments. *Eur. J. Immunol.* **12,** 583–587.
21. Anderson, G., Jenkinson, E. J., Moore, N. C., and Owen, J. J. T. (1993) MHC class II-positive epithelium and mesenchyme cells are both required for T-cell development in the thymus. *Nature* **362,** 70–73.

10

In Vitro Generation of Lymphocytes From Embryonic Stem Cells

Renée F. de Pooter, Sarah K. Cho, and Juan Carlos Zúñiga-Pflücker

Summary

Lymphocytes arise during ontogeny via a series of increasingly restricted intermediates. Initially, the mesoderm gives rise to hemangioblasts, which can differentiate into endothelial precursors, or hematopoietic stem cells (HSCs). HSCs can either self-renew or differentiate into lineage-restricted progenitors and, ultimately, to mature effector cells. This complex process is only beginning to be understood, and the ability to generate lymphocytes from embryonic stem (ES) cells in vitro will facilitate further study by providing a model system in which the effects of genetic and environmental manipulations of ES-cell-derived progenitors can be examined. In this protocol, we describe procedures for generating either B- and NK- or T-lymphocytes from mouse ES cells in vitro.

Key Words: Lymphocyte development; T-cell development; fetal thymic organ culture; reaggregate thymic organ culture; hematopoiesis; B-cell development; hemangioblast; Flk-1; Flt-3L; IL-7; embryonic stem cells; stromal cells.

1. Introduction

Two approaches have successfully generated lymphocytes from embryonic stem (ES) cells. In the first, ES cells were differentiated in vitro into three-dimensional embryoid bodies containing hematopoietic progenitors and then adoptively transferred into recipient hosts to complete their maturation *(1–3)*. This approach is ideal for studying questions of engraftment, homing, and migration in the in vivo context, but it cannot address the identity of the maturational intermediates and, in the case of failure to engraft, cannot distinguish between defects in survival versus homing. Thus, further information could be gained from a system that allows ES cell differentiation into lymphocytes wholly in vitro.

From: *Methods in Molecular Biology, vol. 290: Basic Cell Culture Protocols, Third Edition*
Edited by: C. D. Helgason and C. L. Miller © Humana Press Inc., Totowa, NJ

In vitro differentiation of ES cells into lymphocytes can be achieved by coculturing ES cells with the bone marrow stromal cell line OP9 (ES/OP9 coculture) *(4,5)*, derived from mice deficient in macrophage colony-stimulating factor (M-CSF). The absence of M-CSF prevents macrophages from overwhelming other lineages in the coculture *(6)*. Nakano et al. demonstrated that OP9 stromal cells support the differentiation of ES cells into multiple hematopoietic lineages, including B-cells, although the efficiency of B-cell generation in this system was initially low. However, several cytokines are known to potentiate B-cell production, including stem cell factor (SCF), Flt-3 ligand (Flt-3L), and interleukin (IL)-7 *(7–12)*. In particular, Flt-3L synergizes with IL-7 to promote the growth of lymphoid progenitors *(13)* and Cho et al. demonstrated that the addition of exogenous Flt-3L and IL-7 to ES/OP9 cocultures allowed for the efficient and consistent production of B-lymphocytes *(14)*. Similarly, the addition of IL-15, which has been shown to be involved in NK-cell development *(15)*, enhanced the yield of NK-cells. This system permits detailed molecular studies under various culture conditions and allows the manipulation of ES cells throughout the stages of differentiation from progenitor to mature lymphocyte.

T-cell potential, however, remained elusive. Hypothesizing that the OP9 bone marrow stromal cells might be inducing the commitment of early hematopoietic progenitors to non-T-cell lineages, prehematopoietic Flk-1$^+$ CD45$^-$ cells resembling hemangioblasts were isolated from cocultures and seeded into fetal thymic organ cultures (FTOCs). This modification allowed for the in vitro development of T-cells from ES cells *(16)*. The efficiency of generating T-cells from ES-derived Flk-1$^+$ cells was enhanced by a further modification of this approach, in which progenitors are combined with freshly isolated thymic stroma to create reaggregate thymic organ cultures (RTOCs), which are then deposited as free-standing drops. These drops reform a three-dimensional thymic environment that can support T-cell development, allowing ES cell-derived T-cells to be generated in the RTOC microenvironments *(16)*. These findings demonstrate that T-cells can be generated from Flk-1$^+$ CD45$^-$ ES-derived cells in vitro. In this protocol, we describe procedures for generating either B- and NK- or T-lymphocytes from mouse ES cells in vitro.

2. Materials
2.1. Cellular Components
2.1.1. ES Cells and EF Cells

1. Embryonic stem cells (R1, D3, and E14K derived from 129/Sv mice and ES cells derived from BALB/c and C57BL/6 mice have all been used to generate lymphocytes in vitro).

2. Mouse embryonic fibroblast (EF) cells *(17)*.
3. Fetal bovine serum (FBS). Different sources are required for ES vs OP9/coculture media. Heat-inactivate at 56°C for 30 min and store at 4°C (*see* **Note 1**).
4. High-glucose Dulbecco's modified Eagle's medium (DMEM) (Sigma D-5671). Store at 4°C.
5. 1X Phosphate-buffered saline (PBS) without Ca^{2+}/Mg^{2+} (Gibco 14190-144). Store at room temperature.
6. HEPES, sodium pyruvate, gentamicin (HSG) solution (5 mL HEPES 100X or 1 *M*, Gibco 15630-080; 5 mL sodium pyruvate 100X or 100 m*M*, Gibco 11360-070; 0.5 mL gentamicin 1000X or 50 mg/mL, Gibco 15750-060, aliquoted into 14-mL conical tubes). Store at 4°C; stable for approx 1 yr.
7. Penicillin/streptomycin, Glutamax, 2(β)-mercaptoethanol (PG2) solution (5 mL penicillin/streptomycin 100X or 10000 U/mL penicillin and 10000 µg/mL streptomycin, Gibco 15140-122; 5 mL Glutamax 100X or 200 m*M*, Gibco 35050-061; 0.5 mL β-mercaptoethanol 1000X or 55 m*M*, Gibco 21985-023, aliquoted into 14 mL conical tubes). Store at –20°C; stable for approx 1 yr.
8. ES media: 500 mL of high-glucose DMEM supplemented with 15% heat-inactivated FBS (iFBS), 10.5 mL PGS solution, and 10.5 mL HSG solution (one aliquot each).
9. 2.5% Trypsin (Gibco 15090-046). Dilute with PBS to 0.25% solution as needed and store at 4°C.
10. Mitomycin C solution. Make a 1-mg/mL (100X) mitomycin C (Sigma M-4287) stock solution in PBS. Store in the dark at 4°C; stable for 2 wk.
11. Mouse leukemia inhibitory factor (LIF) (Sigma L5158). Dilute to 7.5 µg/mL (1000X). Aliquot and store at –80°C.
12. Freezing media: 90% iFBS, 10% dimethyl sulfoxide (DMSO).
13. Tissue culture ware, tissue culture treated (suggested suppliers: Sarstedt or Falcon).
14. 70-µm Nylon mesh filter (N70R; BioDesign Inc., Carmel, NY).

2.1.2. OP9 Cells

1. OP9 cells (Riken cell repository; http://www.rtc.riken.go.jp).
2. α-Modified Eagle's medium (αMEM) (Gibco 12561-056). Store at 4°C.
3. OP9 medium: αMEM, supplemented with 20% iFBS and 5 mL of penicillin/streptomycin.

2.2. ES/OP9 Coculture

2.2.1. Coculture

1. Mouse IL-7 (R&D 407-ML). Reconstitute at 1 µg/mL (1000X). Aliquot and store at –80°C.
2. Mouse IL-15 (Peprotech 210-15). Reconstitute at 25 µg/mL (1000X). Aliquot and store at –80°C.
3. Human Flt-3L (R&D 308-FK). Reconstitute at 5 µg/mL (1000X). Aliquot and store at –80°C.

2.2.2. Isolating Flk-1⁺ Cells

1. Anti-Flk-1 phycoerythrin-conjugated antibody (Pharmigen 555308).
2. Magnetic assisted cell sorter (MACS) (Miltenyi Biotech).
3. MACS running buffer: 500 mL Ca/Mg-free Hanks balanced salt solution (HBSS) + 2 mL of 0.5 M EDTA, and 2.5 g BSA.
4. MACS wash buffer: 500 mL Ca/Mg-free HBSS + 2 mL of 0.5 M EDTA.
5. Anti-PE microbeads (Miltenyi Biotech 130-048-801). Microbeads should be stored at 4°C.
6. FTOC medium: 500 mL of high-glucose DMEM supplemented with 15% iFBS and one aliquot each of PGS and HSG solutions.

2.2.3. Preparing Thymic Stroma for RTOC

1. HBSS without Ca^{2+}/Mg^{2+} (Sigma H6648).
2. 0.05% Trypsin, 0.53 mM EDTA (Gibco 25300-054).

2.2.4. Forming RTOCs

1. Gelfoam (Pharmacia and Upjohn 09-0342-01-005), cut to fit six-well plate.
2. Autoclaved 25-µL nonheparinized capillary tube (Fisher Scientific 21-164-2E).
3. 2'-Deoxyguanosine (Sigma D-0901). Reconstitute in DMEM to 11–13.5 mM and store at –20°C.
4. Nucleopore membranes, 13 mm in diameter, 0.8-µm pores (Whatman 110409).

3. Methods

The methods described outline the following: (1) the maintenance of the required cell lines; (2) the coculture of ES cells on OP9 cells for the production of B-cells or NK-cells; and (3) the coculture of ES cells on OP9 cells and the use of RTOCs for the production of T-lymphocytes (*see* **Fig. 1**). It should be noted that all incubations are performed in a standard, humidified, cell culture incubator, at 37°C in 5% CO_2, all tissue culture ware is tissue culture treated, and cells are pelleted by centrifuging for 5 min at 1500 rpm (500g), unless otherwise indicated.

3.1. Cellular Components of Coculture System

3.1.1. ES Cells and EF Cells (see Fig. 2A)

1. The ES cells are maintained as adherent colonies on monolayers of growth-inactivated EF cells (*see* **Fig. 2A**) in ES media. EF inactivation can be performed by either irradiation (3000 cGy) or treatment with mitomycin C. In the case of treatment with mitomycin C, which is light sensitive, EF cells are incubated for 2.5 h in ES media with 10 µg/mL of mitomycin C. Wash three times with PBS and add fresh ES media. EF cells should be used within 5 d of either treatment.
2. For the maintenance of undifferentiated ES cells, Hyclone offers prescreened, characterized lots of FBS (*see* **Note 1**). ES cells should be thawed in a 37°C water bath and transferred to a 14-mL conical tube containing 10 mL of ES media.

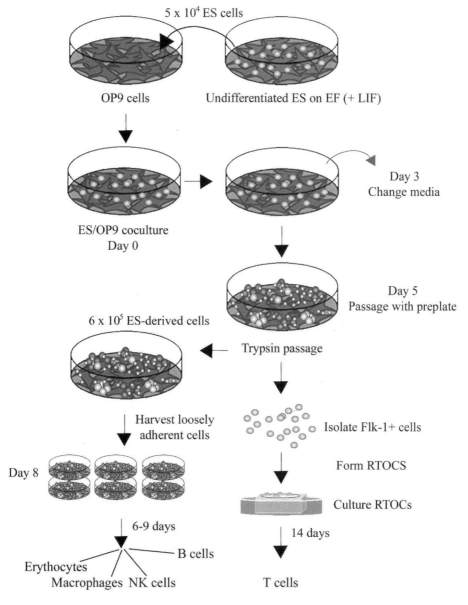

Fig. 1. Schematic overview of the ES/OP9 coculture system (*see* text for details).

Pellet the cells and resuspend in 3 mL of ES media to be plated on a 6-cm dish of approx 80% confluent inactivated EF cells. Add 3 µL of LIF. Change the media the next day, and passage to a fresh plate of inactivated EF the following day (*see* **step 3**), each time adding LIF. Maintain ES by repeating this procedure,

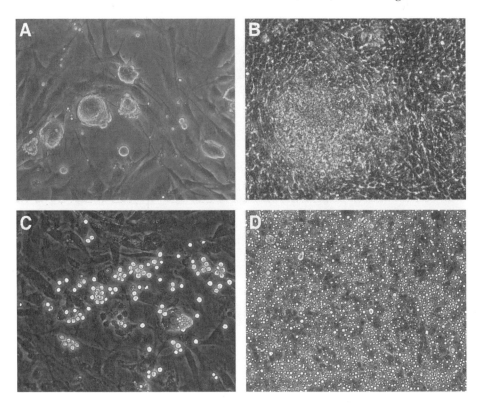

Fig. 2. Photomicrographs of cocultures at various time-points, under ×200 magnification. (**A**) Undifferentiated ES cells on a monolayer of irradiated EF cells; (**B**) mesoderm-like colony at d 5 of ES/OP9 coculture; (**C**) small clusters of hematopoietic cells shown at d 8 of ES/OP9 coculture; and (**D**) hematopoietic and lymphoid cells shown at d 15 of ES/OP9 coculture.

    alternating media changes and passages, and allowing them to become no more than 80% confluent.

3. To passage the ES cells, remove the media and wash the dish gently with 4 mL of PBS. Remove PBS and incubate the plate with 1 mL of 0.25% trypsin for 5 min. Wash the cells from the plate by adding 2 mL of ES media and pipetting vigorously. If the plate has become overconfluent or large colonies with borders of flattened, nonrefractive cells have formed, small, undifferentiated colonies can sometimes be restored by passing the cells through 70-µm Nylon mesh. Pellet cells and resuspend in 3 mL ES media with LIF. Remove the media from a fresh 6-cm dish of 80% confluent inactivated EF cells and add the resuspended ES cells. Gently swish the plate to disperse the cells and LIF.

4. To generate frozen stocks of ES cells, wash with PBS, treat with trypsin, and collect the cells as described in **step 3**. Resuspend the ES cells in ice-cold freez-

ing media and aliquot them into cryovials (two to four vials per confluent 6-cm plate of ES cells). Transfer the vials on ice to a –80°C freezer overnight, and the next day to liquid nitrogen for long-term storage.

3.1.2. OP9 Cells

1. Thaw a vial of OP9 cells as described for ES cells, but substitute OP9 media for ES media. Plate cells in a 10-cm dish with 8–10 mL of fresh OP9 media. Change the media the next day. OP9 cells should not be allowed to become more than 80% confluent and can generally be maintained by splitting 1:4 every 2 d.
2. To passage OP9 cells from a 10-cm plate, remove the media, wash with 6 mL of PBS, remove the PBS, and incubate for 5 min with 4 mL of 0.25% trypsin. Following trypsin treatment, prepare a 50-mL conical tube with 5 mL of OP9 media. Add 4 mL of PBS to the trypsin-treated plate, pipet vigorously, and add the cells to the tube containing media. OP9 cells, especially early-passage cells, are very adherent. Rinse the plate again with 8 mL of PBS and pool this with the first wash. Pellet the cells, resuspend them, and divide them among four 10-cm plates, or four six-well plates. Gently swish the plate to distribute the cells evenly (*see* **Notes 2** and **3**).

3.2. ES/OP9 Coculture (see *Fig. 1*)

The protocol for the differentiation of NK- and B-cells from ES cells is described in **Subheading 3.2.1.** This subsection describes (1) the preparation of the cells for coculture and (2) the production of B-cells and NK-cells. **Subheadings 3.2.2.–3.2.4.** describe (1) the isolation of Flk-1$^+$ prehematopoietic progenitors from cocultures and (2) the preparation of thymic stroma for RTOCs.

3.2.1. Coculture (see *Note 4*)

Day –6 to –2

1. Thaw the ES cells onto inactivated EF cells 4–6 d before beginning the coculture (d –6 to d –4).
2. Maintain undifferentiated ES cells as described in **Subheading 3.1.1.**
3. Thaw OP9 stromal cells at d –4.
4. At d –2, split a confluent plate of OP9 stromal cells onto four 10-cm plates.

Day 0

1. Remove the media from 10-cm dishes of OP9 stromal cells that are no more than 80% confluent and replace with 8 mL fresh OP9 media.
2. Aspirate the media from the ES cells and treat them with trypsin (described in **Subheading 3.1.1.**).
3. Disaggregate the cells by vigorous pipetting and add 6 mL of ES media.
4. Transfer the cells to a new empty 10-cm dish, with no pre-existing EF monolayer.
5. Incubate the cells for 30 min to allow the EF cells to settle and adhere to the plate (plate out).

6. Collect the nonadherent cells from the ES plate and pellet them.
7. Resuspend the ES cells in 3 mL ES media to count.
8. Dilute 5×10^4 ES cells into 2 mL of OP9 media and seed onto a 10-cm dish of 80% confluent OP9 stromal cells from **step 1**.

Day 3

1. Aspirate the coculture media without disturbing the cells or the monolayer.
2. Replace with 10 mL of fresh OP9 media.

Day 5 (*see* **Fig. 2B**)

1. Fifty to one hundred percent of colonies should have mesoderm characteristics (*see* **Fig. 2B**) *(5)* (*see* **Note 5**). Aspirate the media without disturbing the cells or the monolayer.
2. Wash with 10 mL of PBS and remove the PBS.
3. Add 4 mL of 0.25% trypsin to the plates and incubate for 5 min.
4. Disaggregate the cells by vigorous pipetting to create a homogenous suspension.
5. Add 4 mL of OP9 media and incubate the disaggregated cells for 30 min to plate out the OP9 cells.
6. Collect the nonadherent cells and pellet them.
7. Resuspend the cells in 2 mL of fresh OP9 media and count them.
8. Seed 6×10^5 cells per fresh 10-cm plate of 80% confluent OP9 stromal cells. If cells are to be analyzed by flow cytometry at later time-points, a good guideline is to seed one 10-cm plate of OP9 stroma per anticipated time-point (*see* **Note 6**).
9. Add Flt3-L to a final concentration of 5 ng/mL.

Day 8 (*see* **Fig. 2C**)

1. Small clusters of 4–10 round, refractile blastlike cells should be visible (*see* **Fig. 2C**). Transfer all of the culture media into a 50-mL conical tube. Gently wash the surface of the plate using a 10-mL pipet with 8 mL of PBS, attempting to not disrupt the OP9 monolayer. Transfer the wash into the same 50-mL conical tube, passing the wash through a 70-µm filter to exclude pieces of disrupted mono-layer. The object is to collect all round, loosely adherent, blastlike cells. Check by microscope if this has been accomplished.
2. Pellet the collected cells and resuspend them in 2 mL fresh OP9 media.
3. Transfer the cells to fresh six-well plates of 80% confluent OP9 stromal cells: one 10-cm plate's worth of cells is transferred to one well of a six-well plate, in 3 mL of OP9 media.
4. Add Flt3-L to a final concentration of 5 ng/mL.
5. For B-cell differentiation, add IL-7 to a final concentration of 1 ng/mL. For NK- cell differentiation, add IL-15 to a final concentration of 25 ng/mL.

Day 10

1. Change media by collecting culture media into a 14-mL tube and centrifuging.
2. Add 1 mL of fresh OP9 media to the wells to prevent the cells from drying out.

3. Resuspend any pelleted cells with 2 mL fresh OP9 media per well of the six-well plate.
4. Gently pipet the resuspended cells onto the original well without disrupting the monolayer.
5. Add cytokines to the final concentrations described in **Subheading 3.2.1.**, d 8.

Day 12

1. Passage the cells by vigorously pipetting to disrupt the monolayer and pass through a 70-μm mesh into a tube.
2. Pellet the cells and resuspend them in 3 mL per well in fresh OP9 media.
3. Transfer to the same number of wells in fresh six-well plates of 80% confluent OP9 stromal cells, with appropriate cytokines.

Beyond Day 12 (*see* **Fig. 2D**)

To continue the cultures beyond d 12 (*see* **Fig. 2D** for d 15), transfer the cells to fresh OP9 stroma every 4–6 d and change the media every 2–3 d. Alternate the media change and passage protocols described for d 10 and 12, respectively. Although, for efficient hematopoiesis, it is best to leave the cocultures undisturbed as much as possible, overconfluent OP9 monolayers differentiate into adipocytic cells that no longer support hematopoiesis and might begin to detach from the culture dish and roll up from the edges. Also note that B-cells are very sensitive to IL-7 withdrawal and will quickly die should it become exhausted in the culture medium (*see* **Note 6**).

3.2.2. Isolating Flk-1⁺ Cells

1. Follow the protocol, as described in **Subheading 3.2.1.** for the coculture, until d 5. As RTOCs (*see* **Subheading 3.2.4.**) require high numbers of progenitors, it is best to seed 10–15 plates of OP9 stromal cells at 0.
2. Plate out OP9 cells and undifferentiated ES cells as described in **Subheading 3.2.1.**, d 5.
3. Pellet cells and resuspend in 1.2 mL MACS running buffer per plate of cells.
4. Pool pellets and stain with 4 μL phycoerythrin (PE)-conjugated anti-Flk-1 antibody per 1.2 mL of buffer for 30 min, covered, on ice.
5. Rinse with 20 mL running buffer.
6. Reserve a small aliquot for comparison with later Flk-1⁺-enriched population.
7. Stain with anti-PE beads and enrich for Flk-1⁺ cells by MACS (or autoMACS) as described in the manufacturer's protocol.
8. Take a small aliquot for analysis by flow cytometry to ensure that population has been enriched for Flk-1⁺ cells.
9. Pellet cells and resuspend in 200 μL FTOC media. Count the cells.

3.2.3. Preparing Thymic Stroma for RTOC

1. Remove thymic lobes from 15 d postcoitus (dpc) fetal mice.
2. Treat with 1.1–1.35 m*M* deoxyguanosine for 5 d in FTOC on standard FTOC rafts *(18)* (*see* Chapter 9).

3. Transfer thymic lobes to 4 mL Ca^{2+} Mg^{2+}-free HBSS.
4. Irradiate with 1000 cGy.
5. Transfer lobes to an Eppendorf tube containing 600 µL of 0.05% trypsin and 0.53 mM EDTA solution (up to 20 pairs of fetal thymi) and incubate for 30–40 min.
6. Prepare a 5-mL capped polypropylene tube with 600 µL of FTOC media.
7. Transfer lobes using a p1000 pipet to the tube containing media and pipet until thymi are disrupted (3–5 min).
8. Remove any remaining debris by passing through a 70-µm Nylon mesh filter into a fresh 5-mL tube.
9. Pellet cells in a serofuge at high speed (1000g) for 2 min and resuspend in 200 µL of FTOC media to count.

3.2.4. Forming RTOCs

1. Using a razor blade, cut a 10 µL pipet tip until it will just accommodate the diameter of an autoclaved 25-µL nonheparinized capillary tube.
2. Using this altered pipet tip, attach a capillary tube to a 10-µL pipet.
3. Combine approx 1×10^6 each of fetal thymic stroma and Flk-1$^+$-enriched cells in a 1:1.2 to 1:3 ratio in a 1.5-mL eppendorf tube to make 3–4 RTOCs.
4. Pellet the cells and aspirate almost all of the media. The amount of media remaining on top of the pellet should be about half the height of the pellet itself.
5. Resuspend the pellet by gently tapping the tube, to create a thick slurry.
6. Using the adapted pipet–capillary tube hybrid created in **step 1**, deposit 2.5- to 3-µL drops of cell slurry as standing drops onto the nucleopore membranes of the FTOC rafts. The pipet should be held as perpendicular to the surface of the raft as possible.
7. Add Flt3-L and IL-7 to media in the wells containing the rafts, to final concentrations of 5 and 1 ng/mL, respectively.
8. Carefully transfer the RTOCs to a well-humidified incubator.
9. Change the media every 6 d by aspirating the old media without disturbing the RTOC rafts, replacing it with fresh FTOC media and cytokines.
10. Harvest after 12–19 d (*see* **Note 7**).

4. Notes

1. Although Hyclone offers prescreened characterized lots of FBS for the propagation of undifferentiated ES cells, prescreened lots of FBS for ES/OP9 coculture are not yet commercially available. To screen FBS for this purpose, cocultures maintained in OP9 medium supplemented with different lots of heat-inactivated FBS must be run in parallel. The outcome is assessed by the efficiency and cell number of resulting B-cells, coexpressing the cell surface markers CD45R and CD19, at d 16–20 of coculture.
2. For differentiating ES cells, OP9 cells should not be kept in continuous culture for longer than 4 wk. OP9 cells that have been maintained in good condition have a fibroblastic morphology (see background cells in **Fig. 2C**), with short dendritic

protrusions and an overall starlike shape. OP9 cells will lose their ability to induce hematopoiesis from ES cells after prolonged culture, and allowing overconfluency will hasten this. Noticeably increased or decreased rates of division or an increased frequency of adipocytes (rounded OP9 cells with highly refractile fat droplets; notice cell with this phenotype in **Fig. 2C**) are indications of OP9 stroma that might no longer support hematopoiesis from ES cells, but might still support hematopoiesis from fetal liver- or bone marrow-derived progenitors. Older stocks of OP9 cells that might no longer be suitable for initiating an ES/OP9 coculture can still be used at later time-points of a coculture, such as d 8 or d 12. During the course of ES/OP9 cocultures, cells are seeded onto 80% confluent OP9 monolayers, which quickly become overconfluent. Thus, the appearance of some adipocytes during a coculture is normal, but these should not predominate (*see* **Fig. 2C**).

3. To preserve early-passage stocks of OP9 stromal cells (once thawed) OP9 cells grow to 80% confluent, split the 10-cm dish into 4 more dishes and continue subculturing until 16 or 32 plates are 80% confluent. Freeze one 80% confluent plate to one cryovial in freezing media, as described for ES cells. These stocks can be expanded to generate working stocks.

4. During the coculture, hematopoietic cells arise at d 5. This is followed by a wave of erythropoiesis (characterized by TER119$^+$ cells) and myelopoiesis (characterized by CD11b$^+$ cells) that peaks between d 10 and d 14. Lymphopoiesis, producing B-lymphocytes and NK-cells, can begin as early as d 14 and, in the presence of IL-7 or IL-15 respectively should predominate around d 16 and thereafter. When cocultures are maintained past d 20, only lymphoid cells are evident, indicating that multilineage potential does not persist during the coculture.

5. Mesoderm colonies contain tightly packed refractile cells. Early colonies are flat, and the cells can be arranged in somewhat concentric circles. Later colonies acquire pronounced three-dimensional structures and resemble asymmetric wagon wheels, with spokes leading out to the rim from a central hub.

6. The kinetics of the coculture can be assessed by flow cytometry. Hematopoietic cells, defined by the expression of the pan-hematopoietic marker CD45 (leukocyte common antigen, LCA), can be detected as early as d 5 of coculture, but more readily by d 8 *(14)*. Early B-cell progenitors, called pro-B-cells, express CD43. Expression of CD43 is lost as these progenitors mature to pre-B-cells *(19,20)*. Cells restricted to the B-cell lineage can be identified by their expression of CD19 and CD45R (B220) and, later, by upregulation of first surface IgM and then IgD *(21)*. In contrast, NK-cells express the pan-NK-cell integrin DX5 and high levels of CD45, but do not express CD24 (HSA) *(14,22)*. It should be noted that cells differentiated from 129-derived ES cells do not express NK1.1, in keeping with that strain's characteristics. Fluorescently labeled antibodies against the markers described can be purchased from either Pharmingen or eBiosciences.

7. The RTOCs can be harvested by disaggregation through a 70-μm nylon mesh into a standard 5-mL tube (fluorescence activated cell sorting [FACS] tube). A 2.5 × 2.5-cm square of 70-μm Nylon mesh is held across the opening of the

FACS tube and moistened with a drop (approx 20 μL) of 4°C PBS. The RTOC is deposited into the drop and physically disaggregated using the plunger of a 1-cm^3 syringe as a pestle. Following disaggregation, remaining cells adhering to the mesh are collected by washing the mesh with 1 mL of 4°C PBS three to four times, into the same FACS tube. The outcome of the RTOC can be assessed by flow cytometry for T-cell markers such as CD3, CD4, CD8, and the T-cell receptor β chain (TCRβ). The approximate yield of ES-derived T-cells per approx 1 × 10^6 Flk-1$^+$ cells is (6–7) × 10^3 cells.

References

1. Chen, U., Kosco, M., and Staerz, U. (1992) Establishment and characterization of lymphoid and myeloid mixed-cell populations from mouse late embryoid bodies, "embryonic-stem-cell fetuses." *Proc. Natl. Acad. Sci. USA* **89,** 2541–2545.
2. Gutierrez-Ramos, J. C. and Palacios, R. (1992) In vitro differentiation of embryonic stem cells into lymphocyte precursors able to generate T and B lymphocytes in vivo. *Proc. Natl. Acad. Sci. USA* **89,** 9171–9175.
3. Potocnik, A. J., Nielsen, P. J., and Eichmann, K. (1994) In vitro generation of lymphoid precursors from embryonic stem cells. *EMBO J.* **13,** 5274–5283.
4. Nakano, T. (1995) Lymphohematopoietic development from embryonic stem cells in vitro. *Semin. Immunol.* **7,** 197–203.
5. Nakano, T., Kodama, H., and Honjo, T. (1994) Generation of lymphohematopoietic cells from embryonic stem cells in culture. *Science* **265,** 1098–1101.
6. Yoshida, H., Hayashi, S., Kunisada, T., et al. (1990) The murine mutation osteopetrosis is in the coding region of the macrophage colony stimulating factor gene. *Nature* **345,** 442–444.
7. Hirayama, F., Lyman, S. D., Clark, S. C., and Ogawa, M. (1995) The flt3 ligand supports proliferation of lymphohematopoietic progenitors and early B-lymphoid progenitors. *Blood* **85,** 1762–1768.
8. Hudak, S., Hunte, B., Culpepper, J., et al. (1995) FLT3/FLK2 ligand promotes the growth of murine stem cells and the expansion of colony-forming cells and spleen colony-forming units. *Blood* **85,** 2747–2755.
9. Hunte, B. E., Hudak, S., Campbell, D., Xu, Y., and Rennick, D. (1996) *flk2/flt3* ligand is a potent cofactor for the growth of primitive B cell progenitors. *J. Immunol.* **156,** 489–496.
10. Jacobsen, S. E., Okkenhaug, C., Myklebust, J., Veiby, O. P., and Lyman, S. D. (1995) The FLT3 ligand potently and directly stimulates the growth and expansion of primitive murine bone marrow progenitor cells in vitro: synergistic interactions with interleukin (IL) 11, IL-12, and other hematopoietic growth factors. *J. Exp. Med.* **181,** 1357–1363.
11. Lyman, S. D. and Jacobsen, S. E. (1998) c-kit ligand and Flt3 ligand: stem/progenitor cell factors with overlapping yet distinct activities. *Blood* **91,** 1101–1134.
12. Veiby, O. P., Lyman, S. D., and Jacobsen, S. E. W. (1996) Combined signaling through interleukin-7 receptors and flt3 but not c-kit potently and selectively promotes B-cell commitment and differentiation from uncommitted murine bone marrow progenitor cells. *Blood* **88,** 1256–1265.

13. Sitnicka, E., Bryder, D., Theilgaard-Monch, K., Buza-Vidas, N., Adolfsson, J., and Jacobsen, S. E. (2002) Key role of flt3 ligand in regulation of the common lymphoid progenitor but not in maintenance of the hematopoietic stem cell pool. *Immunity* **17,** 463–472.

14. Cho, S. K., Webber, T. D., Carlyle, J. R., Nakano, T., Lewis, S. M., and Zúñiga-Pflücker, J. C. (1999) Functional characterization of B lymphocytes generated in vitro from embryonic stem cells. *Proc. Natl. Acad. Sci. USA* **96,** 9797–9802.

15. Kennedy, M. K., Glaccum, M., Brown, S. N., et al. (2000) Reversible defects in natural killer and memory CD8 T cell lineages in interleukin 15-deficient mice. *J. Exp. Med.* **191,** 771–780.

16. De Pooter, R. F., Cho, S. K., Carlyle, J. R., and Zúñiga-Pflücker, J. C. (2003) In vitro generation of T lymphocytes from embryonic stem cell-derived prehemato-poietic progenitors. *Blood* **102,** 1649–1653.

17. Robertson, E. J. (1997) Derivation and maintenance of embryonic stem cell cultures. *Methods Mol. Biol.* **75,** 173–184.

18. Takahama, Y. (2000) Differentiation of mouse thymocytes in fetal thymus organ culture. *Methods Mol. Biol.* **134,** 37–46.

19. Hardy, R. R., Carmack, C. E., Shinton, S. A., Kemp, J. D., and Hayakawa, K. (1991) Resolution and characterization of pro-B and pre-pro-B cell stages in normal mouse bone marrow. *J. Exp. Med.* **173,** 1213–1225.

20. Li, Y. S., Wasserman, R., Hayakawa, K., and Hardy, R. R. (1996) Identification of the earliest B lineage stage in mouse bone marrow. *Immunity* **5,** 527–535.

21. Melchers, F., Rolink, A., Grawunder, U., et al. (1995) Positive and negative selection events during B lymphopoiesis. *Curr. Opin. Immunol.* **7,** 214–227.

22. Carlyle, J. R., Michie, A. M., Cho, S. K., and Zúñiga-Pflücker, J. C. (1998) Natural killer cell development and function precede alpha beta T cell differentiation in mouse fetal thymic ontogeny. *J. Immunol.* **160,** 744–753.

11

Hematopoietic Development
of Human Embryonic Stem Cells in Culture

Xinghui Tian and Dan S. Kaufman

Summary

The isolation of embryonic stem (ES) cells from human preimplantation blastocysts creates an exciting new starting point to analyze the earliest stages of human blood development. This chapter describes two methods to promote hematopoietic differentiation of human ES cells: stromal cell coculture and embryoid body formation. Better understanding of basic human hematopoiesis through the study of human ES cells will likely have future therapeutic benefits.

Key Words: Embryonic stem cells; embryoid bodies; hematopoietic differentiation; mouse embryonic fibroblast feeders; stromal coculture.

1. Introduction

Human embryonic stem (ES) cells offer many advantages for studies of basic hematopoiesis. Human ES cells can be maintained for months to years in culture as undifferentiated cells, yet retain the ability to form any cell type within the body *(1,2)* (at least this potential is presumed from studies with mouse ES cells, because definitive studies of totipotency cannot be done with human ES cells). Human ES cells are derived from early blastocysts, and development of these cells into specific lineages likely recapitulates events that occur during normal development. Ten million or more human ES cells can be easily grown and sampled at any time-point during differentiation into specific cellular lineages. These numbers should be sufficient for detailed in vitro and in vivo studies. Perhaps most importantly, 20 yr of studies with mouse ES cells clearly demonstrate the value of this model developmental system *(3,4)*. Mouse ES cells have been used to define genetic pathways that regulate blood development (and many other lineages) and this knowledge has been

From: *Methods in Molecular Biology, vol. 290: Basic Cell Culture Protocols, Third Edition*
Edited by: C. D. Helgason and C. L. Miller © Humana Press Inc., Totowa, NJ

shown to translate to human models such as umbilical cord blood or bone marrow-based hematopoietic development. However, only human ES cells allow characterization of the earliest stages of prenatal human hematopoietic development. Specifically, these cells can be used to understand how human hematopoietic stem cells (HSCs) arise from earlier precursors. Although this question has been addressed in murine systems with studies of mouse ES cells and dissection of timed embryos, there are some fundamental differences in mouse and human embryogenesis that suggest that not all developmental pathways will be the same *(5)*. Indeed, mouse and human ES cells have key differences in their phenotype, growth characteristics, and culture requirements that likely translate into unique pathways of development *(1,4,6)*.

This chapter outlines two methods to promote hematopoietic differentiation of human ES cells (*see* **Fig. 1**). Isolation and maintenance of undifferentiated human ES cells will not be discussed in detail. Briefly, human ES cells are routinely maintained in serum-free media either in direct coculture with irradiated mouse embryonic fibroblast (MEF) "feeder" cells or in "feeder-free" conditions by culture on Matrigel or laminin-coated plates *(1,7)*. The "feeder-free" growth still requires ES cells to be grown in medium conditioned by MEFs. Therefore, a requirement for feeder cells remains. Some reports have shown that other (human) feeder cells can be used to maintain undifferentiated human ES cells, thus potentially avoiding some xenogeneic exposures *(8,9)*.

Original studies of hematopoiesis from human ES cells used coculture with stromal cells derived from hematopoietic microenvironments to support or promote development of phenotypical and genotypical blood cells *(10)*. This coculture method is technically straightforward and offers the advantage to potentially characterize and modify the stromal cells to define what components they contribute to the hematopoietic process. Indeed, the finding that nonspecific fibroblasts did not support hematopoietic differentiation strongly suggested that interactions between the differentiating ES cells and the stromal cells were important to support blood development. Therefore, the stromal cells can be easily engineered to understand better what cell-bound and soluble factors they might contribute to hematopoiesis. Because human ES cells were initially thought to be difficult to genetically modify (many methods requiring modifications from methods used for mouse ES cells), the ability to analyze inputs from stromal cells offered an important advantage. However, several recent reports have defined methods to stably express exogenous genes in human ES cells *(11–14)*. Differentiation of human ES cells via embryoid body (EB) formation also offers a suitable method to promote hematopoiesis. EB formation is an important methodology when researchers want to avoid more complex interactions with stromal cells. However, as ES cells differentiate they rapidly become a diverse mixture of cell types. Typically, only a few

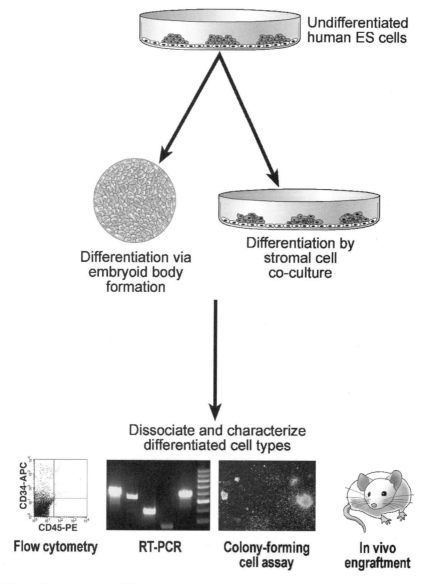

Fig. 1. Schematic of differentiation and analysis of human ES cell-derived hematopoietic cells.

percent are blood cells. Although understanding all components that contribute development of specific lineages will remain a challenge, human ES cells are now an important resource to characterize hematopoietic pathways. Moreover, human ES cells offer the exciting prospect of becoming a suitable source to replace or repair cells, tissues, or organs damaged by disease, trauma, degeneration, or other processes.

In the methods described in this chapter, undifferentiated human embryonic stem (hES) cells were maintained in serum-free media. To promote hematopoietic differentiation via stromal cell coculture, the human ES cells were typically cocultured with the mouse bone marrow stromal cell line S17 *(15)*, although other cell lines derived from hematopoietic microenvironments such as C166 *(16)*, OP9 *(17)*, or primary human bone-marrow-derived stromal cells *(18)* can be used. After a defined number of days, hematopoietic precursor cells derived from human embryonic stem cells were analyzed by fluorescent-activated cell sorting (FACS), colony-forming assay, and reverse transcriptase-polymerase chain reaction (RT-PCR) methods. Other assay systems for early hematopoietic progenitors cells such as long-term culture initiating cell (LTC-IC) assay *(19)* or injection into NOD/SCID mice can be utilized *(20)*.

2. Materials

2.1. Coculture of Human ES Cells and S17 Cells

2.1.1. Cell Culture Media

1. DMEM/F12 with 15% knockout SR media: prepare Dulbecco's modified Eagle's medium/F12 (DMEM/F12) (Invitrogen Corp./Gibco, cat. no. 11330-032) supplemented with 15% knockout SR (Invitrogen Corp./Gibco, cat. no. 10828-028) and 1% MEM nonessential amino acids (NEAA) solution (Invitrogen Corp./Gibco, cat. no. 11140-050). Store at 2–8°C.
2. L-Glutamine/β-mercaptoethanol solution (L-glutamine/β-ME): For culture of undifferentiated hES cells, L-glutamine is routinely made fresh from powder by mixing 0.146 g L-glutamine and 7 μL β-mercaptoethanol in 10 mL phosphate-buffered saline (PBS).
3. Basic fibroblast growth factor (bFGF): Reconstitute the 10-μg vial of recombinant human bFGF powder (Invitrogen, cat. no. 13256-029) in 5 mL of sterile 0.1% fraction V bovine serum albumin (BSA) in PBS. Aliquot 0.5 mL per sterile tube and store at –80°C.
4. hES cell media: DMEM/F12 with 15% knockout SR, 1% NEAA, 2 mM L-glutamine, 4 ng/mL bFGF and 0.1 mM β-ME. To prepare medium, add 2.5 mL L-glutamine/β-ME and 0.5 mL bFGF to DMEM/F12 with 15% knockout SR media in 250 mL total volume. Store at 2–8°C.
5. S17 culture media: RPMI 1640 (Cellgro/Mediatech, cat. no. 10-404-CV) media containing 10% fetal bovine serum (FBS) certified (Invitrogen Corp./Gibco, cat. no. 16000-044), 0.055 mM β-ME (Invitrogen Corp./Gibco, cat. no. 21985-023), 1% NEAA, 1% penicillin–streptomycin (P/S) (Invitrogen Corp./Gibco, cat. no. 15140-122), 2 mM L-glutamine (Cellgro/Mediatech, cat. no. 25-005-CI). Store at 2–8°C.

6. hES/S17 differentiation media: DMEM (Invitrogen Corp./Gibco, cat. no. 11965-092) supplemented with 20% defined FBS (Hyclone, cat. no. SH30070.03), 2 mM L-glutamine, 0.1 mM of β-ME, 1% MEM NEAA, and 1% P/S (same as S17 culture media). Store at 2–8°C.

7. D-10 media used for washing: DMEM supplemented with 10% FBS and 1% P/S.

8. Collagenase split media: DMEM/F12 media containing 1 mg/mL collagenase type IV (Invitrogen Corp./Gibco, cat. no. 17104-019). Collagenase media is filter-sterilized with a 50-mL 0.22-μm membrane Steriflip (Millipore, cat. no. SCGP00525). Store at 2–8°C.

9. Trypsin–EDTA+2% chick serum: 0.05% trypsin/0.53 mM EDTA solution (Cellgro/Mediatech, cat. no. 25-052-CI) with 2% chick serum (Sigma, cat. no. C5405) (*see* **Note 1**).

2.1.2. Cell Culture Supplies

1. Six-well tissue culture plates (NUNC™ Brand Products, Nalgene Nunc cat. no. 152795).

2. Gelatin (Sigma, cat. no. G-1890), 0.1% solution made in water, then autoclave to sterilize.

3. Gelatin-coated six-well tissue culture plates. Add 2–3 mL gelatin per well for a minimum of 1 h. These plates are ready to use after aspirating the extra gelatin solution.

4. Disposable serological pipets (all from VWR Scientific Products): 10 mL (cat. no. 53283-740); 5 mL (cat. no. 53283-738); 1 mL (cat. no. 53283-734) (*see* **Note 2**).

5. 70 μm Cell strainer filter (Becton Dickinson/Falcon, cat. no. 352350).

6. 0.4% Trypan blue stain (Invitrogen Corp./Gibco, cat. no. 15250).

2.2. Embryoid Body Formation From Human ES Cells

2.2.1. Cell Culture Media

1. Stemline hematopoietic stem cell expansion medium (Sigma, cat. no. S-0189) supplemented with 4 mM L-glutamine and 1% P/S. Store according to manufacturer's instructions (*see* **Note 3**).

2. Dispase split media: dissolve 250 mg dispase powder (Invitrogen Corp./Gibco, cat. no. 17105-041) in 50 mL DMEM/F-12 (final 5 mg/mL) and filter-sterilize with a 50-mL 0.22-μm membrane Steriflip (Millipore, cat. no. SCGP00525). Store at 2–8°C.

3. DMEM/FBS: DMEM supplemented with 15% defined FBS (Hyclone, cat. no. SH30070.03), 1% L-glutamine, 1% P/S, 1% MEM NEAA solution, and 0.1 mM of β-mercaptoethanol (Invitrogen Corp., Gibco, cat. no. 21985-023). Store at 2–8°C.

2.2.2. Cell Culture Supplies

1. Poly-2-hydroxyethyl methacrylate, (poly-HEME): Dissolve 1.0 g poly-HEME powder (Sigma, cat. no. P-3932) in 25 mL acetone and 25 mL ethanol in a glass bottle to a final concentration of 2%.

2. Non-tissue-culture-treated T25 flasks (Sarstedt, Ref. no. 83.1810.502) coated with 2% poly-HEME solution. Coat flasks approx 45 min before EB resuspension. Using a glass 1-mL pipet, wash 0.5 mL of 2% poly-HEME solution over the back side of the untreated T25 flask (Sarstedt), which is the bottom surface when placed in the incubator. Aspirate the leftover poly-HEME that will pool in the flask. This can be reused in another flask. Let the flasks sit in a sterile tissue culture hood with the caps completely off for approx 40 min. Wash the poly-HEME-coated surface with 3 mL of desired media before adding the EBs.

3. Blue Max polypropylene 15-mL conical tubes (Becton Dickinson/Falcon, cat. no. 352097).

2.3. Flow Cytometric Analysis

1. FACS wash media: PBS containing 2% FBS and 0.1% sodium azide (Fisher Chemical, cat. no. S227I).

2. 12 × 75-mm Polystyrene round-bottom tube (Becton Dickinson/Falcon, cat. no. 352054).

3. Propidium iodide (Sigma, P4170), 1 mg/mL dissolved in PBS. Store aliquots at 4°C.

2.4. Hematopoietic Colony-Forming Cell Assays

1. MethoCult™ GF⁺ H4435 (StemCell Technologies Inc., Vancouver, BC, cat. no. 04435) consisting of 1% methylcellulose, 30% FBS, 1% BSA, 50 ng/mL stem cell factor, 20 ng/mL granulocyte–macrophage colony-stimulating factor, 20 ng/mL interleukin (IL) 3, 20 ng/mL IL-6, 20 ng/mL granulocyte colony-stimulating factor, and 3 units/mL erythropoietin. This medium is optimized for detection of most primitive colony-forming cells (CFCs). A 100-mL bottle of MethoCult can be aliquoted into 2.5-mL samples. Alternatively, MethoCult can also be purchased prealiquoted into 3-mL samples.

2. I-2 media: Iscove's modified Dulbecco's media (IMDM) (Invitrogen Corp./Gibco, cat. no. 12440-053) containing 2% FBS. Store at 2–8°C.

3. Non-tissue-culture-treated 35-mm Petri dish (Greiner Bio-One, cat. no. 627102) (*see* **Note 4**).

4. 2-mL stripette disposable serological pipet (Corning Inc., cat. no. 4021).

2.5. RNA Isolation

1. TRIzol (Invitrogen, cat. no. 15596-026) (*see* **Note 5**).

2. Diethyl pyrocarbonate (DEPC) (Sigma, cat. no. D5758).

3. 95% Ethanol.

4. Isopropyl alcohol.

3. Methods
3.1. Culture of Undifferentiated hES Cells

Undifferentiated hES cells were cultured as previously described *(1,2)* (*see* **Note 6**). They are maintained in hES cell media by coculture with irradiated

MEF cells or in MEF condition media on Matrigel-coated plates *(7)*. hES cells were fed daily with fresh medium and were passed onto fresh feeder plates or Matrigel-coated six-well plates at approximately weekly intervals to maintain undifferentiated growth.

3.2. Preparation of S17 Feeder Layer

The mouse bone marrow S17 cell line *(10)* is maintained in S17 culture media. To prepare feeder layers, the S17 cells are dissociated with trypsin–EDTA and irradiated with 30 Gy (*see* **Note 7**). Then, 2.5 mL of irradiated S17 cells at 1.0×10^5 cells/mL are plated onto 0.1% gelatin-coated six-well plates (2.5×10^5 cells/well). Feeder layers should be prepared at least 1 day prior to coculture with hES cells and remain suitable for use up to 2 wk when kept in a 37°C, 5% CO_2 incubator. Other stromal cell lines can also be used in a similar manner, although the irradiation dose and cell density might vary.

3.3. Coculture of hES Cells on S17

To improve the viability of human ES cells for hES cells/S17 cocultures, small colonies or clusters of ES cells should be plated onto the S17 cell rather than a single cell-suspension. To maintain small colonies rather than single cells, collagenase type IV is used to harvest ES cells.

1. Warm collagenase split media to 37°C in a water bath.
2. Aspirate media off of hES cells culture and add 1.5 mL/well (six-well plate) of collagenase split media. Place in a 37°C incubator for 5–10 min, observing at approx 5-min intervals. Cells are ready to be harvested when the edges of the colony are rounded up and curled away from the MEFs or from the Matrigel plate.
3. Using a 5-mL pipet, scrape and gently pipet to wash the colonies off of the plate, transfer cell suspension to a 15-mL conical tube, and add another 3–6 mL hES/S17 differentiation media. Centrifuge at 1000 rpm (400*g*) for 5 min. Aspirate media and wash cells with an additional 3–6 mL hES/S17 differentiation media by centrifugation again at 1000 rpm (400*g*) for 5 min.
4. During this last centrifuge step, prepare the S17 feeder layers by aspirating off the S17 media and wash once with 2 mL/well of PBS.
5. Once hES cells are done spinning, aspirate media, resuspend cells in an appropriate volume with hES/S17 differentiation media. Usually, add 1 mL cell suspension per well onto the S17 plate, then add an additional 1.5–2 mL hES/S17 differentiation media to each well. To evenly distribute the cells, gently shake the plate side to side while placing them in a 37°C, 5% CO_2 incubator. This will allow hES cells to attach evenly within the wells. Do not disturb plates for several hours.
6. During differentiation, change the culture medium every 2–3 d. Typical morphology of cells is shown in **Fig. 2**. For the first few days, colonies typically maintain appearance of undifferentiated hES cells. They then show obvious evidence of differentiation, as evidenced by three-dimensional cystic structures and other loosely adherent structures.

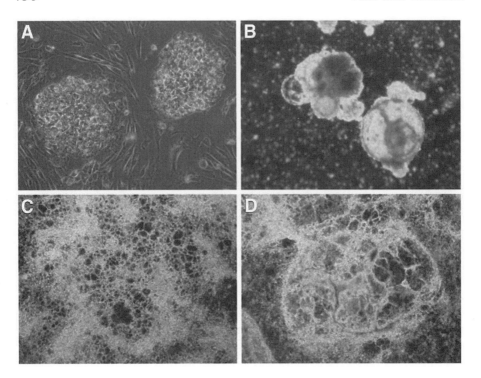

Fig. 2. Hematopoietic differentiation of human ES cells. (**A**) Two colonies of undifferentiated human ES cells grown on MEF feeder layer. These colonies demonstrate uniform morphology with no visible evidence of differentiation. Original magnification: ×100. (**B**) Human ES cells induced to form embryoid bodies in suspension for 14 d. Multiple cell types and cystic regions are evident. Original magnification: ×100. (**C**) Human ES cells allowed to differentiate on S17 stromal cells for 8 d. The majority of cells in this image are derived from a single colony that has differentiated into multiple cell types including thin endothelial-type structures and more densely piled-up regions. Original magnification: ×20. (**D**) Human ES cells allowed to differentiate on S17 cells for 16 d. These cells are now seen to form spherical, cystic structures and a variety of other cell types. Original magnification: ×100. Some aspects of ES cell differentiation on S17 cells begin to resemble structure seen in EBs as seen in (**B**).

3.4. Harvest of Differentiated hES Cells From hES/S17 Cell Cocultures

Optimal time required for differentiation of human ES cells into CD34+ cells and CFCs varies somewhat depending on the hES cell line and stromal cells used. On average, culture for 14–21 d produces the best results for these purposes. A time-course experiment to sample cells every 2–3 d is recommended to find the optimal time-point for specific cells formed. For flow cytometry and colony-forming assays, it is necessary to produce a single-cell suspension of hES cells that have differentiated on S17 or other stromal cells. Because stromal cells are irradiated prior to coculture with ES cells, typically >90% of cells harvested are derived from hES cells.

1. To prepare a single-cell suspension of differentiated hES cell-derived cells from six-well plate cultures, aspirate media and add 1.5 mL collagenase IV per well for 5–10 min until S17 stromal cells can be seen to become more spindle-shaped or break up. Scrape with a 5-mL pipet and transfer hES/S17 cell suspension into a 15-mL conical tube. Add another 6 mL Ca^{2+} and Mg^{2+}-free PBS, break up the colonies by pipetting up and down (vigorously) against the bottom of the tube until there appears to be a fine suspension of cells. Centrifuge cell suspension at 1000 rpm (400g) for 5 min.
2. Remove the supernatant, add 1.5 mL trypsin–EDTA + 2% chick serum solution into the tube. Place at 37°C in a water bath for 5–15 min. Vigorously vortex and observe samples at 3- to 5-min intervals. The single cells suspension is ready when there are minimal clumps of undispersed cells.
3. Add 6 mL DMEM containing 10% FBS (D-10 media) to neutralize the trypsin–EDTA, and pipet up and down to further disperse cells. Centrifuge at 1000 rpm (400g) for 5 min. Resuspend cell pellet with 5 mL D-10 media. Filter the cell suspension with a 70 μm cell strainer filter to remove any remaining clumps of cells. Enumerate cells with a hemocytometer using 0.4% trypan blue to stain the dead cells. From a nearly confluent well, $(1–2) \times 10^6$ hES cell-derived cells can be obtained as single cells.
4. According to the total cell number, aliquot cells as needed for FACS, RNA, protein, and hematopoietic CFC assays. For reasons that should be obvious, doing multiple assays from the same collection of differentiated ES cells will ensure uniformity of results. Depending on the density of the cells, two to three wells can be harvested at a single time-point to collect enough cells for FACS, CFC, and RNA analysis.

3.5. Embryoid Body Formation

This section describes hES cell-derived EB formation in serum-free conditions.

1. Obtain hES cell colonies that have slight to no morphological evidence of differentiated cells. EB formation works best when colonies are neither very large nor very small. One can get a better sense of which colony sizes work with experience. If they are too small, the colonies will dissociate within 2–4 d. If the colonies are too large, they will not dissociate very easily, and in an attempt to break them apart, they could be damaged further. Also, with larger colonies, EB formation might not occur efficiently. An ideal time for hES cells to form EBs is typically 6–7 d after their last passage date. One six-well plate can generate up to six T25 flasks (one well per flask), although EBs can be pooled more densely into flasks as desired.

2. Aspirate the media from each well, without disturbing the adherent hES colonies. Add 1.5 mL/well of 5 mg/mL dispase split media. Incubate at 37°C and 5% CO_2 until approx 50% of the colonies are detached. This usually takes 5–10 min with freshly made dispase and can take up to 15 min with an older dispase solution (over 2 wk). Gently shake the plate until the remaining colonies detach. If they do not, use a 5-mL pipet to wash them off.

3. Add 2 mL of Stemline media to each well and gently separate the colonies by pipetting up and down. Transfer the cell suspension into a 15-mL conical tube and centrifuge at 1000 rpm (400g) for 2 min. Aspirate the supernatant, gently flick the tube and then add 5 mL fresh Stemline media. Repeat for a total of three washes in Stemline media.

4. Aliquot the cell suspension into untreated T25 flasks. Then, add Stemline media to a final volume of 7–8 mL. Incubate at 37°C and 5% CO_2, placing the wide bottom side of the flask horizontal on the shelf.

5. Culture overnight and the following day the cells must be "cleaned up" to remove leftover stromal and dead cells from the suspension.

6. The EBs should be resuspended in fresh media and flasks every 3–4 d to optimize growth and prevent adhesion. To harvest the EBs, resuspend the cells in a 15-mL conical tube. If smaller EBs are desired, or the EBs have "clumped" together overnight, pipet up and down until the EBs are at the desired size. Let EBs settle to the bottom. Gently aspirate the supernatant and try to remove the smaller cells that float in the supernatant. Resuspend the cells in 7–8 mL of Stemline media in new T25 flasks and incubate at 37°C and 5% CO_2.

7 If the EBs are to be generated in DMEM/15% FBS, coat the flask with poly-HEME solution to decrease adherence of EBs to the T25 flasks (*see* **Subheading 2.2.2., item 2**). Culture EBs as described in Materials.

3.6. Harvest of hEB Cells

This subsection describes the isolation of single cells from hEB.

1. Add EBs to a 15-mL conical tube, let them settle by gravity for approx 1 min, and gently aspirate media and floating individualized cells that have not settled out.

2. Wash with 5 mL Ca^{2+} and Mg^{2+}-free PBS, and centrifuge at 1200 rpm (400g) for 3 min.

3. Aspirate supernatant, add 1.5–2 mL trypsin–EDTA with 2% chick serum, vigorously pipet up and down several times, and vortex to break up EBs.

4. Incubate in 37°C water bath for 5 min, and then vortex and pipet vigorously to further dissociate EBs. Repeat steps at 5-min intervals until EBs seem maximally dissociated. This typically takes 10–20 min in total. Some clumps could still remain, but longer incubations usually does not improve this digestion.

5. After the EBs have been maximally digested, add 4 mL D-10 media and centrifuge at 1200 rpm (400g) for 3 min. Wash cells twice more using 5 mL of D10 media for each wash step.

6. Resuspend cells in desired media and filter the cell suspension with a 70 μm cell strainer filter to remove any remaining clumps of cells. Enumerate cells with hemocytometer using 0.4% trypan blue to stain the dead cells.

3.7. Methods for Analyzing hES Hematopoietic Development

This section describes assays for analyzing hematopoietic development from hES/S17 cocultures and hES-derived EBs. The numbers and types of hES-derived hematopoietic cells obtained will vary depending on the ES cell line, method of differentiation, and duration of differentiation cultures. We present protocols for assays that are not specific to analysis of human ES cell-derived blood cells. Many variations are possible for flow cytometric analysis, CFC assays, and RNA isolation. We offer these methods as one example. Differentiated hES/S17 cells are dissociated with collagenase and trypsin/EDTA to make a single-cell suspension as described in **Subheading 3.4.** Alternatively, prepare single-cell suspensions from hES-derived EBs as described in **Subheading 3.6.**

3.7.1. Flow Cytometric Analysis

1. Aliquot approx 2×10^5 cells per tube for each different antibody used. Wash cells one or twice with FACS media before starting staining.
2. Stain with either antigen-specific antibodies or isotype control for at least 15 min on ice. If the first antibodies are unconjugated, the cells should be incubated with conjugated secondary antibodies for another 15–30 min after washing with FACS media between staining steps.
3. Wash one or two times with FACS media. Resuspend cell pellet in 200–500 μL FACS media containing PI. Perform flow cytometric analysis by standard methods. Importantly, to increase specificity, acquire data, or analyze data on PI-negative cells, staining and fixation of cells is not done, as this does not permit PI staining and can increase false-positive events.

3.7.2. Hematopoietic CFC Assay

1. hES/S17 cocultures or EBs are cultured for the appropriate number of days and then single-cell suspensions are prepared. Aliquot 6×10^5 cells into a sterile microfuge tube. Centrifuge at 1500 rpm ($400g$) for 5 min. Resuspend the cell pellet with 100 μL I-2 media. Wash once with I-2 media.
2. Thaw the MethoCult GF+ media to room temperature before starting the colony-forming assay. Add cells to 2.5 mL MethoCult GF$^+$ and vortex until the cells distribute in the media evenly. Place cells in MethoCult GF$^+$ upright and keep at room temperature for approx 15 min to let the bubbles rise and dissipate.
3. Transfer the cells in Methocult GF$^+$ media into sterile Petri dishes. Then, 2.5 mL media should be divided into two 35-mm non-tissue-culture Petri dishes using a wide, blunt 2-mL Stripette (1.1 mL cells = 2.5×10^5 cells per dish). Place these two dishes and another open dish containing water into the 100-mm culture dish

(the additional dish with water helps maintain humidity to prevent drying of methylcellulose-based media).

4. Incubate at 37°C, 5% CO_2 for 2 wk and score for colony-forming units (CFUs) according to standard criteria *(21)*.

3.7.3. RNA Isolation

1. Spin down an aliquoted single-cell suspension, remove the media, and then homogenize in TRIzol reagent by repetitive pipetting (1 mL Trizol per [5–10] × 10^6 cells). Incubate the samples for 5 min at room temperature (15–30°C). Washing cells before addition of TRIzol should be avoided, as this will increase the possibility of mRNA degradation.
2. Add 0.2 mL chloroform per 1 mL TRIzol into the aqueous phase. Cap sample tube, shake tubes vigorously by hand for 15 s, and then incubate at room temperature for 2–3 min. Centrifuge at a maximum of 12,000g for 15 min at 2–8°C.
3. Transfer the colorless upper aqueous phase into a fresh tube and save the lower red organic phase. Add 0.5 mL of isopropyl alcohol per 1 mL TRIzol to precipitate the RNA by incubating the samples at room temperature for 10 min. Centrifuge at a maximum of 12,000g for 10 min at 2–8°C.
4. Remove the supernatant. Wash the RNA pellet once with 75% ethanol, adding at least 1 mL of 75% ethanol per 1 mL of TRIzol reagent used for the initial homogenization. Mix the sample by vortexing and centrifuge at a maximum of 7500g for 5 min at 2–8°C.
5. Dry the RNA pellet (5–10 min) at room temperature; do not dry the RNA by centrifugation under vacuum. Dissolve RNA in RNase-free (DEPC treated) water or 0.5% sodium dodecyl sulfate (SDS) solution by passing the solution a few times through a pipet tip and incubating for 10 min at 55–60°C.
6. The purified RNA can be stored at –20°C or –80°C for an extended time and for use in RT-PCR analysis. The protein obtained from the red organic phase can be used for Western blot analysis.

4. Notes

1. Chick serum is added to trypsin–EDTA solution to add proteins that improve cell viability, but unlike FBS, chick serum does not contain trypsin inhibitors. Trypsin–EDTA + 2% chick serum should be warmed to 37°C before use.
2. Based on standard materials and methods used for embryo culture, disposable glass pipets are used for the culture of undifferentiated human ES cells. Some researchers feel that these pipets help maintain the ES cells in an undifferentiated state by minimizing exposure to plastics (which can vary between lots) or detergents (possible toxins when used to clean reusable glass pipets). Use of these disposable glass pipets for cultures of differentiated ES cells is probably optional.
3. Other serum-free media can be used for culture of embryoid bodies. We typically prefer to culture the dispase-harvested ES cell colonies overnight in serum-free media to best allow formation of EBs with minimal adherence to the plastic cul-

ture ware. Adherence often occurs with serum containing media, even when tissue culture dishes pretreated with poly-HEME are used.

4. We have tested several brands of non-tissue-culture-treated dishes and only Greiner dishes showed no adherent cells when this complex mixture of cells was plated in the CFC assay. If cells do adhere and grow, these proliferating cells will likely interfere with results.

5. Other means of RNA and/or protein isolation are also available. For a small amount of cells, the RNeasy Mini Kit (Qiagen, cat. no. 74104) is suitable, especially if protein samples are not desired.

6. Detailed protocols for the maintenance of human ES cell lines are beyond the scope of this chapter. The reader is encouraged to consult the references cited, other published literature, and various websites (i.e., http://stemcells.nih.gov/registry/index.asp.)

7. Reports using S17 cells to support hematopoietic differentiation of rhesus monkey ES cells did not irradiate or otherwise mitotically inactive the S17 stromal cells *(22)*. These authors felt that growth inhibition of S17 cells when confluent was sufficient to prevent overgrowth when cocultured with ES cells. Although this does seem to be the case, we prefer irradiation of stromal cells to prevent subsequent growth and proliferation that could complicate interpretation of subsequent assays such as the CFC assay.

Acknowledgments

We thank Julie Morris, Rachel Lewis, and Dong Chen for assistance with embryoid body protocols.

References

1. Thomson, J. A., Itskovitz-Eldor, J., Shapiro, S. S., et al. (1998) Embryonic stem cell lines derived from human blastocysts. *Science* **282,** 1145–1147.

2. Odorico, J. A., Kaufman, D. S., and Thomson, J. A. (2001) Multilineage differentiation from human embryonic stem cell lines. *Stem Cells* **19,** 193–204.

3. Keller, G., Kennedy, M., Papayannopoulou, T., and Wiles, M. V. (1993) Hematopoietic commitment during embryonic stem cell differentiation in culture. *Mol. Cell. Biol.* **13,** 473–486.

4. Smith, A. G. (2001) Embryo-derived stem cells: of mice and men. *Annu. Rev. Cell. Dev. Biol.* **17,** 435–462.

5. Palis, J. and Yoder, M. C. (2001) Yolk-sac hematopoiesis: the first blood cells of mouse and man. *Exp. Hematol.* **29,** 927–936.

6. Thomson, J. A. and Marshall, V. S. (1998) Primate embryonic stem cells. *Curr. Topics Dev. Biol.* **38,** 133–165.

7. Xu, C., Inokuma, M. S., Denham, J., et al. (2001) Feeder-free growth of undifferentiated human embryonic stem cells. *Nature Biotechnol.* **19,** 971–974.

8. Richards, M., Fong, C. Y., Chan, W. K., Wong, P. C., and Bongso, A. (2002) Human feeders support prolonged undifferentiated growth of human inner cell masses and embryonic stem cells. *Nature Biotechnol.* **20,** 933–936.

9. Cheng, L., Hammond, H., Ye, Z., Zhan, X., and Dravid, G. (2003) Human adult marrow cells support prolonged expansion of human embryonic stem cells in culture. *Stem Cells* **21,** 131–142.

10. Kaufman, D. S., Hanson, E. T., Lewis, R. L., Auerbach, R., and Thomson, J. A. (2001) Hematopoietic colony-forming cells derived from human embryonic stem cells. *Proc. Natl. Acad. Sci. USA* **98,** 10,716–10,721.

11. Ma, Y., Ramezani, A., Lewis, R., Hawley, R. G., and Thomson, J. A. (2003) High-level sustained transgene expression in human embryonic stem cells using lentiviral vectors. *Stem Cells* **21,** 111–117.

12. Zwaka, T. P. and Thomson, J. A. (2003) Homologous recombination in human embryonic stem cells. *Nature Biotechnol.* **21,** 319–321.

13. Pfeifer, A., Ikawa, M., Dayn, Y., and Verma, I. M. (2002) Transgenesis by lentiviral vectors: lack of gene silencing in mammalian embryonic stem cells and preimplantation embryos. *Proc. Natl. Acad. Sci. USA* **99,** 2140–2145.

14. Gropp, M., Itsykson, P., Singer, O., et al. (2003) Stable genetic modification of human embryonic stem cells by lentiviral vectors. *Mol. Ther.* **7,** 281–287.

15. Collins, L. S. and Dorshkind, K. (1987) A stromal cell line from myeloid long-term bone marrow cultures can support myelopoiesis and B lymphopoiesis. *J. Immunol.* **138,** 1082–1087.

16. Wang, S. J., Greer, P., and Auerbach, R. (1996) Isolation and propagation of yolk-sac-derived endothelial cells from a hypervascular transgenic mouse expressing a gain-of-function fps/fes proto-oncogene. *In Vitro Cell. Biol. Anim.* **32,** 292–299.

17. Nakano, T., Kodama, H., and Honjo, T. (1994) Generation of lymphohemato-poietic cells from embryonic stem cells in culture. *Science* **265,** 1098–1101.

18. Simmons, P. J. and Torok-Storb, B. (1991) Identification of stromal cell precursors in human bone marrow by a novel monoclonal antibody, STRO-1. *Blood* **78,** 55–62.

19. Sutherland, H. J., Lansdorp, P. M., Henkelman, D. H., Eaves, A. C., and Eaves, C. J. (1990) Functional characterization of individual human hematopoietic stem cells cultured at limiting dilution on supportive marrow stromal layers. *Proc. Natl. Acad. Sci. USA* **87,** 3584–3588.

20. Dick, J. E., Bhatia, M., Gan, O., Kapp, U., and Wang, J. C. (1997) Assay of human stem cells by repopulation of NOD/SCID mice. *Stem Cells* **15(Suppl. 1),** 199–203.

21. Eaves, C. and Lambie, K. (eds.) (1995) *Atlas of Human Hematopoietic Colonies.* StemCell Technologies Inc., Vancouver, BC.

22. Li, F., Lu, S., Vida, L., Thomson, J. A., and Honig, G. R. (2001) Bone morphogenetic protein 4 induces efficient hematopoietic differentiation of rhesus monkey embryonic stem cells in vitro. *Blood* **98,** 335–342.

12

Generation of Murine Stromal Cell Lines

*Models for the Microenvironment
of the Embryonic Mouse Aorta–Gonads–Mesonephros Region*

Robert A. J. Oostendorp, Kirsty Harvey, and Elaine A. Dzierzak

Summary

We describe a method to derive cell lines and clones from cells of the murine midgestation aorta–gonads–mesonephros (AGM) microenvironment. We start from subdissected AGM regions in "explant-" or "single-cell suspension"-type cultures from embryos transgenic for *tsA58*, a temperature-sensitive mutant of the SV40 T antigen gene. The number of cells in such cultures initially expand, but in most cases, this expansion phase is followed by a stable or even decline in cell number. After this so-called crisis phase, cell proliferation is noticeable in more than 90% of the cultures. Stromal cell clones can be isolated from these cultures, some of which have been cultured for more than 50 population doublings. These stromal cell clones are valuable tools for the study of the regulation of hematopoietic stem and progenitor cells in the midgestation mouse embryo.

Key Words: Aorta–gonads–mesonephros; AGM; hematopoietic stem cells; stromal cell lines; *tsA58* mutants.

1. Introduction

The differentiation of progenitor and stem cells of many tissues depend on their interactions with mesenchymal and other cells of the microenvironment. Our understanding of the molecular mechanisms governing development and differentiation of stem cells has improved over many years through the widespread use of cell lines. Such cells have been isolated from already existing tumors, from spontaneous immortalized variants of normal cells, or from primary isolates transduced with genes facilitating unlimited growth (immortalizing genes) *(1–3)*. Central to the use of cell lines in the study of cellular differentiation and development is the assumption that they are representative

From: *Methods in Molecular Biology, vol. 290: Basic Cell Culture Protocols, Third Edition*
Edited by: C. D. Helgason and C. L. Miller © Humana Press Inc., Totowa, NJ

of cells that function within the normal cellular physiology of the organism. However, the methodology required for their isolation and growth necessitates extended cultivation periods or cultivation conditions that could alter them *(4)*.

1.1. Immortalizing Genes

The most commonly used immortalizing gene to generate cell lines is that encoding the SV40 large T-antigen (*TAg*). In addition, investigators have used ectopic expression of the catalytic component of telomerase gene (*TERT*) *(5)* and *p53*-deficient cells *(6,7)* to generate cell lines. It is important to note that expression of one immortalizing gene does not suffice to transform cells, but that additional gene mutations are required *(8)*. The conditionally active form of the *TAg* gene, *tsA58*, produces a thermolabile protein that is active at 33°C *(9)*. Most often, the *TAg* or the *tsA58* gene has been introduced into cells via retroviral-mediated transduction by the cocultivation of target cells with virus-producing feeder layers *(10)*. However, this method of gene transduction requires that the cells of interest be dividing in order to achieve the integration of the provirus and immortalizing gene DNA sequences into the cellular genome and subsequent gene expression. The extended cultivation period necessary to allow cell proliferation, integration, and drug selection of the transduced cells might alter or exclude the physiologically relevant cells. Hence, we and others have generated transgenic mice expressing the immortalizing genes (*TAg*, *tsA58*, *hTERT*) or deleted *p53* to alleviate such problems by allowing for the immediate expression upon plating cells in vitro, without requirement for previous proliferation or selection steps. An additional advantage of using temperature-sensitive mutants such as *tsA58* is that they proliferate at the activating temperature (33°C) and usually stop proliferation and differentiation at the nonpermissive temperature (37–39°C) *(11,12)*.

1.2. Hematopoietic Microenvironment

Investigators are interested in the influence of the microenvironment of hematopoietic tissues on the development, expansion, and differentiation of hematopoietic stem cells. This microenvironment is composed of stromal cells *(13,14)* that interact and regulate the hierarchy of hematopoietic stem cells, progenitors, committed cells, and functional circulating blood cells *(15,16)*. Stromal cells within the context of the bone marrow and fetal liver are thought to maintain and support hematopoiesis throughout adult and fetal stages, respectively *(17,18)*. Recent developmental studies suggest that during early to midgestation, unique prefetal liver microenvironments in the yolk sac and the aorta–gonads–mesonephros (AGM) region play an important role in the differentiation, generation, maintenance, and perhaps even the expansion of the first hematopoietic cells in the mouse embryo *(19)*.

To facilitate the isolation of cell lines representative of the in vivo hematopoietic microenvironments present in the midgestation embryo, as well as to isolate cell lines from other tissues of the embryo and adult, we generated transgenic mouse lines that express the thermolabile *tsA58* gene in a constitutive and ubiquitous manner *(20)*. In this chapter, we describe methods to derive stromal cell lines and clones from cells of the murine midgestation AGM microenvironment.

Using the protocols described, cell lines from wild-type and transgene-expressing mouse bone marrow, spleen, liver, and thymus tissue, as well as embryonic liver, gastrointestinal tissue, and androgen-responsive vas deferens cell lines *(21)* have been generated.

2. Materials

1. 0.1% Gelatin. A suspension of 0.4 g gelatin powder (Sigma, cat. no. G-9391) in 400 mL distilled water in a 500-mL bottle (loose cap) is autoclaved. The gelatin is now dissolved and sterile. Store this solution at 4°C or room temperature. If you do not culture cells often, make smaller aliquots (100 or 200 mL).
2. 0.25% Trypsin. This is obtained from Gibco–Invitrogen (cat. no. 25050-014). Alternatively, 0.05% trypsin/0.53 mM EDTA (Gibco–Invitrogen, cat. no. 25300-054) gives similar results. Store in aliquots at –20°C.
3. Alpha-MEM. Alpha-MEM is from Gibco–Invitrogen with added Glutamax I (cat. no. 32571-028). Store at 2–8°C.
4. Long-term culture medium: MyeloCult™ M5300 (StemCell Technologies, Vancouver, BC, Canada). Store in aliquots at –20°C.
5. Stroma medium: The stroma medium contains 50% long-term culture medium (M5300, Stem Cell), 15% fetal bovine serum (FBS) (*see* **Note 1**), 35% alpha-MEM (Gibco–Invitrogen), antibiotics (penicillin and streptomycin; Gibco–Invitrogen, cat. no. 15140-122), Glutamax I (Gibco–Invitrogen, cat. no. 35050-038), and 10 µM β-mercaptoethanol (Sigma, cat. no. M-7522). Filter medium using 0.2-µm filters (Millipore, SCGPT05RE bottle-top filters) to remove debris and other particles that could stimulate phagocytosis and promote stromal cell differentiation and senescence. Store in aliquots at –20°C.
6. Conditioned medium (CM): prepared from each passage of the developing cell lines. The CM is collected in conical tubes and spun at 3500 rpm (2500g) for 7 min to remove debris and contaminating cells. Larger samples (>1 mL) and CM used for cloning is additionally 0.2-µm-filtered using a syringe or bottle-top filter (e.g., Millipore Millex-GV filters). Store at 4°C for up to 2 wk or for longer periods in aliquots at –20°C.
7. 0.4% Trypan blue for viable cell counts
8. Freeze medium: 90% FBS and 10% dimethyl sulfoxide (DMSO) (Sigma, cat. no. D-5879). Prepare just before use.
9. Cultureware (Costar): 94-mm, 60-mm, and 35-mm tissue-culture-treated dishes and 48- and 24-well tissue-culture-treated plates.

3. Methods

The methods outline the different steps in establishing a new cell line: (1) choice of mouse strain, (2) isolation of primary cells, (3) growth of primary cells until growth crisis, (4) growth of cells after growth crisis and cloning, (5) characterization of isolated stromal cell clones. The latter phase in cell line development usually involves screening the newly established lines for a particular desired functional behavior. Functional screening methods are beyond the scope of this chapter and will not be described.

3.1. Choice of Mouse Strain

To generate cell lines, we developed transgenic mouse strains expressing *tsA58* under the control of the β-actin (TAg05) and phosphoglycerate kinase-1 (TAg11) promotors *(20)*. Cell lines can also be generated from other mouse strains (*see* **Note 2**). Animals should be housed according to institutional guidelines, with free access to food and water. Animal procedures should be carried out in compliance with the Standards for Humane Care and Use of Laboratory Animals.

3.2. Isolation of Primary Cells

AGM and subdissected tissues were obtained from E10 and E11 embryos as described in detail elsewhere in this series *(22)*. A full description of this methodology is beyond the scope of the present chapter.

3.3. Explant and Single-Cell Cultures

Throughout this procedure, cells are cultured on 0.1% gelatin-coated tissue culture plates. Culture vessels are coated with 0.1% gelatin (100 µL/cm^2) either at 37°C (for at least 1 h) or at 4°C (overnight) with similar results. The plates can be stored at 4°C for up to 1 wk. Prior to cell seeding, the excess 0.1% gelatin solution is washed off and the vessel washed once with PBS. Once washed, these vessels should be used immediately.

Because optimal growth conditions were unknown for AGM stromal cells (*see* **Note 3**), we chose to culture the subdissected tissues on 0.1% gelatin-coated 24-well plates in either long-term culture medium or in stroma medium at 33°C (permissive temperature for *tsA58*), 5% CO_2, and greater than 95% humidity (*see* **Note 4**) using both explant and single-cell culture methods. Both the "explant" and the "single-cell suspension" methods yield stromal cell lines.

3.3.1. Explant Cultures

In this type of culture, the tissue of interest is cultured as a whole and the stromal cells are allowed to migrate and grow out of the tissue. Isolated tissues are cultured at the air–medium interface on 24-well plates (one tissue piece per

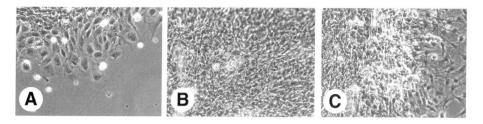

Fig. 1. Outgrowth of fibroblastoid cells 4 d after the start of primary cell culture. Cells from midgestation embryonic tissues were cultured using the "explant" method on 0.1% gelatin-coated culture dishes. Shown are explants of embryonic liver EL17 (**A**), aorta–mesenchyme AM20 (**B**), and urogenital ridges UG26 (**C**).

well) with a minimal amount of Stroma medium (100 µL/cm² of culture area). Thus, the tissue is in contact with the gelatin-coated cultureware. Tissues will attach to the plastic cultureware surface and, at the same, time, fibroblastoid cells can be seen to migrate out of the tissue (*see* **Fig. 1**).

3.3.2. Single-Cell Cultures

Spin the isolated tissues at 400g for 5 min and then wash the tissues once in serum-free Alpha-MEM. Then, subject tissues to a 15-min incubation with 0.25% trypsin and gently spin at 400g for 10 min at room temperature. Resuspend in a small volume of Stroma medium and vigorously pipet to dissociate remaining cell clumps and obtain a single-cell suspension. Count the number of viable cells prior to plating using the trypan blue exclusion and a Neubauer cytometer. The cell suspension is cultured on 24-well plates in 300 µL Stroma medium at a density of 10^5 cells per well or, if less cells are available per tissue, one tissue per well. After 1 d, single cells can be observed to be attached to the cultureware. As an alternative to establishing cell lines from single embryos, tissues from several embryos can be pooled, treated in the same manner, and cultured in six-well plates.

3.3.3. Cell Culture Until Growth Crisis

Cultures are incubated at 33°C, 5% CO_2, and greater than 95% humidity (*see* **Note 4**).

1. After 1 or 2 d, the first fibroblastoid cells can be seen to grow out of the explanted tissues (*see* **Fig. 1**). The explantlike cultures are now topped off to a total volume of stroma medium of 300 µL/cm² of cultureware area.
2. After 2–3 more days, the culture supernatant is collected as described in **Subheading 2.**, **item 6**. The adherent cells (from explant and single-cell cultures) are washed once in Alpha-MEM (no serum) and harvested by brief trypsin exposure (not more than 10 min). Detached cells are collected in polypropylene 15-mL tubes. The cells are then replated at a density of 5×10^4 cells/cm².

3. Because the growth factor requirements of the derived cell line is often not known (*see* **Note 3**), the stroma medium is supplemented with 20% 0.2-μm-filtered CM from its own previous passage as a source of autocrine growth factors for all the subsequent culture steps

4. In the first few passages, the total cell number will increase. This is usually followed by a period of passages in which the number of cells harvested is stable and then begins to be lower than the number cells initially seeded (growth crisis). During this phase, the cells are seeded in consecutively smaller culture vessels (94-mm dish [70 cm^2, 10 mL] → 60-mm dish [28 cm^2, 4 mL] → 35-mm dish/ 6-well plate [both 10 cm^2, 2 mL] → 24-well plate [2 cm^2, 300 μL] → 96 well [0.8 cm^2, 100 μL]) to maintain the number of cells at around 5×10^4 cells/cm^2. This procedure facilitates cell–cell contact and allows for the sufficiently high production of autocrine growth factors. Always add the 20% CM obtained from the previous passage. Alternatively, if sufficient CM is not available, a 0.22-μm-filtered CM from a semiconfluent cell line from the same tissue can be used as the growth supplement.

5. This procedure is repeated each week (regardless of whether cell proliferation is observed) until a consistent increase in cells is notable (*see* **Note 5**).

3.4. Culture of Cells After Growth Crisis: Cloning

The crisis period of cell senescence is usually followed (in 32 of 36 cases in our hands) by outgrowth of cells. As soon as a cell line shows consistent growth (*see* **Note 5**), cells are cloned at a density of 1 cell per 300 μL per well in 0.1% gelatin-coated 24- or 48-well plates. Cultures are incubated at 33°C, 5% CO_2, and greater than 95% humidity (*see* **Note 4**).

1. Conditioned medium is prepared from the parental cell line. The clones are grown on 0.1% gelatin-coated wells in stroma medium supplemented with 30% 0.2-μm-filtered CM of the parental cells. The cloning was more efficient when using 30% instead of the usual 20% CM.

2. After 3 d, the wells are supplemented with 300 μL stroma medium supplemented with 30% 0.2-μm-filtered CM of the parental cell line

3. The clones are maintained for 2–3 wk with medium changes every 3 or 4 d.

4. When individual wells are subconfluent (*see* **Note 6**), clones are harvested by trypsin treatment (first passage) and expanded in larger culture vessels (100-mm dishes).

5. When these larger vessels are subconfluent (i.e., range: 50–80%), again the cells are harvested by trypsin treatment and an aliquot of the clones should be frozen as (3–5) $\times 10^5$ cells per vial in freeze medium. It is important to freeze cells at the earliest stage, to ensure availability of low passage cells for future use (*see* **Notes 7** and **8**).

6. The clones are propagated as 5×10^4 cells per 100-mm dish and passaged once a week, or more often if cells reach subconfluence more quickly.

7. Clones generated in this manner can usually be cultured for more than 50 passage doublings without any sign of cellular senescence (*see* **Fig. 2** and **Note 9**).

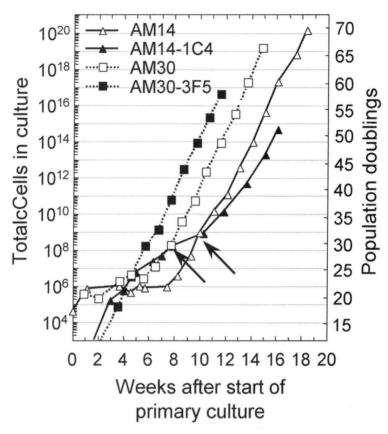

Fig. 2. Growth curves of the aorta–mesenchyme (AM)-derived AM14 and AM30 cell lines and clones thereof. AM30 (open squares) was derived from a pool of eight embryos of a TAg11 litter, whereas AM14 (open triangles) was derived from a "control" litter that did not express the immortalizing *tsA58* gene. Please note that the AM14 crisis period lasted for about 8 wk, whereas AM30 did not seem to show signs of a proliferation crisis. AM14 and AM30 were cloned after nine and seven passages, respectively (arrows). Two of the clones generated were followed for more than 50 population doublings after cloning (AM14-1C4 [closed triangles] and AM30-3F5 [closed squares]) without any sign of cellular senescence.

4. Notes

1. Select an FBS batch that gives good performance of the primary cells in the assays that you wish to perform. If such a batch is not available in your laboratory, please try to obtain such a batch from your colleagues performing similar assays. This batch will serve as a "positive" control. To obtain your own batch, it is prudent to test at least 10 different batches of FBS in this same assay. The assay you will use to test FBS batches is, however, up to you.

2. We found that it was possible to generate cell lines from early midgestation embryos from "normal" mice (the *lacZ* transgenic BL1b strain) as well as mice expressing *tsA58 (20)*. Thus, the expression of an immortalizing is not required for cell line generation. By direct comparison, however, twofold more lines were isolated from the *tsA58* transgenic embryos than from the control *lacZ* transgenic embryos. Furthermore, the presence of the *tsA58* gene allowed for a threefold to fourfold greater cloning efficiency compared to the control *lacZ* marker transgenics. Although the *tsA58* gene had an enhancing effect on the growth of the liver, urogenital ridge, and gastrointestinal-derived cell lines, no enhancing effect was observed with the aorta–mesenchyme-derived lines *(20)*.

3. It is important to know under which culture conditions the primary cells you are interested in will grow. Issues you should resolve prior to generating cell lines are as follows: (1) Which medium do the primary cells require (with or without serum)? (2) Do you need CM or have growth factor requirements been established? (3) Do the primary cells require anchoring? Using gelatin, fibronectin, laminin, or other coatings can drastically alter the cell type that will grow out of your culture. The methods described can be used to generate cell lines from different types of tissue and to grow different types of adherent cell. We have not tried to derive nonadherent, suspension-type cell lines by the method described here.

4. It is very important to regularly check the temperature, CO_2 levels, and humidity of the incubator used and make sure the incubator is level. Humidity is checked by weighing a 100-mm dish and adding exactly 10 mL of water (10 g, weigh again). One week later, weigh the dish. Water dissipation should not exceed 10% (1.0 g); 5% or less water loss is optimal. In particular, cloning efficiency depends on optimal levels of CO_2 and humidity.

5. The generation of cell lines is a time-consuming and labor-intensive process. We found that optimal results were obtained when cells are passaged weekly or prior to reaching confluence. Do not keep cells unpassaged for more than 1 wk. In our hands, it appeared that failure to passage cells regularly favored cell senescence. In some cases, the crisis period can last for weeks, sometimes for more than 3 mo *(20)*. Thus, it is important to keep culturing and passaging the cells, even when no cell proliferation is apparent. In our hands, cell lines eventually grew out of 30 of 32 (94%) of primary cell cultures of embryonic tissues.

6. The density of your cultures should be monitored daily. Always passage your cultures prior to reaching confluence (i.e., between 40% and 80%). Especially in the case of contact inhibition, a sizable proportion of cells will cease to be reactivated once proliferation has stopped by contact inhibition.

7. It is important to freeze samples of newly established cell lines (precloning and postcloning) at low passage numbers. This will ensure that there are low passage cells to go back to in case certain functional phenotypes are revealed only at these passages or when disaster strikes (contamination, CO_2 failure, etc.). In addition, once the cell lines have been characterized, cells with suitable passage numbers can be shared with collaborators.

8. The cell lines that will be generated will differ in growth characteristics and requirements: Some will be contact inhibited and some of the generated cell lines will be growth factor dependent. Because after thawing no fresh CM will be available, a mixture of CMs from semiconfluent cells of different tissues (either per tissue or all tissues together) can be prepared, filtered (0.2-μm bottle-top filter), and stored at 4°C. We stored this CM not more than 6 mo. This CM mix can then be used as a growth factor supplement for the stromal cells until the first passage after thawing. After this first passage, cells will produce their own CM for the next passage, which can be collected as described in **Subheading 2., item 6**.

9. Cell lines generated from *tsA58* transgenic mice showed a stable functional phenotype up to 50–60 population doublings after cloning *(20)*. It is known that expression of the immortalizing SV40 large T gene is, by itself, not sufficient to immortalize cells. Rather, a secondary event, such as activation of TERT, is required to produce a stable phenotype of cells for more than 150 population doublings *(8)*. Thus, it is likely that culturing cell lines beyond 60 population doublings will select for transformed cells. This should be kept in mind when early-passage cells are compared with late-passage cells (>60 population doublings).

Acknowledgment

We gratefully acknowledge Jessyca Maltman (Terry Fox Laboratory, BC Cancer Agency, Vancouver, BC, Canada) for her help and experience in the culture of mouse marrow cells and introducing me (RAJO) to the art of deriving stromal cell lines from adult tissues.

References

1. Santerre, R. F., Cook, R. A., Crisel, R. M., et al. (1981) Insulin synthesis in a clonal cell line of simian virus 40-transformed hamster pancreatic beta cells. *Proc. Natl. Acad. Sci. USA* **78**, 4339–4343.
2. Bayley, S. A., Stones, A. J., and Smith, C. G. (1988) Immortalization of rat keratinocytes by transfection with polyomavirus large T gene. *Exp. Cell Res.* **177**, 232–236.
3. Jat, P. S. and Sharp, P. A. (1986) Large T antigens of simian virus 40 and polyomavirus efficiently establish primary fibroblasts. *J. Virol.* **59**, 746–750.
4. Ridley, A. J., Paterson, H. F., Noble, M., and Land, H. (1988) Ras-mediated cell cycle arrest is altered by nuclear oncogenes to induce Schwann cell transformation. *EMBO J.* **7**, 1635–1645.
5. Morales, C. P., Holt, S. E., Ouelette, M., et al. (1999) Absence of cancer associated changes in human fibroblasts immortalized with telomerase. *Nature Genet.* **21**, 115–118.
6. Ohsawa, K., Imai, Y., Nakajima, K., and Kohsaka, S. (1997) Generation and characterization of a microglial cell line, MG5, derived from a p53-deficient mouse. *Glia* **21**, 285–298.

7. Thompson, D. L., Lum, K. D., Nygaard, S. C., et al. (1998) The derivation and characterization of stromal cell lines from the bone marrow of p53–/– mice: new insights into osteoblast and adipocyte differentiation. *J. Bone Miner. Res.* **13,** 195–204.

8. O'Hare, M. J., Bond, J., Clarke, C., et al. (2001) Conditional immortalization of freshly isolated human mammary fibroblasts and endothelial cells. *Proc. Natl. Acad. Sci. USA* **98,** 646–651.

9. Tegtmeyer, P. (1975) Function of simian virus 40 gene A in transforming infection. *J. Virol.* **15,** 613–618.

10. Jat, P. S., Cepko, C. L., Mulligan, R. C., and Sharp, P. A. (1986) Recombinant retroviruses encoding simian virus 40 large T antigen and polyomavirus large and middle T antigens. *Mol. Cell Biol.* **6,** 1204–1217.

11. Morgan, J. E., Beauchamp, J. R., Pagel, C. N., et al. (1994) Myogenic cell lines derived from transgenic mice carrying a thermolabile T antigen: a model system for the derivation of tissue-specific and mutation-specific cell lines. *Dev. Biol.* **162,** 486–498.

12. Okuyama, R., Yanai, N., and Obinata, M. (1995) Differentiation capacity toward mesenchymal cell lineages of bone marrow stromal cells established from temperature-sensitive SV40 T- antigen gene transgenic mouse. *Exp. Cell Res.* **218,** 424–429.

13. Lord, B. I., Testa, N. G., and Hendry, J. H. (1975) The relative spatial distributions of CFUs and CFUc in the normal mouse femur. *Blood* **46,** 65–72.

14. Ogawa, M. (1993) Differentiation and proliferation of hematopoietic stem cells. *Blood* **81,** 2844–2853.

15. Metcalf, D. (1988) *The Molecular Control of Blood Cells*, Harvard University Press, Cambridge, MA.

16. Lemischka, I. R. (1991) Clonal, in vivo behavior of the totipotent hematopoietic stem cell. *Semin. Immunol.* **3,** 349–355.

17. Moore, M. A. S. and Metcalf, D. (1970) Ontogeny of the haemopoietic system: yolk sac origin of in vivo and in vitro colony forming cells in the developing mouse embryo. *Br. J. Haematol.* **18,** 279–296.

18. Jordan, C. T. and Lemischka, I. R. (1990) Clonal and systemic analysis of long-term hematopoiesis in the mouse. *Genes Dev.* **4,** 220–232.

19. Dzierzak, E., Medvinsky, A., and de Bruijn, M. (1998) Qualitative and quantitative aspects of haemopoietic cell development in the mammalian embryo. *Immunol. Today* **19,** 228–236.

20. Oostendorp, R. A. J., Medvinsky, A. J., Kusadasi, N., et al. (2002) Embryonal subregion-derived stromal cell lines from novel temperature-sensitive SV40 T antigen transgenic mice support hematopoiesis. *J. Cell Sci.* **115,** 2099–2108.

21. Umar, A., Luider, T. M., Berrevoets, C. A., Grootegoed, J. A., and Brinkmann, A. O. (2003) Proteomic analysis of androgen-regulated protein expression in a mouse fetal vas deferens cell line. *Endocrinology* **144,** 1147–1154.

22. Dzierzak, E. and de Bruijn, M. (2002) Isolation and analysis of hematopoietic stem cells from mouse embryos In *Hematopoietic Stem Cell Protocols* (Klug, C. A. and Jordan, C. T., eds.), Humana, Totowa, NJ, pp. 1–14.

13

Culture of Human and Mouse Mesenchymal Cells

Emer Clarke

Summary

Normal human and mouse bone marrow is composed of hematopoietic and non-hematopoietic cells. The latter have also been termed stromal cells, microenvironment cells, colony-forming-unit fibroblasts (CFU-F), and mesenchymal cells. These cells were originally thought to provide an appropriate matrix for hematopoietic cell development, but recent examination of these cell populations suggests a much broader spectrum of activity, including the generation of bone, cartilage, muscle, tendon, and fat. In the future, these mesenchymal cell populations could be used for the treatment of specific diseases and to enhance the engraftment of hematopoietic cells. This chapter describes methods for the human CFU-F assay, culture and expansion of mesenchymal cells, as well as their differentiation to adipocytes. In addition, this chapter describes the mouse CFU-F assay.

Key Words: Marrow stroma; microenvironment; colony-forming unit; fibroblast; CFU-F; stromal progenitors; mesenchymal cells; differentiation; adipocytes.

1. Introduction

The bone marrow stroma was originally thought to function mainly as a structural framework for the hematopoietic stem and progenitor cells present in the bone marrow. It has been established that the stroma consists of a heterogeneous population of cells, including endothelial cells, fibroblasts, adipocytes, and osteogenic cells, a subset of which exerts both positive and negative regulatory effects on the proliferation and differentiation of hematopoietic cells *(1,2)*. The adherent stromal cell population is also believed to contain other nonhematopoietic cells that are capable of both self-renewal and differentiation into bone, cartilage, muscle, tendon, and fat *(3–5)*. Characterization of the stromal cells was initiated many years ago where the morphology as well as the cytochemical characterization of the cultured cells was described (sudan black[+ve], alkaline phosphatase [+ve], esterase [−ve], collagen IV [+ve], fibronectin [+ve])

From: *Methods in Molecular Biology, vol. 290: Basic Cell Culture Protocols, Third Edition*
Edited by: C. D. Helgason and C. L. Miller © Humana Press Inc., Totowa, NJ

(6,7). A number of years later, Simmons and Torok-Storb described the first antibody (Stro-1) that targeted the stromal precursor in human bone marrow *(8)*.

The colony-forming-unit fibroblast (CFU-F) assay has been used by many investigators as a functional method to quantify the stromal progenitors *(9,10)*. Abnormal function of these precursors has been implicated in several diseases *(11,12)*. Transplantation of unprocessed bone marrow cells can restore microenvironment function, suggesting unprocessed bone marrow contains both the stromal precursor as well as the hematopoietic precursors. Studies by Gallatto et al. confirm that these microenvironment precursor cells, as measured by the CFU-F assay, are susceptible to damage following chemotherapy or radiation and they remain at a significantly reduced frequency for a considerable time following transplantation *(13)*.

There has been a resurgence of interest in the stromal cells and their function in both the tissue engineering and stem cell plasticity fields. This interest was fueled by the observation that cultured stromal cell populations were capable of both self-renewal and differentiation, characteristics typically associated with stem cells. These traits have led many researchers to refer to these cultured stromal cells as mesenchymal stem cells (MSCs). Cultured mesenchymal cells have been characterized using panels of antibodies and are defined as $CD45^{-ve}$, $CD34^{-ve}$, $SH2^{+ve}$ (CD105), $SH3^{+ve}$, and $SH4^{+ve}$ (CD73) cells *(5)*. The isolation and enrichment of human mesenchymal cells have utilized some of their simple characteristics like adherence as well as cell-separation strategies using cocktails of antibodies that deplete the bone marrow of specific cell populations *(14,15)*. Despite these advances, the exact phenotype of the stromal (mesenchymal) precursor cell in human bone marrow (i.e., the cell phenotype prior to culture) is still debated.

Enrichment of mouse CFU-F has been described by Short and Simmons who identified the femoral bone itself as a richer source of progenitors than the marrow plug within it *(16)*. Using a number of physical and enzymatic treatments of the bone to generate a single-cell suspension followed by depletion of cells expressing the lineage (Lin) antigens CD3, CD4, CD5, CD8, CD11b, and GR1, they could enrich the CFU-F significantly. Further cell-sorting experiments using flow cytometry identified the stromal (mesenchymal) mouse precursor as Lin^{-ve}, $CD45^{-ve}$, $CD31^{-ve}$, and Sca^{+ve}.

Cultured mesenchymal cells have been shown to exhibit some unique properties that challenge the dogma that stem cells derived from adult tissue produce only the cell lineages characteristic of tissues wherein they reside. Studies published by Verfaillie's group have demonstrated the ability of cultured MSCs to differentiate into neural cells, skeletal cells, cardiomyocytes, endothelial cells, and smooth muscle cells *(17)*. The expanding knowledge of the biology of specific cell populations might be the foundation for future therapies in many areas outside of hematology and oncology.

This chapter describes methodology for the human CFU-F assay, culture and expansion of human mesenchymal cells, and differentiation to adipocytes, as well as the mouse CFU-F assay.

2. Materials

2.1. Human CFU-F Assay

1. Ammonium chloride buffer (StemCell Technologies Inc., Vancouver, BC, Canada).
2. Ficoll–Hypaque (density 1.077 g/mL) (Sigma–Aldrich, cat. no. F8636; StemCell, cat. no. 07907). Store in a sterile manner in the dark, at room temperature.
3. Phosphate-buffered saline (PBS) (StemCell, cat. no. 37350).
4. PBS + 2% fetal bovine serum (FBS) (StemCell, cat. no. 07905).
5. MesenCult™ basal medium (human) (StemCell, cat. no. 05401), modified McCoy's 5A medium (StemCell, cat. no. 36350), or Dulbecco's Modified Eagle's medium (DMEM) (StemCell, cat. no. 36253). Store liquid media in the dark at 4°C.
6. FBS. Each batch must be pretested for its ability to support human CFU-F (*see* **Note 1**). Pre-tested batches are available from StemCell (06471).
7. L-Glutamine or 200 mM L-glutamine solution (StemCell, cat. no. 07100). L-Glutamine solution is stable for 2 yr at –20°C or 1 mo at 4°C (*see* **Note 2**).
8. Complete medium. Prepare 10% prescreened FBS in MesenCult basal medium, modified McCoy's 5A medium, or DMEM. Add L-glutamine to give a final 2 mM concentration. Store medium at 4°C for up to 1 mo.
9. Automated cell counter (Coulter) or 3% glacial acetic acid (StemCell, cat. no. 07060) and Neubauer counting chamber for manual cell counts.
10. Tissue culture materials; T-25 tissue-culture-treated flasks (Falcon, cat. no. 353108), 1-mL and 10-mL sterile pipets, hand pipettors, and 20-µL, 200-µL, and 1000-µL tips.
11. Water-jacketed incubator calibrated to 37°C, 5% CO_2 in air and >95% humidity
12. Inverted microscope equipped with ×2, ×4, and ×10 objectives.
13. Methanol ACS (BDH, cat. no. ACS531).
14. Giemsa staining solution (EM Science, cat. no. RO3055/76)

2.2. Human Mesenchymal Cell Culture and Expansion

1. 0.25% Trypsin–EDTA (StemCell, cat. no. 07901)
2. Materials described in **Subheading 2.1.**, **items 1–11**.

2.3. Differentiation of Human Mesenchymal Cells to Adipocytes

1. 0.25% Trypsin–EDTA (StemCell, cat. no. 07901).
2. PBS + 2% FBS (StemCell, cat. no. 07905).

3. MesenCult basal medium (human) (StemCell, cat. no. 05401), modified McCoy's 5A medium (StemCell, cat. no. 36350), or DMEM (StemCell, cat. no. 36253). Store liquid media in the dark at 4°C.
4. Adipogenic supplements (StemCell, cat. no. 05403). Store at –20°C. Stable for 2 yr at –20°C or 1 mo at 4°C (*see* **Note 3**).
5. Adipogenic differentiation medium. Add 50 mL of adipogenic supplement to 450 mL of MesenCult basal medium, modified McCoys 5A medium, or DMEM. Store complete medium at 4°C for up to 1 mo.
6. Tissue culture materials, incubator, and microscope are described in **Subheading 2.1.**, **items 10–12**.

2.4. Mouse CFU-F Assay

1. PBS + 2% FBS (StemCell, cat. no. 07905).
2. 3% Glacial acetic acid (StemCell, cat. no. 07060) and Neubauer counting chamber for manual cell counts.
3. MesenCult basal medium mouse (StemCell, cat. no. 05401) or modified McCoy's 5A medium (StemCell, cat. no. 36350). Store liquid media in the dark at 4°C.
4. FBS and horse serum (HS). Batches of FBS and HS must be pretested in combination for their ability to support mouse CFU-F (*see* **Note 4**). Pretested batches are available from StemCell (cat. no. 05502).
5. Mouse CFU-F medium. Add 10% prescreened FBS and 10% prescreened HS to MesenCult basal medium or modified McCoy's 5A medium. Add L-glutamine to give a final 2 mM concentration. Store at 4°C for up to 1 mo.
6. Tissue culture materials; six-well tissue-culture-treated plates (Falcon, cat. no. 353502; Corning, cat. no. 3506), 1-mL and 5-mL sterile pipets, hand pipettors, 20-µL, 200-µL, and 1000-µL tips, 1-cm^3 syringes (Becton Dickinson), 21-gage needle (Becton Dickinson), and 23 gage needles (Becton Dickinson).
7. Materials described in **Subheading 2.1.**, **items 11–14**.

3. Methods

All culture steps should be done using a sterile technique and performed in a certified biosafety cabinet.

3.1. Human CFU-F Assay

3.1.1. Preparation of Human Bone Marrow Cells

Two suitable methods are described in **Subheadings 3.1.1.1.** and **3.1.1.2.** for preparation of human bone marrow cells.

3.1.1.1. RED BLOOD CELL-DEPLETED BONE MARROW

1. Collect bone marrow samples using heparin as the anticoagulant.
2. Dilute the unprocessed bone marrow with nine times the volume of the ammonium chloride buffer (e.g., 5 mL of bone marrow and 45 mL of buffer) and mix well.

3. Incubate for 5 min at room temperature and centrifuge at 1200 rpm (330g) for 10 min.
4. Discard supernatant, resuspend cell pellet, and wash once using PBS + 2% FBS.

3.1.1.2. Mononuclear Bone Marrow Cells

1. Collect bone marrow samples using heparin as the anticoagulant.
2. Dilute the unprocessed bone marrow with an equal volume PBS + 2% FBS.
3. Carefully layer 20 mL of diluted cells on 15 mL of Ficoll–Hypague in each 50-mL conical tube. Centrifuge at 1200 rpm (330g) for 25 min with the brake set to the "off" position.
4. Carefully harvest the mononuclear cells from the buffy layer located at the interface between the medium and ficoll using a 5-mL pipet.
5. Dilute cells with a minimum 5X volume of PBS + 2% FBS and centrifuge at 1200 rpm (330g) for 10 min.
6. Discard supernatant, resuspend cell pellet, and wash once using PBS+2% FBS.

3.1.2. Performing the Human CFU-F Assay

1. Count nucleated cells using an automatic cell counter, or a hemacytometer and a light microscope. Resuspend cells in complete medium at 10^7 cells/mL.
2. Place 10 mL of complete medium into each of four T-25 tissue culture-treated flasks and then add 300 μL of the stock cell solution to one (3×10^6 cells per flask), 200 μL of stock to the second (2×10^6 cells per flask), 100 μL of stock to the third (1×10^6 cells per flask), and 50 μL to the fourth (0.5×10^6 cells per flask) (*see* **Note 5**).
3. Place the cap onto the flask following the addition of cells and swirl the flask gently to ensure equal distribution of the cells. Avoid getting any medium into the neck of the flask as this could promote contamination in the culture (*see* **Note 6**).
4. Incubate for 14 d. Maximum colony size and numbers are typically observed at this time.
5. Evaluate the culture microscopically using the ×2 objective prior to staining.
6. Remove the medium from the tissue culture flasks and discard appropriately. Rinse the culture flasks with PBS (without FBS) to remove any remaining medium and discard.
7. Add 5 mL of methanol to a T-25 flask for 5 min at room temperature to fix the cells to the tissue culture flasks. Discard the methanol and allow flasks to air-dry at room temperature.
8. Add 5 mL of Giemsa to a T-25 flask for 5 min at room temperature. Remove the Giemsa solution and rinse thoroughly with water (tap water can be used).
9. Discard the water and air-dry, because enumeration of CFU-F is simpler once the plate has dried.
10. Count human CFU-F macroscopically and determine CFU-F frequency (*see* **Fig. 1** and **Notes 5** and **7**).

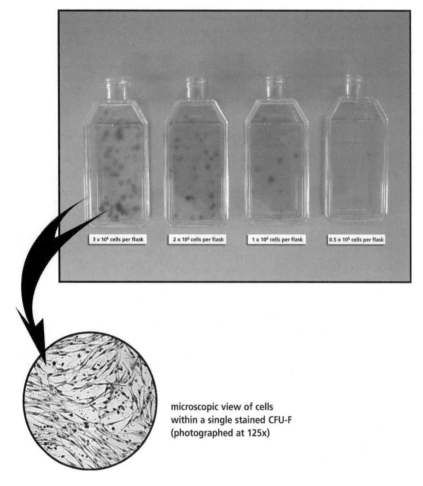

Fig. 1. Dose–response curve of CFU-F in T-25 flasks using various concentrations of Ficolled human bone marrow cells.

3.2. Human Mesenchymal Cell Culture and Expansion

1. Prepare a mononuclear cell population as described in **Subheading 3.1.1.2.** and dilute the cells at 10^7 cells/mL in complete medium.
2. Place 9 mL of complete medium per T-25 tissue culture treated flask and add 1 mL of the stock cell solution (10^7 cells per flask) (*see* **Note 8**).
3. Place the cap onto the flask following the addition of cells and swirl the flask gently so as to ensure equal distribution of the cells throughout. Avoid getting any medium into the neck of the flask as this could promote contamination in the culture.

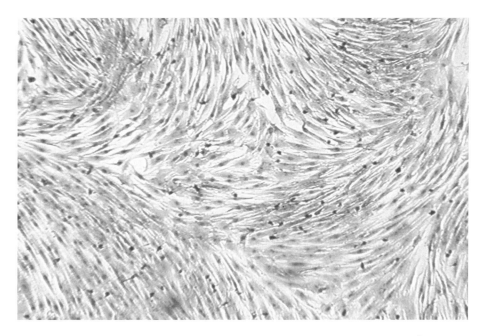

Fig. 2. Confluent mesenchymal cell layer at passage 2, generate from 10^7 Ficolled mononuclear human bone marrow cells (photographed at ×50).

4. Incubate cultures at 37°C for 14 d. Typically at this time, there is a confluent layer of cells as well as some round nonadherent cells floating throughout the medium.
5. Discard the medium (and nonadherent cells). Rinse the culture flasks with PBS (without FBS) to remove any remaining medium and FBS and discard appropriately.
6. Add 5.0 mL of trypsin–EDTA and place the T-25 flask in a 37°C incubator until the adherent cells begin to lift off (approx 3–5 min).
7. Add 5.0 mL of the complete medium to the T-25 flask, rinse flask surface using the 5-mL pipet, and transfer cells and medium into a 15-mL tube (*see* **Note 9**).
8. Spin the tube at 1200 rpm (330*g*) for 7 min with the break set at the "high" position. Discard the supernatant and resuspend the cells in 1–2 mL of complete medium.
9. Perform a cell count if cell expansion assessment is required. Alternatively, divide the contents of one T-25 flask into four new T-25 tissue culture-treated flasks in a total volume of 10 mL of complete medium per flask.
10. Incubate cultures for approx 5 d or until the cells become confluent (*see* **Fig. 2** and **Note 10**).
11. This procedure of passaging and expanding the mesenchymal cells can be repeated for 8–10 passages with normal human bone marrow (*see* **Note 11**).

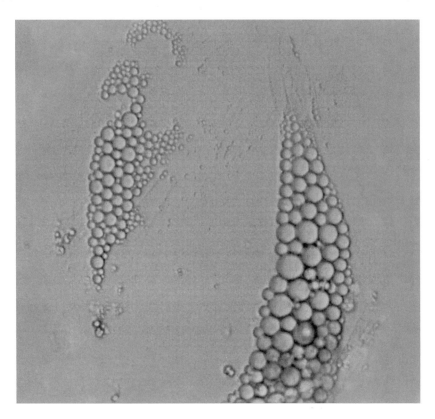

Fig. 3. Differentiation of cultured mesenchymal cells to adipocytes (photographed at ×125 magnification).

3.3. Differentiation of Human Mesenchymal Cells to Adipocytes

1. Generate cultured mesenchymal cells and harvest cells as described in **Subheading 3.2.** (*see* **Note 12**).
2. Place 9 mL of adipogenic differentiating medium per T-25 flask. Add 25% (1/4 flask equivalent) of the total cells harvested from one T-25 flask of cultured mesenchymal cells.
3. Place the cap onto the flask and swirl the flask gently so as to ensure equal distribution of the cells throughout. Avoid getting any medium into the neck of the flask, as this could promote contamination in the culture.
4. Incubate cultures for 14 d. Typically at this time, there is an abundance of adipocytes (*see* **Fig. 3**).
5. Evaluate numbers and size of adipocytes using an inverted microscope with ×4 and ×10 objectives.

3.4. Mouse CFU-F Assay

Mice should be maintained and sacrificed using procedures approved by your institution.

3.4.1. Extraction of Mouse Bone Marrow Cells

1. Place the mouse on a flat surface on its back and wet the pelt thoroughly with 70% isopropyl alcohol.
2. Cut pelt and peel back to expose the hind limbs.
3. Using sterile sharp scissors (to avoid splitting the bone), cut the knee joint in the center, and remove ligaments and excess tissues. Remove the femur and tibia by severing them from the animal at the hip and ankle joints, respectively.
4. Trim the ends of the long bones to expose the interior of the marrow shaft. Flush the marrow from the femurs using 1–2 mL PBS + 2% FBS and 21-gage needle attached to a 1-cm^3 syringe. A smaller needle (23-gage) is more efficient at marrow cell removal from the tibia.
5. Prepare a single-cell suspension by gently aspirating several times using the same needle and syringe.
6. Perform a cell count by diluting a cell aliquot 1:50 or 1:100 in 3% glacial acetic acid and count the nucleated cells using a hemacytometer and a light microscope. The expected cell recovery is $(1–2) \times 10^7$ cells per femur and 6×10^6 cell per tibia.

3.4.2. Performing the Mouse CFU-F Assay

1. Dilute the mouse bone marrow cells to 1×10^7 cells/mL in mouse CFU-F medium.
2. Place 5.85 and 5.7 mL of mouse CFU-F medium into two 13-mL polystyrene tubes and then add 150 μL of the stock cell solution to one (final concentration of 5×10^5 cells per well) and 300 μL of stock to the other (to obtain a final concentration of 1×10^6 cells per well).
3. Vortex the tubes to ensure a well-mixed cell suspension, and plate 2.0 mL of medium containing cells into three replicate wells in a six-well tissue-culture-treated plate (*see* **Note 13**).
4. Place the lid on the plate and carefully swirl the plate gently so as to ensure equal distribution of the cells throughout.
5. Incubate cultures for 10 d. Maximum colony size and numbers are typically observed at this time (*see* **Fig. 4**).
6. Evaluate the culture microscopically using the ×4 and ×10 objective prior to staining.
7. Remove the medium from each well of the six-well plate using a 2-mL pipet and discard appropriately. Gently rinse each well with PBS (without FBS) to remove any remaining medium and discard appropriately.
8. Add 2 mL of methanol to each well for 5 min at room temperature to fix the cells to the tissue culture flasks. Discard the methanol and allow wells to air-dry at room temperature.
9. Add 2 mL of Giemsa to each well for 5 min at room temperature. Remove the Giemsa solution and rinse thoroughly with water (tap water can be used).

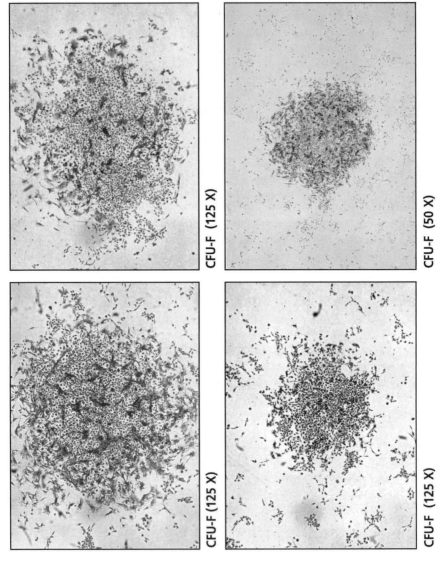

CFU-F (125 X)

CFU-F (50 X)

CFU-F (125 X)

CFU-F (125 X)

Fig. 4. Stained mouse CFU-F colonies (photographed at magnification indicated).

10. Discard the water and allow the plate to air-dry, because enumeration of mouse CFU-F is simpler in the absence of water droplets.
11. Count the mouse CFU-F microscopically with an inverted microscope with ×4 and ×10 objectives (*see* **Note 14**).

4. Notes

1. It is necessary to test different batches of FBS to select one that gives the greatest number of CFU-Fs and optimal colony morphology. Although there are differences between individuals, the average frequency of CFU-F is 1:100,000 in Ficolled normal bone marrow cells and the average size of a CFU-F is 3 mm in diameter.
2. Although many basal media formulations contain L-glutamine, this amino acid is stable for only 1 mo at 4°C. Therefore, it is recommended that L-glutamine be added each time that "complete medium" is prepared.
3. Adipogenic supplements contain prescreened serum, hydrocortisone, and dexamethasone.
4. For optimal growth of mouse CFU-F, a mixture of FBS and HS must be used and, therefore, these sera must screened individually and in combination. Batches of serum that support human CFU-Fs do not typically support mouse CFU-Fs. With optimal batches of FBS and HS, the frequency of mouse CFU-Fs is approx 1:40,000 and the size of the colony is 0.5–1.0 mm in diameter.
5. It is essential to use tissue-culture-treated flasks for this assay, as the mesenchymal progenitors (CFU-Fs) must adhere to promote growth and replication. Because there are differences in the proliferative potential of human marrow, the CFU-F assay is initiated at four distinct cell concentrations. It is anticipated that three of these four cultures will generate appropriate data. When one doubles the cells plated, one would anticipate a twofold increase in CFU-Fs (i.e., 20 colonies from a culture initiated with 2×10^6 as compared to 10 colonies from a culture initiated with 1×10^6 cells). However, this does not always happen. Sometimes, when there is poor proliferation (which can be associated with increasing age of a donor or disease status), there could be few, if any, colonies at the lowest cell concentrations plated. Determining a frequency from such data could be erroneous. In addition, if the marrow is a highly proliferative one (very young donor), cultures initiated at 3×10^6 cells per flask could be overplated and the resulting CFU-F number could be an underestimate of the true value. To generate the frequency of the CFU-Fs, determine which cell concentrations are appropriate. This can be done by plotting a graph with CFU-F numbers on the *y*-axis and cell concentration on the *x*-axis and drawing a line of best fit. The line should go through the origin (if there are no cells added to the medium, there are no CFU-F). It might be necessary to exclude data points that deviate significantly. The frequency is calculated by dividing the total CFU-Fs generated by the total number of cells in the assay. An example is given in **Table 1**.
6. Contamination can be minimized by preventing medium lying in the neck of the flask. This could be achieved by adding the cell suspensions to the complete

Table 1
Representative Human CFU-F Numbers

Ficolled BM cells cultured	CFU-F enumerated	Comment
0.5×10^6	10	In linear range
1.0×10^6	22	In linear range
2.0×10^6	45	In linear range
3.0×10^6	52	Not in linear range (overplated, therefore underestimated)

Note: Using the data provided in Table 1, the frequency of the CFU-F = $(10 + 22 + 45)/(0.5 \times 10^6 + 1 \times 10^6 + 2 \times 10^6)$

$= (77)/(3.5 \times 10^6)$

$= 1{:}45{,}454$

medium in a tube and then transferring the entire contents of the tube using a 10-cm^3 pipet and placing the pipet tip at the bottom of the flask. Care must be used in removing the pipet so that medium is not inadvertently dropped in the flask neck.

7. Human CFU-Fs are large enough to see with the naked eye, and following staining with Giemsa, they are very easy to score. We recommend taking a felt-tip pen and marking each CFU-F on the flask when counted. This prevents counting colonies more than once. Having determined the CFU-F number per cell concentration plated, one can determine the frequency (*see* **Note 5**).

8. Cell density is critical in establishing either CFU-Fs or MSCs. At low cell concentrations CFU-Fs result, but when approx 10^7 cells are plated, discreet colonies do not form and instead a monolayer of mesenchymal cells is established.

9. Trypsin is used to detach the adherent cell populations from the plastic surface; however, the action of trypsin will continue to act on the cells (and eventually reduce viability) until neutralized. Serum inhibits the activity of the trypsin; therefore, the addition of 5 mL of complete medium (containing 10 to 20% serum) will inhibit any further trypsin activity. We recommend that you evaluate the culture after 3 min with trypsin and see if the majority of the cells are nonadherent and floating. If many cells are adherent, return the flask to the incubator for an additional 2 min. Once you have established that most of the cells are nonadherent, add the complete medium immediately.

10. Depending on the cell number plated as well as the proliferative capacity of the marrow, the cells could become almost confluent between 3 and 7 d of culture. In order to maintain a healthy cell population, it is advisable to passage the cells when they are about 80 to 85% confluent. These culture cells have been characterized and have been shown to lack expression of CD45 and CD34. They express CD105 (SH2) and CD73 (SH3, SH4) (**5**).

11. Ficolled bone marrow from normal donors can be passaged 8–10 times; however, there is variability from donor to donor and the ability to passage cells might be limited if the culture expanded mesenchymal cells are allowed to sit in a confluent state for a number of days. Typically, cells should be passaged when the cells are 80 to 85% confluent.

12. Although many investigators use passaged mesenchymal cells as their cell source for adipogenic differentiation, adipocytes can be generated from human bone marrow by plating 10^7 Ficolled bone marrow cells in complete adipogenic medium.

13. It is essential to use tissue-culture-treated plates for this assay, as the mouse mesenchymal progenitors (CFU-Fs) need to adhere before they can replicate and generate colonies.

14. Mouse CFU-Fs are smaller than the human counterparts, so microscopic evaluation is essential. The colony, typically 0.5–1.0 mm in diameter, contains two distinct cell types: one fibroblastlike cell and another a rectangle-shaped cell (which, because of its shape, has been referred to as a blanket cell). We have tested the marrow from a number of different strains of mice, including CD1, N/M mice, and Balb/C mice, and the frequency of CFU-Fs was similar.

References

1. Dexter, T. M., Allen, T. D., and Lajtha, L. G. (1977) Conditions controlling the proliferating of hematopoietic cells in vitro. *J. Cell Physiol.* **91,** 335–344.
2. Verfaillie, C. M. (1993) Soluble factor(s) produced by human bone marrow stroma increase cytokine-induced proliferation and maturation of primitive hematopoietic progenitors while preventing their terminal differentiation. *Blood* **82,** 2045–2053.
3. Bruder, S. P., Jaiswal, N., and Haynesworth, S. E. (1997) Growth kinetics, self-renewal, and osteogenic potential of purified human mesenchymal stem cells during extensive subcultivation and following cryopreservation. *J. Cell Biochem.* **64,** 278–294.
4. Mackay, A. M., Beck, S. C., Murphy, J. M., Barry, F. P., Chichester, C. O., and Pittenger, M. F. (1998) Chondrogenic differentiation of cultured human mesenchymal stem cells from marrow. *Tissue Eng.* **4,** 415–428.
5. Pittenger, M. F., Mackay, A. M., Beck, S. C., et al. (1999) Multilineage potential of adult human mesenchymal stem cells. *Science* **284,** 143–147.
6. Friedenstein, A. J. (1980) Stromal mechanisms of bone marrow: cloning in vitro and transplantation in vivo. *Hematol Bluttransfus.* **25,** 19–29.
7. Castro-Malaspina, H., Gay, R. E., Resnick, G., et al. (1980) Characterization of human bone marrow fibroblast colony forming cells (CFU-F) and their progeny. *Blood* **56,** 289–301.
8. Simmons, P. G. and Torok-Storb, B. (1991) Identification of stromal cells in human bone marrow by a novel monoclonal antibody Stro-1. *Blood* **78,** 55–62.
9. Friedenstein, A. J., Chailakhjan, R. K., and Lalykina, K. S. (1970) The development of fibroblast colonies in monolayer cultures of guinea-pig bone marrow and spleen cells. *Cell Tissue Kinet.* **3,** 393–403.

10. Clarke, E. and McCann, S. R. (1989) Age dependent *in vitro* stromal growth. *Bone Marrow Transplant.* **4,** 596–597.

11. Minguell, J. J. and Martinez, J. (1983) Growth pattern and function of bone marrow fibroblasts from normal and acute lymphoblastic leukemia patients. *Exp. Hematol.* **11,** 522–526.

12. Scopes, J., Ismail, M., Marks, J. K., et al. (2001) Correction of Stromal cell defect after bone marrow transplantation in aplastic anaemia. *Br. J. Haematol.* **115,** 642–652.

13. Galotto, M., Berisso, G., Delfino, L., et al. (1999) Stromal damage as consequence of high-dose chemo/radiotherapy in bone marrow transplant recipients. *Exp. Hematol.* **27,** 1460–1466.

14. Clarke, E., Wognum, A. W., Marciniak, R., and Eaves, A. C. (2001) Mesenchymal cell precursors from human bone marrow have a phenotype that is distinct from cultured mesenchymal cells and are exclusively present in a small subset of CD45lo, SH2+ cells. *Blood* **98,** 355a.

15. Reyes, M., Lund, T., Lenvik, T., Aguiar, D., Koodie, L., and Verfaillie, C. M. (2001) Purification and ex vivo expansion of postnatal human marrow mesodermal progenitor cells. *Blood* **98,** 2615–2625.

16. Short, B. J., Brouard, N., and Simmons, P. J. (2002) Purification of MSC from mouse compact bone. *Blood* **100,** 62a.

17. Jiang, Y., Jahagirdar, B. N., Rheinhardt, R. L., et al. (2002) Pluripotency of mesenchymal stem cells derived from adult marrow. *Nature* **418,** 41–49.

14

Isolation, Purification, and Cultivation of Murine and Human Keratinocytes

Frizell L. Vaughan and Ludmila I. Bernstam

Summary

The architecture of mammalian skin incorporates an outer layer of stratified epithelium. This enables the organism to conserve internal homeostasis and maintain protection from adverse environmental exposure. The keratinocyte is the cell primarily responsible for this structure. Isolation and in vitro cultivation of this cell type is widely used in dermatological and other investigations as opposed to using whole animals. However, this cell is very fastidious as compared to other skin cells (fibroblasts, etc.) and thus requires special procedures to obtain successful in vitro cultivation. This chapter describes the methodology required to isolate, purify, and cultivate keratinocytes to produce both monolayer and stratified cultures. The methodologies for producing cultures of keratinocytes obtained from rat skin and from human skin are described.

Key Words: Keratinocytes; basal; cultivation; primary; rat; murine; human; monolayer; multilayer; in vitro; epidermis.

1. Introduction

The methodology of in vitro cultivation of both rat and human keratinocytes is outlined. Depending on an investigator's goals, each source might offer an advantage. Experiments with inbred, syngeneic strains of rats might offer comparatively superior statistical data. However, using tissue obtained from humans could reduce extrapolation problems and bypass animal rights concerns. Rat keratinocytes are obtained from the skins of syngeneic newborn albino rats. Human keratinocytes are obtained from skin biopsies resulting from various surgical procedures. Both tissues are first processed to (1) minimize microbial contamination and (2) remove subcutaneous elements. The skin is then treated physically and chemically to obtain separation of the dermis from the epidermis. The separation procedure used results in the splitting of the skin

From: *Methods in Molecular Biology, vol. 290: Basic Cell Culture Protocols, Third Edition*
Edited by: C. D. Helgason and C. L. Miller © Humana Press Inc., Totowa, NJ

at the junction where the basal layer of keratinocytes join with the underlining dermis. The resulting basal keratinocytes are then removed, suspended in a special isotonic solution, and purified via centrifugation on a density gradient. The purpose of this procedure is to remove the fibroblasts and cellular debris resulting from previous procedures. The purified keratinocytes are then resuspended and quantified. Precise amounts of keratinocytes are plated on various substrata depending on whether the purpose is to obtain a monolayer culture or a stratified, differentiated culture. Both culture types are maintained at 35°C, 5% CO_2 in a humidified environment. Growth of monolayer cultures is monitored using an inverted phase-contrast microscope or by standard histological procedures if grown on opaque substrata. The production of stratified, differentiated keratinocytes requires further steps and special substrata to obtain the desired epithelium. The cells are first incubated on a membrane submerged in growth medium until they form a confluent monolayer. They are then incubated at the air–liquid interface in order to encourage stratification to form an epidermal-like structure. This resulting culture can be examined using histological procedures for light and transmission electron microscopy.

The methodology involved in culturing mammalian cells is very complex. The demand for purity of materials that are used, the necessity of maintaining strict aseptic conditions, and the requirement of specific incubation procedures must all be met for successful cell cultivation. Any attempt to discuss all of these specific procedures in this chapter would be inadequate at best. However, a thorough knowledge of this methodology is essential for all investigators involved in mammalian cell cultivation. Fortunately, there are a number of manuals that have been published in which this methodology is discussed in detail. The one that these authors have used for many years is authored by R. Ian Freshney, the latest edition published in 2000 *(1)*. Explicit descriptions of necessary and useful equipment, aseptic techniques required, and valuable tips designed to improve performance are contained in such manuals. For investigators not familiar with these procedures, the information contained in manuals of this type is a necessity. For seasoned investigators, such manuals remain very helpful.

2. Materials

1. Human full-thickness skin, surgically removed. Process immediately to obtain and cultivate keratinocytes.
2. Syngeneic albino rats, 2–3 d old. Process immediately to obtain and cultivate keratinocytes.
3. Biosafety cabinet equipped with a HEPA filter.
4. CO_2 Incubator with atmosphere controls.
5. Inverted phase-contrast microscope (IPCM).
6. Sterilizing equipment.

7. Water purification equipment (if needed).
8. Hemocytometer.
9. Reagent-grade chemicals only (alcohol, NaCl, buffers, etc.).
10. Bard/Parker surgical scalpel (no. 22), disposable, sterile (Baxter).
11. Plastics, sterile (tissue culture flasks, covered multiwells, dishes, etc.) (*see* **Note 1**).
12. Sterile centrifuge tubes (*see* **Note 1**).
13. Filter units for sterilization of chemicals and biologicals (Nalgene).
14. Sterile disposable pipets (*see* **Note 1**).
15. Trypsin, crude, 1:250, unsterile powder. Use working solutions immediately (BD-Difco Laboratories).
16. Trypsin, porcine pancreas, cell culture tested. Store at 4°C or –20°C (*see* **Note 1**).
17. Trypsin inhibitor, soybean, cell culture tested. Store at 4°C or –20°C (*see* **Note 1**).
18. Earle's balanced salt solution (EBSS). Store at 4°C or –20°C (*see* **Note 1**).
19. Phosphate-buffered saline (PBS), Ca^{2+} and Mg^{2+}-free.
20. Ethylenediaminetetracetic acid (EDTA) (*see* **Note 1**).
21. Trypan blue; cell culture tested (*see* **Note 1**).
22. Percoll™, sterile solution (Amersham Biosciences). Store at ambient temperature.
23. Percoll density marker beads (Amersham Biosciences).
24. Polycarbonate centrifuge tubes (Nalgene).
25. Minimum essential medium (MEM). Refrigerate or freeze working solutions until used (1 mo maximum) (*see* **Note 1**).
26. L-Glutamine, 200 mM. Store at –20°C. Use immediately after thawing (*see* **Note 1**).
27. Fetal bovine serum (FBS). Store at –20°C. Use immediately after thawing (*see* **Note 1**).
28. Insulin (IN), solution, from bovine pancreas, cell culture tested. Store at 4°C (*see* **Note 1**).
29. Hydrocortisone (HC)–cortisol. Cell culture tested. Store at 4°C (*see* **Note 1**).
30. Antibiotic/antimycotic solutions (penicillin, streptomycin, gentamycin, amphotericine B, neosporin, etc.). Store solutions at –20°C (*see* **Note 1**).
31. Epidermal growth factor (EGF), mouse natural, cell culture tested. Store working solutions at 4°C or –20°C (*see* **Note 1**).
32. Bovine pituitary extract (BPE). Store at –20°C. Use immediately after thawing.
33. Collagen, calf skin type I; cell culture tested; powder or solution. Store solutions at 4°C or –20°C (*see* **Note 1**).
34. Laminin (LMN). Engelbreth–Holm–Swarm rat sarcoma (basement membrane); cell culture tested. Store at –70°C (*see* **Note 1**).
35. Porous, inert membrane (13 mm in diameter) (Pall-Gelman, Millipore).
36. Glass fiber filter (44 mm in diameter) (Pall-Gelman).
37. Sable hair brush sterilized with 70% ethanol.

3. Methods

The methods described in this section include (1) the preparation of solutions and biologicals necessary for successful isolation and cultivation of mammalian cells, (2) the initial procedures for processing full-thickness epithelium received from rat and human skin to obtain viable keratinocytes, (3) procedures necessary to produce splitting of full-thickness skin into epidermis and dermis, (4) steps to remove mostly basal keratinocytes from the epidermis and dermis, (5) procedures for purification and enumeration of suspended keratinocytes, (6) cultivation of keratinocytes to produce monolayer cultures, (7) subcultivation of confluent monolayer cultures, and (8) cultivation of cells to produce multilayered differentiated cultures. The progress of the methodology used in obtaining viable keratinocytes for cultivation can be monitored using microscopic and/or histological examinations.

Strict aseptic conditions must be maintained in all procedures for preparing keratinocytes for eventual in vitro cultivation.

3.1. Preparation of Chemical and Biological Solutions Necessary for the Processing and Subsequent In Vitro Cultivation of Basal Keratinocytes

A number of important solutions must be carefully prepared in order to successfully process the skin for isolation, purification, and subsequent cultivation of basal keratinocytes. Because of the delicate nature of living mammalian cells, they must be suspended in solutions with the proper osmolality in order to maintain cell membrane integrity. Fortunately, most of these solutions can be purchased fully prepared for immediate use. Some can be obtained at higher concentrations (10X, 100X) for better storage and handling. If sterilization is necessary, it can be accomplished using filters of various sizes and configurations depending on the solution, its volume, and its characteristics (*1*). The websites of the companies listed in **Subheading 2.** (*see* **Note 1**) usually contain descriptions of the products that they sell.

3.1.1. Isotonic Solutions Used in Washing Cells and in Dissolving Solid Substances

These materials can be purchased as sterile solutions in various concentrations or as powders to be dissolved in purified water and sterilized. The exact formulations can be found in cell culture manuals or in descriptions supplied by the company.

1. EBSS. This solution is specially formulated to maintain cell membrane integrity. It has various uses in producing and maintaining cell suspensions for short periods prior to actual cultivation.
2. PBS (1–10X strength). This isotonic solution is used mostly as a solvent for dissolving various substances to be used in cell preparation procedures prior to cultivation. It is not used to store cells for extended periods. It may be necessary to include glucose (0.01% [v/v]) in this solution (PBSG).

3.1.2. Special-Purpose Solutions Necessary
for Suspending, Counting, and Dissociating Keratinocytes

1. Trypan blue. This is the dye most frequently used in the dye exclusion test when it is necessary to determine the viability of isolated mammalian cells. The powder is usually dissolved in PBS to obtain a 0.1% solution. Small samples of cell suspensions are then added and the test performed without asepsis.
2. EDTA. This chemical is used to dissociate cell clumps into single cells and to dislodge cultivated cells from the substratum to result in a suspension. A concentration of 0.02 % in PBS is usually employed for those purposes. It can also be included in solutions containing other chemicals when appropriate.
3. Trypsin. This is the enzyme of choice for isolating keratinocytes from full-thickness skin. Solutions in EBSS or PBS ranging from 0.1% to 0.25% are used for epidermal–dermal separation. Also, it can be dissolved in EBSS or PBS to obtain solutions of 0.03% to be included with 0.02% EDTA in detaching monolayer cultures for subcultivation.
4. Trypsin inhibitor. It might be necessary to inhibit further trypsin enzymatic activity after a selected incubation period. A solution of soybean trypsin inhibitor at a concentration of 1 mg/mL dissolved in PBS or EBSS is used for this purpose. Complete growth medium, described in **Subheading 3.1.3.** can also be used to inhibit trypsin activity.

3.1.3. Basal Growth Medium Specifically Formulated
for the Growth of Basal Keratinocytes

Because of the fastidious characteristics of basal keratinocytes, careful attention must be made in selecting the proper medium and specific supplements in order to promote both attachment and growth of these cells in vitro. The growth medium is the same for both rat and human cultures. It contains supplements shown to be required for optimal cell viability and proliferation of basal keratinocytes. The basal medium is MEM, which contains an exacting balance of amino acids, vitamins, inorganic salts, and glucose *(1)*. Supplements are added to it to obtain optimal cultivation.

1. Hormonal supplements have been shown to affect growth control of cells in culture *(2)*. The combination of HC and IN was shown to support proliferation of keratinocytes *(3)*. Thus, basal medium is supplemented with both HC and IN at 10 µg/mL.
2. Both EGF *(4)* and BPE *(5)* have been identified as stimulating keratinocyte proliferation in culture. EGF at 10 ng/mL and BPE at approx 60 µg/mL is added to the basal medium.
3. Antibiotics and antimycotics are necessary supplements that control microbial contamination during long-term cultivation. Penicillin (100 units/mL) and streptomycin (100 µg/mL) are used as antibiotics in MEM. These reagents are unstable and, thus, freshly prepared solutions of these reagents should be added to growth

Table 1
Ingredients Needed for 500 mL of Complete Minimum Essential Medium

Ingredient	State	Solvent	Stock solution	mL per 500 mL CMEM
MEM	Liquid	H_2O	1X	436.5
L-Glutamine	Liquid	H_2O	100X	5.0
FBS	Liquid	None	100%	50.0
HC	Powder	EtOH–H_2O	10 mg/mL	0.5
IN	Powder	HCl–H_2O	10 mg/mL	0.5
EGF	Powder	H_2O	10 µg/mL	0.5
BPE	Powder	MEM	14 mg/mL	2.0
Antibiotic/Antimycotic	Liquid	Saline	100X	5.0
Total				500.0

Note: This includes the original form, the solvent, the stock solution, and the amount of the stock solution added to prepare 500 mL of the growth medium for keratinocyte cultivation. EtOH, absolute ethanol.

medium weekly. Either fungizone (0.25 µg/mL) or amphotericin B (25 µg/mL) are used as antimycotics. These components form the basal MEM that support keratinocyte cultivation. It does not include animal serum supplements.

4. The unstable amino acid L-glutamine (200 m*M*) at a concentration of 0.29 mg/mL is added to the basal medium and is necessary for medium stabilization. Fresh medium containing this reagent should be used for no more than 1 wk after preparation. Thawed L-glutamine can be added to older medium at the prescribed concentration at weekly intervals.

5. To obtain complete growth medium (CMEM), FBS is added to the MEM with supplements to obtain a 10% (v/v) solution. It is necessary to pretest each lot of serum purchased from suppliers to confirm its effectiveness in supporting keratinocyte cultivation. Although there have been reports of successful growth of keratinocytes in vitro using serum-free medium, we find that basal keratinocytes, especially of human origin, require serum-supplemented medium for optimal attachment and proliferation. However, media can be purchased from suppliers that affirm optimal growth of keratinocytes without any animal components (Cascade Biologics; *see* **Note 1**). The preparation of 500 mL of MEM is outlined in **Table 1**.

3.1.4. Growth Factors That Enhance the Attachment and Thus Subsequent Growth of Keratinocytes on Various Substrata

The surface of commercial tissue culture vessels are prepared to promote attachment and growth of mammalian cells in vitro. However, keratinocytes are more exacting in their requirement for optimal proliferation. Two substances included in the basement membrane of stratified epithelium have been shown to enhance attachment of keratinocytes (i.e., collagen *[6]* and laminin *[7]*). Various substrata used in culturing basal keratinocytes, such as plastic and glass culture vessels or synthetic membranes, should be precoated with these substances.

1. We have experienced best results with LMN at 1 mg/cm^2 of surface. Aliquots of the purchased stock solution of LMN, containing 1 mg/mL in buffered NaCl, are added to the culture surface and are allowed to evaporate to dryness in a biosafety cabinet. Coated culture vessels should be used within 5–7 d after evaporation of NaCl.
2. Calf skin collagen type 1 is applied at 6–10 μg/cm^2 of surface *(8)*. A purchased solution of collagen (1 mg/mL) is diluted to 50 μg/mL with 0.1 *N* acetic acid. The required amount is plated onto the surface and incubated for 1 h at room temperature. The surface is then rinsed with PBS to remove the acid and air-dried in a biosafety cabinet. Coated culture vessels should be used within 5–7 d after drying is complete.

3.2. Preparation of Full-Thickness Skin for Basal Keratinocyte Isolation

This subsection details steps necessary to (1) minimize microbial contamination of the tissues obtained from rat and human biopsies and (2) remove as much subcutaneous and dermal tissue as possible so as to facilitate epidermal–dermal separation. Because to the considerable differences in the anatomical structure of newborn rat skin as compared to human skin, procedures for the preparation, and treatment of the two skin types to obtain basal keratinocytes must be described separately. Also, keratinocytes obtained from individual human skins, both neonatal and adult, cannot be successfully pooled for subsequent cultivation as can skin from syngeneic rat littermates.

3.2.1. Preparation of Full-Thickness Skin From Inbred Laboratory Rats for Basal Keratinocyte Isolation (see **Note 2**)

The inbred CFN albino rat was found to be a convenient and accessible source of rat keratinocytes. The skins from animals 24–36 h old were found to give the best results in subsequent epidermal–dermal separation. One litter usually consists of 10–15 animals.

1. The animals are killed by cervical dislocation and the total body cleansed using cotton soaked with 70% ethanol.
2. To minimize variations in skin thickness, only the backs of the animals are used. With surgical scissors, the back skin from the nape of the neck to the beginning of the tail is dissected. The resulting skin tissues measure approx (2–3) × (4–5) cm, depending on the age of the animals.

3. Any loose subcutaneous elements are carefully removed with a scalpel and discarded. The processed skins can be placed in a dish containing EBSS or PBS before proceeding to the next step.

3.2.2. Human Skin Obtained From Surgical Procedures

Human skin, handled aseptically, can be obtained from various sources, including neonatal foreskin and skins resulting from cosmetic surgery. Most of the adult tissues received are the result of breast reduction and the removal of abdominal skin after weight loss.

1. The skins are handled aseptically in a biological safety cabinet with filtered positive air pressure. The objective is to eliminate possible microbial contamination resulting from preparation procedures.
2. All subcutaneous elements, containing mostly adipose tissue, are removed and discarded.
3. The full-thickness skin is placed, dermal side down, in a 150-mm plastic dish and covered with PBS (minus Ca^{2+} and Mg^{2+}) containing the antibiotics described in **Subheading 3.1.3., step 3**. They remain in the dish for 30–40 min before the solution is removed.
4. An attempt is then made to remove as much of the underside of the dermis as possible to facilitate dermal–epidermal separation. To accomplish this, the skins are placed, epidermal side down, in a 150-mm dish cover and the surface of the dermis scraped with a scalpel. The purpose is to reduce the thickness of the dermis to approx 3–5 mm.
5. The processed tissue is then cut into sections approx 4–5 mm wide and 3–4 cm long using a scalpel.

3.3. Separation of the Dermal and Epidermal Layers of the Skin for Subsequent Harvesting of Basal Cell Keratinocytes

This procedure is designed to promote separation of the two anatomical layers where basal keratinocytes in the epidermis are attached to the surface of the dermis. The objective is to obtain as many viable basal cells as possible. Experimental studies indicate that only these cells are capable of proliferation in vitro *(9)*. Both physical and chemical steps are incorporated in producing epidermal–dermal separation. It has been reported that stretching skin biopsies promotes this separation *(10)* and that trypsinization at lower temperatures results in less cell damage *(11)*. This subsection describes epidermal-separation procedures for both rat and human keratinocytes. These steps and all subsequent steps are performed under strict aseptic conditions in biological safety cabinets.

3.3.1. Treatment of Rat Skin Samples
to Obtain Dermal–Epidermal Separation

1. The prepared skin samples are washed with fresh EBSS chilled to 4°C. Three individual samples are then placed, stratum corneum down, on the surface of 100-mm plastic tissue culture dishes.
2. Each sample is held down with forceps and, using a scalpel, scraped laterally from the center with force in order to cause the skin to stretch and adhere to the bottom of the dish.
3. The adhering samples are then chilled to 4°C in the refrigerator and a sterile stock solution of cold 1:250 trypsin at 0.25% (w/v) in EBSS or PBS carefully added to completely cover the skins. Incubation in the trypsin is continued at 4°C for 14–16 h depending on the age of the litter (*see* **Note 3**).

3.3.2. Treatment of Human Skin Samples
to Obtain Dermal–Epidermal Separation

1. The narrow skin strips described in **Subheading 3.2.2.** are placed dermal side down in a large plastic tissue culture dish (100–150 mm) prechilled to 4°C to promote adhesion of the skin to the bottom of the dish.
2. Powdered trypsin is dissolved at room temperature in EBSS to obtain a 0.1–0.13% solution. It is then sterilized using a 0.2-µm filter unit (*1*) and chilled to 4°C in the refrigerator.
3. Approximately 50 mL of the chilled trypsin solution is added to the adhering skins in a 150-mm dish (less for smaller dishes). Incubation of the tissue in the enzyme is continued at 4°C for 12–18 h or until the epidermis and dermis can be physically separated (*see* **Note 4**).

3.4. Obtaining Keratinocytes From the Separated Epidermis and Dermis

This subsection describes the methodology required to harvest basal keratinocytes from the skin after splitting its two main layers. Steps must be taken to protect the viability of the cells that have been subjected to very harsh physical and chemical procedures. As the process of isolating and suspending cells progresses, resulting samples of cell suspensions can be monitored microscopically using IPCM.

3.4.1. Suspension of Rat Basal Keratinocytes Obtained
From the Separated Epidermis and Dermis into Stabilizing Medium

1. After the enzymatic treatment, the trypsin solution is aspirated from the 100-mm dish and the tissues washed two to three times with approx 10 mL of EBSS to remove residual trypsin.
2. The enzymatic activity of the trypsin is neutralized by adding 10–20 mL of CMEM to each dish and allowing it to remain for 2–5 min before removal.

3. Using two sterile forceps, individual samples are removed and placed, stratum corneum down, in a plastic tissue culture dish. The dermis is then removed by grasping its edges with the forceps, lifting it from the epidermis, and placing it, epidermal side up, next to the exposed epidermis.
4. Fresh CMEM is then added to the dishes to cover the surfaces of both skin layers, and keratinocytes are carefully liberated into the medium by applying delicate strokes with a sterile fine sable hair brush across the tops of the skin layers. The sable brush was sterilized by immersion in 70% ethanol for 15–30 min, followed by washing with EBSS or PBS to remove the ethanol.
5. The resulting cells suspended in CMEM from all dishes are combined in one container to be processed further.

3.4.2. Suspension of Human Basal Keratinocytes Obtained From the Separated Dermis into Stabilizing Medium

The procedure for collecting human cells is identical to the one described in **Subheading 3.4.1.** for rat cells.

3.5. Purification and Enumeration of Basal Cell Keratinocytes Collected From the Surface of the Dermis and Epidermis

The methodology described in this subsection applies to both rat and human basal keratinocytes. Purification and enumeration must be accomplished before cell cultivation can proceed.

3.5.1. Purification of Basal Keratinocytes to Remove Fibroblasts and Cellular Debris

The cells harvested from the separated skin elements include various cell types and cellular debris resulting from the physical and chemical procedures described.

1. The cells collected from the tissue samples are first consolidated in conical centrifuge tubes and centrifuged at $30g$ for 5 min at 4°C for initial purification. This removes most of the tissue and cellular debris.
2. The cell pellet is gently resuspended in 5 mL of EBSS.
3. Added to a Nalgene centrifuge tube are 6.6 mL EBSS, 0.8 mL of 10X PBS, and 7.6 mL Percoll. The 5-mL cell suspension is then added to the tube and the contents mixed thoroughly by inverting the tube two or more times.
4. A continuous gradient is formed by centrifuging the resulting 38% Percoll at $30,000g$ for 15 min at 4°C *(12,13)*. *See* **Fig. 1** for the location of density marker beads and the various cell layers in the density gradient. The identities of the cells in the bands can be determined by viewing inocula on slides and cover slips via phase-contrast microscopy (*see* **Fig. 2** and **Note 5**).
5. With the aid of a sterile Pasteur pipet connected to a vacuum, all of the Percoll and cellular components above the lower band of the gradient (containing the basal cells) are aspirated and discarded.

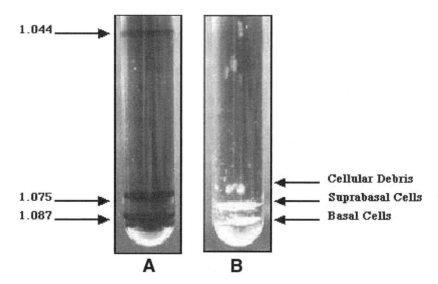

Fig. 1. Percoll density gradient compartmentalization of cellular components resulting after keratinocyte isolation from skin samples. (**A**) Nalgene tube showing the location of three marker beads resulting after gradient formation; (**B**) Nalgene tube showing the location of cellular components resulting after gradient formation. Differentiated cells (spinous, granular, etc.), fibroblasts, and debris are located between densities of 1.075 and 1.087 g/cm³. Basal keratinocytes are located proximally at the 1.087-g/cm³ density.

6. The basal cells in the lower band are collected using a 5.0-mL pipet attached to a controlled pipetting device, suspended in CMEM (8–9 mL) and centrifuged in a 10-mL graduated centrifuge at 16*g* for 10 min at 4°C.
7. CMEM is then added to the pellet to result in exactly 10 mL of packed cells and medium.
8. The cells are carefully and uniformly suspended into the CMEM by slow, repeated filling and emptying a 10-mL pipet.
9. The resuspended suspension should be enumerated immediately.

3.5.2. Enumeration of Purified Basal Keratinocytes

The percent of viable cells in the total cell number in a suspension can be determined visually using an IPCM and a hemocytometer (*1*). Only viable cells are able to attach and proliferate in vitro. Therefore, the percent of viable cells in the total cell count must be determined in the attempt to produce consistent seeding inocula. The trypan blue exclusion test is employed for this purpose. The blue dye will not stain cells that are actively metabolizing and have intact membranes.

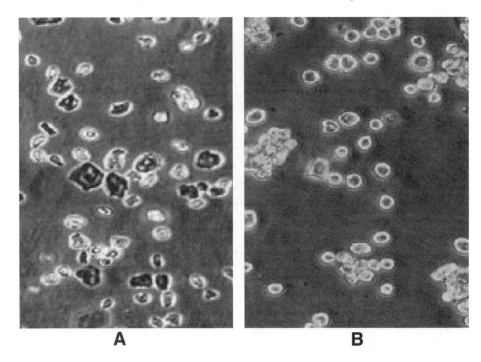

Fig. 2. Phase-contrast micrographs of cellular components separated in a Percoll density gradient and resuspended in a supporting medium. **(A)** Cells resuspended from the 1.075- to 1.087-g/cm^3 density. Most of the cells are identified as differentiated cells and fibroblasts plus some basal cells. **(B)** Cells collected from the 1.087-g/cm^3 density are almost exclusively basal keratinocytes (×240). Cultivation of cell suspensions from the two bands in CMEM can be used to confirm that the rounded cells in the upper band are predominantly fibroblasts, whereas the lower band contains basal cells almost exclusively.

1. In a small tube, add 200 µL of the cell suspension to 600 µL of PBS and then add 200 µL of the 0.1% trypan blue solution and mix gently but thoroughly by inverting the tube three to five times.
2. With a cover slip in place on each side of the hemocytometer, add a small amount of the cell suspension to fill each chamber using a Pasteur pipet, filling via capillary action (do not overfill).
3. Count the entire chamber on both sides and determine the average cell number.
4. The following formula can be used for determining the total number basal cells in a suspension *(1)*: average cell number per grid (mm^2) × 10^4 × dilution factor (5) = number of cells per milliliter in the suspension. Approximately (1–1.5) × 10^8 cells can be obtained from the skins of a litter (10–15) of rats. Approximately (1.0–1.2) × 10^8 cells can be obtained from one human skin sample 25 cm^2 in size, whereas approx (2.2–6.5) × 10^5 cells can be obtained from one foreskin.

5. The percent viability of the cell suspension is expressed as the number of cells unstained by Trypan blue per 100 cells counted.

3.6. Establishing Monolayer Cultures of Basal Keratinocytes

Successful in vitro cultivation of isolated, purified basal keratinocytes depends on specific procedures and environmental conditions. Such considerations include (1) substratum for initial attachment, (2) initial plating density, (3) temperature, and (4) atmosphere.

3.6.1. Establishing Monolayer Cultures of Rat Basal Cells

1. The purified and enumerated cells are centrifuged at $16g$ for 10 min at 4°C and resuspended in CMEM to obtain a cell suspension containing approx 5 × 10^5 cells/mL. We have observed that a 0.2% suspension (v/v) of rat basal cells (pellet) will result in a suspension of similar composition.
2. Various culture vessels are seeded with this inoculum in amounts that satisfy the working volume and surface area (*see* **Note 6**).

3.6.2. Establishing Monolayer Cultures of Human Basal Cells

The purified and quantified basal keratinocytes described in **Subheading 3.5.** are used to produce monolayer and multilayer basal keratinocyte cultures.

1. Experimental results in our laboratory have shown that plating approx 2 × 10^5 purified, viable cells per square centimeter of cultivation surface results in optimal attachment and proliferation of human keratinocytes. Other investigators may prefer different initial seeding densities to obtain desired results.
2. The cell suspension in CMEM is pipetted into culture vessels in amounts that satisfy the working volume and surface area. For examples of steps in seeding various culture vessels with the desired amount of cells (*see* **Note 6**).

3.6.3. Incubation of Rat and Human Basal Cells to Establish Monolayers

1. The seeded cells are placed in an incubator maintained at 35°C with the atmosphere set at 95% air–5% CO_2 with a humidity of 95%. Rat and human cells respond similarly using this procedure.
2. Incubation is allowed to proceed for 18–24 h to allow attachment of the cells to the surface.
3. The original medium, containing unattached cells and possible debris, is aspirated and discarded.
4. An equal volume of fresh CMEM, warmed to 35°C in a water bath, is then added and incubation continued.
5. Every 2 d, the old medium is removed from the culture and fresh medium (warmed to 35°C) added. Cell attachment and growth is monitored daily using an IPCM. Typical cultures at various stages are shown in **Fig. 3**.

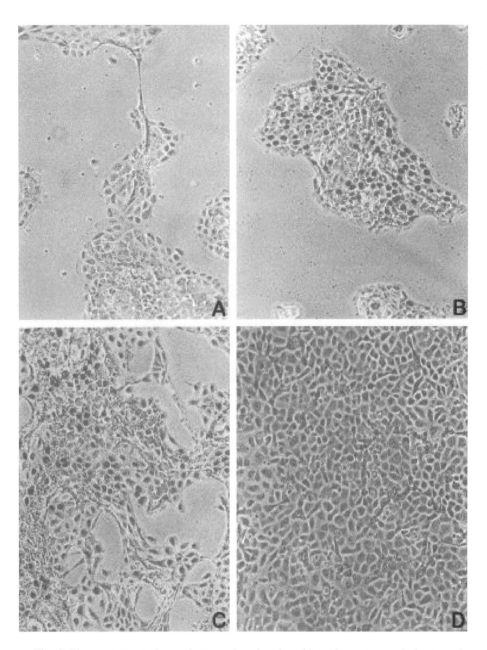

Fig. 3. Phase-contrast photomicrographs of rat basal keratinocytes seeded onto collagen-coated plastic substrata. Soon after plating (18–24 h), the cells have attached to the substratum as shown in (**A**). Also shown in this figure are the results of cell proliferation in 2 d (**B**) and 3 d (**C**). Cells firmly connect to each other and form a continuous sheet constituting a confluent monolayer in 5–7 d (**D**) (original magnification: ×235).

3.7. Subcultivation of Primary Monolayers of Keratinocytes

Primary cultures of human basal keratinocytes can be subcultured for use in experimentation. This is best accomplished when the primary culture reaches 70 to 80% confluence. Primary cells obtained from one culture vessel can be subcultured into two vessels of similar surface area.

1. The growth medium (CMEM) is first removed from the selected culture, which is then washed with a solution containing 0.02% EDTA in PBS minus Ca^{2+} and Mg^{2+}.
2. Next, a solution containing 0.03% trypsin and 0.01 % EDTA in PBS is added to the culture vessel to cover the cells.
3. Incubation proceeds at 37°C for approx 2 min or until the cells detach from the surface of the culture vessel as determined via microscopic examination.
4. Further enzymatic activity is inhibited by adding an equal or greater volume of CMEM or an equal volume of soybean trypsin inhibitor dissolved in PBS (1 mg/mL).
5. The detached cell suspension is transferred to a centrifuge tube and centrifuged at 16g for 5 min at room temperature.
6. The cells are resuspended in fresh CMEM resulting in an amount that will produce the 1:2 split. Cultivation of the seeded, passed cells follows the described procedures for keratinocyte growth in vitro.

3.8. Methodology for Constructing a Differentiated, Stratified Keratinocyte Culture In Vitro

A stratified differentiated keratinocyte culture developed in vitro using isolated basal cells might be advantageous in some experimental designs *(8,14,15)*. Such a culture is produced by incubating cultured keratinocytes at the air–liquid interface *(16)*.

1. The initial seeding, as described in **Subheading 3.6.**, is on a porous, inert membrane that is autoclave sterilized. We have used the Puropore membrane supplied by Gelman (Ann Arbor, MI) with a pore size of 0.2 μm and a diameter of 13 mm (*see* **Note 7**).
2. To enhance attachment, the membrane should be pre-coated with laminin or collagen using the procedure for coating culture vessels (*see* **Subheading 3.1.4.**).
3. The coated membranes are then placed in 24-well plastic tissue culture vessels and seeded with 1×10^6 cells/cm^2. At least two coated wells are seeded in the absence of a membrane so that the attachment and proliferation of the culture can be monitored using IPCM.
4. The cell inoculum seeded on the coated membranes is incubated submerged in CMEM for 4–5 d as described for monolayer cell cultivation (*see* **Note 8**). Verification of the attachment and growth of the cells on the membrane can be accomplished via histological staining procedures and light microscopy *(1)*. A stained 5-d-old culture of human keratinocytes on a membrane is shown in **Fig. 4**.

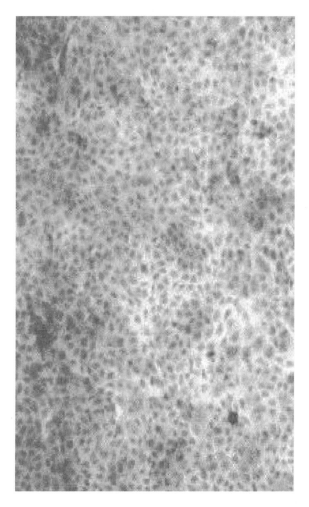

Fig. 4. Light photomicrograph of a 5-d culture of human basal keratinocytes on a collagen-coated porous membrane. The membrane containing the cells was stained with hemotoxylin and eosin using histological procedures (original magnification: ×150).

The next step is to raise the porous membrane to the air–liquid interface.

5. Glass fiber filters 44 mm in diameter, sterilized via autoclave, are placed in 60-mm culture dishes and CMEM is added to accomplish complete saturation of the filters without excess fluid.

6. Up to five Nylon membranes (two is optimal) previously coated, seeded, and incubated for the 4- to 5-d period are transferred to the surface of the saturated filters. They are positioned so that the surface containing the cultured keratinocytes is exposed to the atmosphere rather than covered with medium and fed with CMEM via contact with the saturated filter below (*see* **ref. 9**).

7. Incubation of the lifted cultures is continued for an additional 10–15 d while being fed fresh medium three times per week. This is done by aspirating as much of the old medium as possible and adding fresh CMEM to resaturate the filter.

8. The resulting stratification and differentiation can be observed using histological procedures for preparing sections for light microscopy and for transmission electron microscopy (TEM). Such observations have verified the development of a stratified epithelium complete with stratum corneum. Ultrastructural markers characteristic of normal mammalian skin also develop in the lifted culture. Using TEM methodology, these structures have been observed in both rat *(7)* and human *(13)* lifted cultures.

4. Notes

1. These companies are the prime suppliers of mammalian cell culture products. Most of the products listed can be purchased from more than one source. Also, many of the reagents listed can be purchased in various forms, including dry and lyophilized powders, frozen in ampules, and as ready-to-use or concentrated solutions. They can also be purchased in various combinations to fit specific cultivation procedures. Descriptions of these products can be obtained at the company's websites or in brochures: Sigma–Aldrich (www.sigmaaldrich.com/); Invitrogen (www.invitrogen.com/); BD Biosciences (www.bdbiosciences.com/); BD (www.bd.com/ds/); Cellgro (www.cellgro.com/); Falcon Labware (www. bacto.com.au/falcon.htm); Corning (www.corning.com/); ICN Biomedicals (www.icnbiomed.com/); Cascade Biologics (www.cascadebiologics.com).

2. The newborn CFN albino rats from our breeding pens average 10–15 animals per litter. Smaller litters (i.e., those with less than 10 animals) tend to produce larger, more developed animals and are not used. Newborn rats can also be obtained from laboratory animal suppliers such as Charles River Laboratories (Wilmington, MA). Litters of animals purchased from suppliers can be shipped along with the nursing doe or shipped under other arrangements.

3. We began cultivating rat keratinocytes in 1971. At that time, the trypsin available was the crude powder (1:250) supplied by Difco Laboratories (Detroit, MI). A number of purified trypsin enzymes are now available and much less toxic. Investigators should explore these substances, which will most likely give better results. Because of our ongoing experimentation using rat keratinocytes, we did not introduce such a change in our procedure, which might make comparisons with earlier experimental data invalid.

4. We use the same procedure for splitting human skin and rat skin. However, because of the advanced development of human skin, results similar to those of newborn rat skin cannot be duplicated. First, it is very difficult to obtain a comparable degree of stretching with the firm human skin and, second, enzymatic treatment under cold conditions does not produce the ease of epidermal–dermal separation observed with newborn rats. As a result, skin splitting of most human skin samples following cold enzymatic treatment is often difficult to obtain. The strength of the trypsin solution and the incubation time selected for enzy-

matic treatment depend on the thickness of the individual skin specimens. There are, undoubtedly, other variations in individual human skin samples that could impact on the enzymatic separation. Skin characteristics differ depending on the age, gender, and physical condition of the donor. Also, the postoperation time of skin storage before being processed must be considered. These variations might necessitate adjustments in the trypsinization procedure for dermal–epidermal separation. An increase in the strength and incubation time of enzymatic treatment might obtain better separation results. However, this also produces an adverse affect on cell viability. The degree of separation should be checked prior to removing the enzyme from the dish. If, with the use of forceps, the epidermis can be lifted from the dermis but is still connected at the center of the skin strips, then, in our experience, this results in an optimal separation. However, if the separation is already complete and the epidermis has floated from the dermis, there has been overtrypsinization. In the former case, the resulting quantity of isolated basal cells might be less but the quality as observed in cell viability, cellular attachment, and proliferation will be more satisfactory. Experience obtained from processing different human skins is the only solution to this problem.

5. It might not be possible to remove all fibroblasts from suspensions of basal keratinocytes using the Percoll gradient. However, this purification step is necessary in order to obtain a successful cultivation of the latter cell type. Comparatively small numbers of fibroblasts remaining in suspensions can eventually outgrow the keratinocytes in areas, dislodging them from the culture. Therefore, the more fibroblasts that can be removed, the more uniform the resulting monolayer of keratinocytes will be.

6. The proper seeding of the purified and quantified basal cell keratinocytes to obtain the desired number of cells per square centimeter of substratum surface can be accomplished in a number of ways. One example in seeding cells at 2×10^5 cells/cm^2 of culture surface could include the following steps in the procedure: First, if a total of 1×10^8 cells were found to be obtained from a 25-cm^2 human skin sample (or from the skins of 12 newborn rats) following purification and enumeration, they could be resuspended in 10 mL of CMEM to result in a stock suspension containing 1×10^7 cells/mL. The exact amount of cells could then be seeded into vessels of various surface sizes. Because a T75 flask has 75 cm^2 of culture surface and a working volume of 15 mL, a total of $(2 \times 10^5)(75)$ or 150×10^5 cells are needed (i.e., 1.5 mL of the stock suspension). Pipetting 1.5 mL of the stock suspension into the flask plus 14.5 mL of additional CMEM completes the procedure. Other culture vessels could be seeded with this stock suspension depending on their culture surface and working volume. Another example would be more appropriate if only one vessel type is to be seeded. T75 flasks can be seeded with 15 mL of a stock suspension containing 2×10^6 cells. It must be pointed out that the culture surface and working volume of vessels obtained from various suppliers could differ.

7. We have used Gelman's nylon membrane (Puropor-200), which is no longer available, almost exclusively. Other membranes of similar pore size might produce similar results. Glass fiber filters were used because of their ability to become saturated with medium. Other such filters of similar inert structure could be just as effective. Selections can be made from companies manufacturing porous membranes (Gelman, Millipore, etc.).

8. The methodology developed for producing stratified cultures of human keratinocytes results in variations not observed with rat keratinocytes. This is most likely the result of the differences in human samples as described in **Note 4**. We have been able to produce somewhat better results by modifying the CMEM supplements. The FBS is increased to 15% and BPE doubled to approx 120 µg/mL.

Acknowledgments

The authors take this opportunity to acknowledge the invaluable contribution of the late Isadore A. Bernstein in making our scientific endeavors conceivable. His distinctive vision concerning the possible use of mammalian cells in dermatological investigations plus his ability and untiring quest for funding made our efforts possible. Appreciation also goes to Anna Vaughan for proofreading the manuscript and her photography expertise in preparing the figures.

References

1. Freshney, R. I. (ed.) (2000) *Culture of Animal Cells. A Manual of Basic Techniques*, 2nd ed., Wiley, New York.
2. Hayashi, I., Larher, J., and Sato, G. (1978) Hormonal growth control of cells in culture. *In Vitro* **14,** 23–30.
3. Vaughan, F. L., Kass, L. L., and Uzman, A. (1981) Requirement of hydrocortisone and insulin for extended proliferation and passage of rat keratinocytes. *In Vitro* **17,** 941–946.
4. Rheinwald, J. G. and Green, H. (1977) Epidermal growth factor and the multiplication of cultured human keratinocytes. *Nature* **265(5593),** 421–424.
5. Boyce, S. and Ham, R. (1983) Normal human epidermal keratinocytes, in *In Vitro Models for Cancer Research* (Weber, M. M. and Sekely, L., eds.), CRC, Boca Raton, FL, pp. 245–274.
6. Karasek, M. and Charlton, M. F. (1971) Growth of post-embryonic skin epithelial cells on collagen gels. *J. Invest. Dermatol.* **56,** 205–210.
7. Vaughan, F. L., Gray, R. H., and Bernstein, I. A. (1986) Growth and differentiation of primary rat keratinocytes on synthetic membranes. *In Vitro* **22,** 141–149.
8. Bernstam, L., Lan, C.-H., Lee, J., and Nriagu, J. O. (2002) Effects of arsenic on human keratinocytes: Morphological, physiological, and precursor incorporation studies. *Environ. Res.* **89,** 220–235.
9. Vaughan, F. L. and Bernstein, I. A. (1971) Studies of Proliferative capabilities in isolated epidermal basal and differentiated cells. *J. Invest. Dermatol.* **56,** 454–466.

10. Van Scott, E. J. (1952) Mechanical separation of the epidermis form the corneum. *J. Invest. Dermatol.* **18,** 377–379.

11. Szabo, G. (1955) Modification of the technique of "skin splitting" with trypsin. *J. Pathol. Bacteriol.* **70,** 545.

12. Fischer, S. M., Nelson, K. D., Reiners, J. J., Jr., Viage, A., Pelling, J. S., and Slaga, T. J. (1982) Separation of epidermal cells by density centrifugation: a new technique for studies on normal and pathological differentiation. *J. Cutan. Pathol.* **9,** 43–49.

13. Bernstam. L. I., Vaughan, F. L., and Bernstein, I. A. (1990) Stratified cornified primary cultures of human keratinocytes grown on microporous membranes at the air-liquid interface. *J. Dermatol. Sci.* **1,** 173–181.

14. Scavarelli-Karantsavelos, R. M., Zaman-Saroya, S., Vaughan, F. L., and Bernstein, I. A. (1990) Pseudoepidermis, constructed in vitro, for use in toxicological and pharmacological studies. *Skin Pharmacol.* **3,** 115–125.

15. Vaughan, F. L. (1994) The pseudoepidermis: An in vitro model for dermatological investigations. *Am. Biotechnol. Lab.* **12,** 26–28.

16. Prunieras, M. D., Regnier, M., and Woodley, D. (1983) Methods of cultivation of keratinocytes with an air–liquid interface. *J. Invest. Dermatol.* **8,** 28–33.

15

Isolation and Culture of Primary Human Hepatocytes

Edward L. LeCluyse, Eliane Alexandre,
Geraldine A. Hamilton, Catherine Viollon-Abadie,
D. James Coon, Summer Jolley, and Lysiane Richert

Summary

As our knowledge of the species differences in drug metabolism and drug-induced hepatotoxicity has expanded significantly, the need for human-relevant in vitro hepatic model systems has become more apparent than ever before. Human hepatocytes have become the "gold standard" for evaluating hepatic metabolism and toxicity of drugs and other xenobiotics in vitro. In addition, they are becoming utilized more extensively for many kinds of biomedical research, including a variety of biological, pharmacological, and toxicological studies. This chapter describes methods for the isolation of primary human hepatocytes from liver tissue obtained from an encapsulated end wedge removed from patients undergoing resection for removal of liver tumors or resected segments from whole livers obtained from multiorgan donors. The maintenance of normal cellular physiology and intercellular contacts in vitro is of particular importance for optimal phenotypic gene expression and response to drugs and other xenobiotics. As such, methods are described for culturing primary hepatocytes under various matrix compositions and geometries. Differential expression of liver-selective properties occurs over time in primary hepatocytes dependent on the culture and study conditions. Overall, improved isolation and cultivation methods have allowed for exciting advances in our understanding of the pathology, biochemistry, and cellular and molecular biology of human hepatocytes.

Key Words: Primary human hepatocytes; in vitro hepatic model systems; cell isolation methods; sandwich culture.

1. Introduction

The liver serves as the primary site of detoxification of natural and synthetic compounds in the systemic circulation. Other biological and physiological functions include the production and secretion of critical blood and bile components, such as albumin, bile salts, and cholesterol. The liver is also involved

From: *Methods in Molecular Biology, vol. 290: Basic Cell Culture Protocols, Third Edition*
Edited by: C. D. Helgason and C. L. Miller © Humana Press Inc., Totowa, NJ

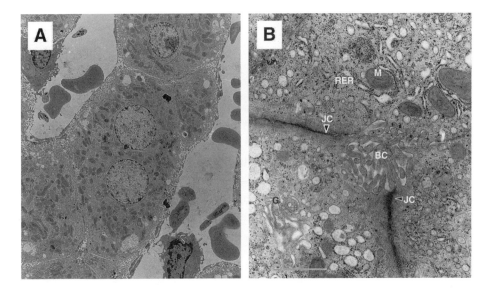

Fig. 1. Electron micrographs of the whole liver illustrating the structural complexity of hepatocytes. **(A)** Hepatocytes that line the sinusoids as cell plates exhibit a complex cytoplasm that features both polarity of organelles and the plasma membrane. **(B)** At higher magnification, typical ultrastructure of the apical (canalicular) domain of the hepatocyte plasma membrane can be observed, including junctional complexes (JC) and microvilli. Note the polar distribution of the Golgi apparatus (G) near the bile canaliculus (BC), which is typically found in hepatocytes. M, mitochondria; RER, rough endoplasmic reticulum. Bar = 1 μ*M*.

in the protein, steroid, and fat metabolism, as well as vitamin, iron, and sugar storage. The parenchymal cells or hepatocytes are highly differentiated epithelial cells that perform many of the functions attributed to the liver. Much of their functional diversity is revealed in the complexity of the cytological features of the cells (*see* **Fig. 1**). Hepatocytes are highly polarized cells that are dependent on the maintenance of two distinct membrane domains. The sinusoidal and canalicular membrane domains are separated by tight junctions and exhibit striking ultrastructural, compositional, and functional differences. The maintenance of a polarized cell and membrane architecture is essential for maintaining normal biliary excretion and xenobiotic elimination.

One of the most complex functions specific to the liver is its ability to metabolize an enormous range of xenobiotics. Many drugs present in the blood are taken up by hepatocytes, where they can be metabolized by phase I and II biotransformation reactions. Much remains to be learned about the biochemical and molecular factors that control the expression and regulation of normal hepatocyte structure and function in humans. Because of these issues, the use of in vitro and in vivo systems to evaluate hepatic drug uptake and metabolism, cytochrome P450 (CYP450) induction, drug interactions affecting hepatic metabolism, hepatotoxicity, and cholestasis is an essential part of toxicology and pharmacology *(1–9)*.

Within the literature, one can find a number of different approaches that have been applied successfully for the isolation and cultivation of primary human hepatocytes *(1,2,10–21)*. However, for the novice who is attempting to identify those methods and conditions that are most appropriate for a particular type of study, this task might appear overwhelming initially. Likewise, there are few sources available for obtaining detailed information needed to perform in vitro studies utilizing primary human hepatocytes. This chapter describes the isolation and culture of human hepatocytes from liver tissue obtained from one of two sources: an encapsulated end wedge removed from patients undergoing resection for removal of liver tumors or from resected tissue from whole livers obtained from multiorgan donors. This procedure is essentially a modification of the two-stage perfusion and digestion described by MacDonald et al. *(20)* and has been adopted by an interlaboratory consortium sponsored by the European Centre for the Validation of Alternative Methods (ECVAM) for the isolation and cultivation of primary human hepatocytes for testing the potential of new drugs to induce liver enzyme expression. This chapter attempts to address some of the more important issues and caveats that must be considered when utilizing primary cultures of human hepatocytes for drug evaluation, especially for long-term studies of gene expression (e.g., induction or suppression). The effects of different culture conditions on the restoration and maintenance of normal hepatic structure and function in vitro also are presented, especially as they relate to testing the potential of new drugs to alter liver enzyme expression.

2. Materials

2.1. Human Liver Tissue

Adult human liver tissue suitable for the isolation of hepatocytes is either from donors undergoing surgical liver resection for the removal of metastatic tumors or from brain-dead-but-beating-heart donors, inasmuch as liver tissue is exquisitely sensitive to ischemia and deteriorates rapidly after death. Rejected livers are shunted to agencies such as the National Disease Research Interchange (NDRI) (Philadelphia, PA), Tissue Transformation Technologies (T-Cubed) (Edison, NJ), or NIH contract organizations that are part of the Liver Tissue Procurement and Distribution System (LTPADS) (*see* **Note 1**) to be distributed to academic and industrial researchers. These livers, ranging in weight from 1500 to 2500 g, are rarely sent as whole livers but, rather, are carved up by agency staff members to maximize the number of researchers

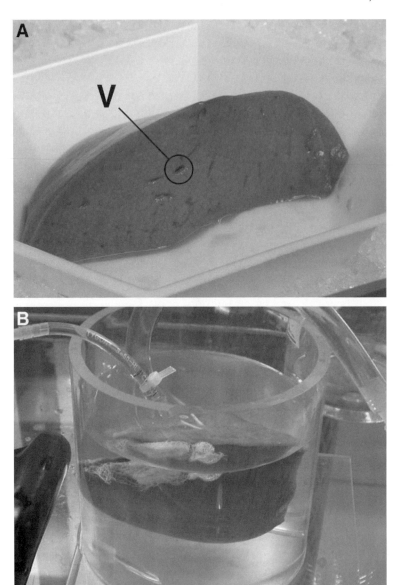

Fig. 2. Resected human liver tissue (**A**) prior to cannulation, illustrating a candidate vessel (**V**) for placing the cannula, and (**B**) during perfusion. Note that the resection is entirely submerged and floating in the perfusion buffer.

receiving samples. Each researcher receives a piece that is usually about 100–200 g and that must be perfused through cut blood vessels exposed on the surface of the sample (*see* **Fig. 2A**). The sample is shipped to the investigator as quickly as possible, but it often arrives late in the evening, meaning that the initial work on human liver samples is often overnight. The triaging of the liver from donor to either recipient or to investigators takes about 12–24 h. The conditions prior to death and the cold ischemia of the transport conditions can result in the deterioration of the sample. Thus, the quality of the starting material is extremely variable. The samples arrive flushed with cold-preservation buffer, most commonly University of Wisconsin solution ("UW" solution or Viaspan®), bagged and on ice.

For donor organs, it is generally accepted that the overall organ integrity and function begins to deteriorate after 18 h of cold storage and will not be used for transplant after this time. In our experience, the quality of the cells prepared from donor organs that have been procured more than 18–20 h reflect this general phenomenon, and lower yields and viability of the polyploidal cell populations are observed compared with fresher organs or tissue. We have also observed that, in general, organs received more than 24 h after clamp time often do not yield cells of adequate quality nor are the cells able to efficiently attach to culture substrata *(21)*. However, the time threshold after which a particular organ cannot produce cells of adequate quality is affected by several factors, including age of the donor, proficiency of organ preservation, the quality of the tissue perfusion, and disease state of the organ (e.g., extent of cirrhosis and steatosis) *(22)*. For the most part, organs should be a uniform tan or light brown color when received; organs that appear "bleached" or dark brown should not be used and generally yield only nonviable or CYP450-depleted cells. Medium containing phenol red with hepatocytes isolated from these organs often has a characteristic pink color, especially when mixed with Percoll®, which is believed to be reflective of the depletion of certain macromolecules from the damaged cells.

Normal remnants from partial hepatectomy can represent an alternative source of tissue for the isolation of primary hepatocytes, especially for many European and Asian countries because of legal and ethical considerations. In our experience, fresh surgical waste tissue often yields better preparations of cells, especially when prolonged warm and cold ischemia times are avoided. In a retrospective examination of the influence of human donor, surgical, and postoperative characteristics on the outcome of hepatocyte isolation obtained from liver surgical waste following hepatectomy from 149 patients, we showed that neither donor disease nor mild steatosis has a detrimental effect on the yield, viability, or attachment rate of the cells *(22)*. However, it was concluded that biopsy tissue weight (>100 g) and warm ischemia longer than 60 min effected the total yield and overall viability of the preparations. Recently, a multilaboratory study examined the effects of liver source, preflushing conditions, tissue transport time, and specific hepatocyte isolation conditions and concluded that (1) surgical liver resections are preferable to tissue from rejected donor organs, (2) preflushing is only necessary if transport time from the surgical suite is greater than 1 h, (3) preflushed tissue is stable during transport for at least 5 h, and (4) ideally digestion times not longer than 20 min should be used *(21)*.

2.2. Collection of Liver Samples

Based on the above discussion and depending on the source of the donated adult human liver specimen, one of two protocols should be followed when transporting tissue directly from procurement centers.

1. For livers obtained from centers where they can be transported from source to the laboratory in less than 60 min, the lobe should be placed in ice-cold medium (e.g., Dulbecco's modified essential medium [DMEM]).
2. For livers obtained from remote locations, where transport will take 2–6 h, samples should be preperfused with UW solution (Viaspan) or Soltran (Baxters) and transported in this solution on melting ice.

2.3. Supplies and Equipment

1. Suitable apparatus to include platform for liver undergoing perfusion and digestion, peristaltic pumps to ensure flow of appropriate buffers, heater unit to maintain temperature of system at a constant 34–35°C, and variable-sized tanks to accommodate liver tissue (*see* **Note 2**).
2. Water bath at 36–37°C.
3. Class II safety cabinet.
4. Suitable surgical instruments, including tissue and hemoclip forceps.
5. Sterile gauze and cotton-tipped applicators.
6. Disposable pipets.
7. Silk (3-0) and needle.
8. Suitable apparatus for size separation (850–1000 μm, 500 μm, and 100 μm).
9. Microcentrifuge tubes, 1.5 mL.
10. Polyethersulphone 0.2-μm filters.
11. Suitable refrigerated centrifuge for cell sedimentation.
12. Centrifuge tubes (50–250 mL sterile).
13. Cannulas—14–22G or equivalent. Flexibility is required to address the wide variety of vessel sizes.
14. Masterflex® biocompatible tubing (size 14–16), joints, and suitable connectors for cannulas.
15. Suitable disinfectant for surfaces and instruments.
16. Suitable sterile containers including trays and glassware.
17. Protective gear: safety glasses, surgical mask, lab coat, and protective sleeves.
18. Teflon mesh filters (Spectra Labs, Inc., Tacoma, WA): 850–1000, 400–500, and 80–100 μm mesh sizes.
19. Tissue culture-treated dishes (NUNC Permanox® 60-mm dishes; Naperville, IN), multiwell plates and flasks (Biocoat®; BD Biosciences, Palo Alto, CA).

2.4. Reagents

1. Instant medical adhesive (Loctite® 4013; Loctite Corp., cat. no. 20268).
2. Percoll (Sigma, cat. no. P-4937).
3. Phosphate-buffered saline (PBS), 10X (Gibco, cat. no. 14080).
4. Trypan blue (Sigma, cat. no. T-8154).
5. Ethyleneglycol-bis(2-aminoethylether)N,N,N,N-tetraacetic acid (EGTA), tetra-sodium salt (Sigma, cat. no. E-8145).
6. DMEM with HEPES and 4.5 g/L glucose, without phenol red (Gibco, cat. no. 21063). If medium is kept longer than a period of 1 mo, add 1 mL of L-glutamine 100X (Gibco, cat. no. 25030) or 1 mL of Glutamax® I 100X (stable L-glutamine) (Gibco, cat. no. 35050) to 100 mL DMEM
7. Insulin: prepare bovine insulin (Gibco, cat. no. 13007-018) at 4 mg/mL. Store at 4°C.
8. Collagenase Type IV (Sigma, cat. no. C-5138), preferred activity 400–600 units/mg (*see* **Note 3**).
9. Fetal bovine serum (FBS) (Gibco, cat. no. 16000).
10. Penicillin–streptomycin 100X solution (Gibco, cat. no. 15140).
11. Dexamethasone (Sigma, cat. no. D-4902; cell culture tested): dissolve 3.925 mg in 1 mL DMSO to prepare 10 mM solution and store aliquots of 100 μL at –20°C. Use at a final concentration of 1 μM (dilution: 1/10,000) (*see* **Note 4**).
12. Dimethyl sulfoxide (DMSO) (Sigma, χατ. vo. D-5879)
13. Hanks' balanced salt solution (HBSS): Ca^{2+}- and Mg^{2+}-free, without phenol red (Gibco, cat. no. 14175).
14. Bovine solution albumin (BSA) Fraction V (Sigma, cat. no. A-3059).
15. Wash buffer (P1 medium). Prepare 0.5 mM EDTA (208.1 mg/L), 0.5% (w/v), BSA and 50 μg/mL ascorbic acid in Ca^{2+}- and Mg^{2+}-free HBSS. Filter-sterilize using a 0.2-μm polyethersulphone filter. Store at 4°C for up to 4 wk.
16. Digestion medium (P2 medium). Prepare 0.03–0.05% (w/v) Collagenase Type IV (300–500 mg/L) and 0.5% (w/v) BSA in DMEM. Filter-sterilize using 0.2-μm polyethersulphone filter. Store at 4°C for up to 4 wk.
17. Suspension and attachment medium. Prepare 5% FBS, and penicillin–streptomycin (100 U/mL and 100 μg/mL, respectively) in DMEM. Filter-sterilize and store at 4°C for 4 wk. Complete medium by adding insulin (4 μg/mL, 1/1000 of stock) and 1 μM dexamethasone (1/10,000 of stock) just before use. Complete suspension medium may be stored for up to 3 d at 4°C.
18. Percoll (90% isotonic solution). Prepare fresh on each occasion. Mix 45 mL of Percoll and 5 mL PBS (10X). Ensure well mixed before use. Store at 4°C until use.
19. Rat-tail collagen (BD Biosciences, Palo Alto, CA) at 4 mg/mL.
20. DMEM 10X (Sigma, cat. no. D-2429).
21. 0.2 N NaOH.
22. Matrigel® (BD Biosciences, Palo Alto, CA).
23. Cell harvest and homogenization buffer: 50 mM Tris-HCl (Sigma, cat. no. T-3253), 150 mM KCl (Sigma, cat. no. P-9333), 2 mM EDTA (Sigma, cat. no. E-6511), pH 7.4.

3. Methods

The following procedure describes the isolation of human hepatocytes from liver tissue obtained from one of two sources: an encapsulated end wedge removed from patients undergoing resection for removal of liver tumors or resected segments from whole livers obtained from multiorgan donors.

3.1. Preparation for Liver Perfusion

1. Place P2 medium (100 mL/10 g liver) in water bath at 34–35°C.
2. Keep 100–200 mL of P1 medium at 4°C for initial preperfusion of liver segment.
3. Set up perfusion apparatus, rinse perfusion lines with plenty of 70% ethanol and reagent-grade water, and ensure temperature of system is slightly hypothermic at 34–35°C (*see* **Note 2**) *(17)*.
4. Set up culture materials, switch on class II cabinet, disinfect surfaces, prepare suspension and attachment medium and sufficient 90% Percoll solution and place in water bath.
5. Place the appropriate size tank in the perfusion apparatus and fill with P1 medium. Purge all lines and bubble trap of air prior to initiating perfusion.

3.2. Perfusion of Resected Liver Tissue

3.2.1. Preparation and Cannulation of Tissue

1. Weigh the piece of liver tissue and record weight (*see* **Note 5**).
2. Using a Teflon cannula attached to a 60-mL syringe, flush the liver tissue with ice-cold P1 medium using several blood vessels on the cut surface. This will clear any excess blood from the liver and help to determine the vessel(s) that will offer optimal perfusion of the tissue.
3. Using a sterile gauze pad, dab dry the cut surface of the liver.
4. Cannulate the chosen vessel(s) (one to two cannulas is generally sufficient, but up to four might be required) using one of the following:
   a. A 200-µL pipet tip will be suitable in most cases (cut off end of pipet tip to obtain optimal size to match vessel opening).
   b. A 16–22-gage Teflon cannula (must remove the needle). This is best for very small pieces.
   c. Plastic serological pipet; most useful with larger pieces and lobes (1- to 10-mL pipets scored and broken off to the appropriate length can be used as required).
5. Make a collar around the periphery of the cannula with medical adhesive at the point where it will join the tissue on the cut surface; then, insert the cannula into vessel opening. Secure the cannula in place by adding more adhesive around the cannula–tissue interface.
6. Seal all other openings on the cut surface using medical adhesive. For the larger openings, it might be necessary to seal them using hemoclip forceps or a cotton-tipped applicator. The wooden dowel from the cotton-tipped applicator can be

used, or the cotton tip can be reduced to fit in the opening size (cut the wooden dowel to a small size so that no more than 1 cm is sticking out). Secure the cotton tip or wooden dowel in place by making a collar around the edge with medical adhesive.

7. Once again, dab dry the cut surface of the liver and cover with a thin layer of adhesive; apply using a cotton-tipped applicator.

8. In some cases, there might be a cut or tear on the outer capsule of the liver tissue (Glisson's capsule) or there might be more than one cut surface. These must be sealed to ensure optimal perfusion of the tissue.

9. Allow the medical adhesive to dry sufficiently before initiating the perfusion.

3.2.2. Perfusion of Resected Liver Tissue

1. Once the adhesive has dried adequately, place the liver into a weigh boat and connect perfusion tubing to the cannula(s) and slowly start the perfusion (do not exceed 10–15 mL/min initially). If no overt leaks are observed, place two or three small incisions along the encapsulated edge of the tissue and carefully place the liver into the tank containing P1 medium inside the perfusion unit prewarmed to 34–35°C (*see* **Fig. 2B**). If a self-contained, temperature-regulated unit is not utilized, then place the tank containing the liver resection and P1 medium into a water bath at 35–37°C (*17*).

2. Slowly increase the flow rate for the P1 medium until residual blood and perfusate are observed flowing from the incisions and/or extreme edges of the cut face. The flow rate will vary with the size of the tissue and how well it is sealed. On average, flow rates vary between 15 and 30 mL/min for resections weighing between 20 and 100 g.

3. While P1 medium is perfusing throughout the liver, prepare P2 medium with collagenase. See **Subheading 2.4.** for additional details. For most normal pieces of liver, use 60–100 mg collagenase per 100 mL of P2 medium, and for cirrhotic or steatotic (>40% fat) piece of liver, use 100–120 mg collagenase per 100 mL of P2 medium. Depending on the size of the tissue, the volume of P2 should be approx 100 mL/10 g liver tissue, and, therefore, the amount of collagenase will vary accordingly.

4. After 10–15 min, stop the pump and carefully drain the tank as completely as possible of P1 medium and then add a similar volume of prewarmed P2 medium containing collagenase.

5. Perfuse for approx 15–25 min; the time will vary depending on the activity of the collagenase and the size of the liver resection. Indications of a complete digestion are softening and enlargement of the tissue. Complete digestion is generally achieved within the specified timeframe if a proper batch and concentration of collagenase has been chosen. However, it is important not to overextend the perfusion time, as this will lead to excessive cell damage and a significant loss in viability.

6. When the perfusion is complete, remove the liver from the tank and place in a covered sterile bowl/dish and then proceed to a biosafety cabinet for hepatocyte isolation.

3.3. Isolation of Hepatocytes

1. Add a sufficient volume (approx 1–2 mL/g tissue) of ice-cold suspension medium (DMEM supplemented with 5% FBS and hormones; *see* **Subheading 2.4.**) to the dish containing the digested liver tissue.
2. Using tissue forceps and scissors, remove the glue and gently tear open the Glisson's capsule. With the aid of the tissue forceps, release the hepatocytes into the medium by gently shaking and passing the tissue between the tissue forceps, leaving behind the connective tissue and any undigested material.
3. Add additional suspension medium (final volume: approx 5 mL/g tissue) and filter the digested material through a series of Teflon or stainless steel mesh filters using further cold (4°C) media (up to 1 L) to aid this process as appropriate:

   850- to 1000-μm mesh → 400- to 500-μm mesh → 90- to 100-μm mesh

   Use large funnels and filter into sterile beakers. It might be necessary at the initial stage to use a syringe plunger to carefully encourage filtering.
4. The resulting cell suspension is then divided equally into sterile centrifuge bottles (ensure that the suspension is not too dense [approx 5–10 mL/g total liver]) and washed by low-speed centrifugation (75*g* for 5 min). The size of the centrifuge tubes will vary according to the amount of material (50–200 mL).
5. Discard or retain the supernatant (*see* **Note 6**) and gently resuspend each pellet in approx 5–10 mL of suspension medium and combine. Subject to pellet size, cells are resuspended in suspension medium using roughly a 1:8-fold dilution. At this stage, cells should be counted and viability assessed using trypan blue. If the viability is greater than 75%, then a Percoll wash should be utilized as described in **steps 6–8**. If viability is less than 75%, then a more stringent Percoll wash should be utilized (*see* **Note 7**).
6. Switch to 50-mL sterile centrifuge tubes if larger tubes were used for the spin in **step 4**. Resuspend the pellets in suspension medium and 90% isotonic Percoll; the ratio of volumes should be 3 parts cell suspension to 1 part isotonic Percoll (see **Subheading 2.4.** for details on Percoll preparation) (e.g., 37.5 mL of cells in DMEM + 12.5 mL of 90% isotonic Percoll). Sample tubes should be loaded with a maximum of 500×10^6 total cells per 50-mL tube. (If the liver has a high fat content [≥40%], then see **Note 8**.)
7. Centrifuge at 100*g* for 5 min.
8. Carefully remove the top layer of the supernatant that contains dead cells and other debris; care should be taken not to disrupt the pellet(s) or contaminate it with the contents from the top layer of debris. Gently resuspend the pellet(s) in suspension medium, combine into one or two 50-mL tubes and centrifuge for a final time at 75*g* for 5 min.
9. Gently resuspend the final cell pellet in 40 mL of suspension medium and place on ice. (If the final pellet volume is greater than approx 8 mL, then resuspend in 80 mL, and if the pellet is less than 4 mL, then resuspend in 20 mL.)

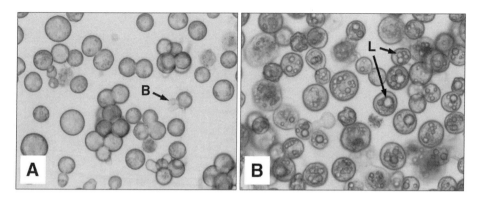

Fig. 3. Primary human hepatocytes from two separate donor organs viewed under bright field optics. (**A**) Hepatocytes exhibiting mostly normal morphology with clear cytoplasms and intact, well-delineated plasma membranes. Note that some cells possess surface blebs (B), which are caused by either physical or chemical damage and/or oxidative stress. (**B**) Hepatocytes isolated from a donor organ with high fat content. Note the presence of large lipid droplets within the cytoplasm of most cells. Although the presence of lipid changes the centrifugation characteristics of the hepatocytes considerably, after several days in culture they generally function much in the same manner as hepatocytes from normal, healthy liver tissue.

3.4. Cell Count and Viability Assessment

1. Perform a cell count and viability assessment by trypan blue exclusion using a hemocytometer. Prepare eight parts Suspension medium, one part trypan blue, one part cell suspension (v/v/v) and invert tube gently to ensure a uniform cell suspension.
2. Add 10 μL of cell suspension to the hemocytometer and count at least four of the mm^2 quadrants with an average of 80–120 cells per quadrant (approx 400 cells total).
3. Record total cell yield, viability, and cell morphology (*see* **Fig. 3**).
4. Remove sufficient cells for d 0 biochemical assessments (*see* **Note 9**).

3.5. Monolayer Culture of Primary Human Hepatocytes

3.5.1. Plating Hepatocytes

Human hepatocytes derived from the two-step liver digestion method described in the previous subsections can be cultured for a variety of biochemical, cellular, and molecular studies. This subsection describes the seeding, maintenance, and harvest of primary cultures of human hepatocytes.

1. Dilute the cell suspension with attachment medium (see comments in **Subheading 2.4.**) to give the required final cell density (*see* **Table 1** and **Note 10**). Check the cell density under the microscope and adjust if necessary.

Table 1
Determination of Seeding Density for Different Types of Tissue-Culture-Treated Vessels

Type of dish or multiwell plate	Seeding density	Volume/dish or well	Total no. of viable cells
100-mm dish	1.5×10^6–1.75×10^6 viable cells/mL	6 mL	9×10^6–10.5×10^6
60-mm dish	1×10^6–1.33×10^6 viable cells/mL	3 mL	3×10^6–4×10^6
6-Well plate	5×10^5–7.5×10^5 viable cells/mL	2 mL	1×10^6–1.5×10^6
12-Well plate	5×10^5–7.5×10^5 viable cells/mL	1 mL	5×10^5–7.5×10^5
24-Well plate	5×10^5–7.5×10^5 viable cells/mL	0.5 mL	2.5×10^5–3.75×10^5
96-Well plate	5×10^5 viable cells/mL	125 µL	6.25×10^4

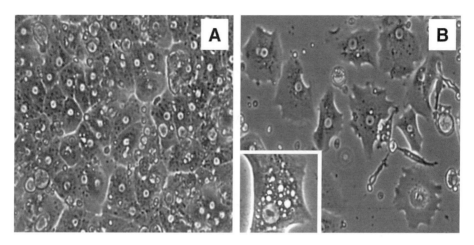

Fig. 4. Light micrographs of hepatocyte monolayers at normal (**A**) and low (**B**) seeding density. Note the difference in the confluence of the monolayer and the corresponding changes in the morphology of both the cytoplasm and nucleus of most cells. Inset: Increased vacuole formation over time is often observed in hepatocytes at low plating densities.

2. Add the appropriate volume of cell suspension to each well or dish (*see* **Table 1** and **Note 11**). Swirl the bottle of cells gently before seeding each multiwell plate or stack of dishes to ensure that the suspension remains homogenous (*see* **Note 12**).
3. Place the stack of dishes or plates in a 95%/5% air/CO_2 incubator at 37°C.
4. In order to ensure formation of uniform monolayers, gently swirl the dishes or plates in a figure-of-8 pattern when placing them in the incubator. In the case of 24- to 48-well plates, make a cross-shape (⇔, ⇕) while shaking the plates.
5. Allow hepatocytes to attach for 4–12 h at 37°C in the incubator.
6. Assess attachment efficiency by gently swirling the culture vessels and counting cells in the aspirated medium from two to three dishes or wells (attachment efficiency of ≥75% is required for optimal monolayer formation). Observe the cells under the microscope to confirm confluence (should be ≥80%) (*see* **Fig. 4**).
7. After attachment, cultures should be swirled adequately to remove unattached cells and debris, and the attachment medium carefully aspirated and replaced with the appropriate medium, depending on the specific studies to be performed (*see* **Note 13**). In some cases, the cells can be overlaid with either Matrigel or collagen gels to enhance the development of a more histotypic architecture (*see* **Subheading 3.6.** and **Fig. 5**).

3.5.2. Maintenance and Dosing of Hepatocyte Cultures

1. Generally, medium is replaced on a daily basis and hepatocytes are maintained for 36–48 h prior to treatment with drugs or other agents intended or expected to alter the gene expression profiles (*see* **Note 14**). Dosing with test compounds

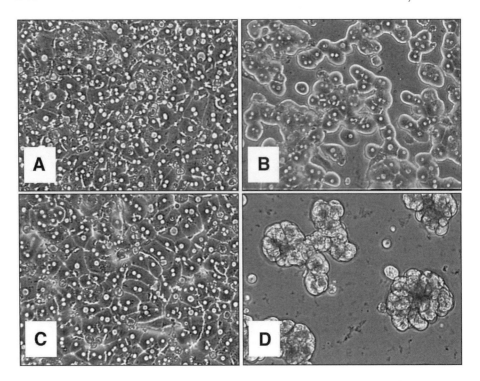

Fig. 5. Human hepatocytes cultured under different matrix conditions for 72 h.
(**A**) Freshly isolated hepatocytes on a rigid collagen substratum and overlaid with
medium alone. (**B**) Hepatocytes maintained between two layers of gelled collagen,
type I. Hepatocytes maintained in the "sandwich" configuration form trabeculae or
cordlike arrays throughout the monolayers. (**C**) Hepatocyte cultures on a rigid collagen substratum with a top layer of Matrigel. (**D**) Hepatocytes maintained on a substratum of Matrigel. Human hepatocytes maintained on a gelled layer of Matrigel
aggregate together to form clusters or colonies of cells that become more three-dimensional over time in culture. All cultures were maintained in modified Chee's medium
supplemented with insulin (6 μg/mL) and dexamethasone (0.1 μ*M*).

generally is started 48 or 72 h postplating. Dosing solutions containing drugs and
xenobiotics that modulate liver enzymes are renewed typically every 24 h for
3–5 d depending on the purpose and end point of the studies (*5*).
2. Stock solutions of drugs are prepared in a compatible solvent, such as
DMSO or methanol, at 1000-fold higher concentrations as those required for
experimentation.
3. Dosing tubes are prepared prior to the first dosing day and labeled according to
the dosing groups. Plates or dishes are labeled and arranged in stacks according
to dosing groups.

4. At the end of the treatment period, monolayers can be harvested for biochemical assessment (*see* **Subheading 3.5.3.**), fixed for microscopic evaluation and immunostaining *(23)*, or treated with substrates directly to assess inherent enzyme activities *(24)*.

3.5.3. Harvest of Plated Cells

1. After the dosing period, cells should be harvested into appropriate solutions depending on the biochemical or molecular tests to be conducted, such as homogenization buffer or appropriate RNA preservation reagent (e.g., TRIzol, RNAeasy) *(5,9)*, and stored at –80°C. This procedure need not be performed under sterile conditions; however, standard precautions should be observed when handing samples for isolation of RNA to minimize RNase contamination and loss of sample integrity.
2. Place homogenization buffer and HBSS on ice. Label 5- to 10-mL tubes according to the treatment groups and place on ice.
3. Gently rinse each culture dish or well twice with ice-cold HBSS, taking care not to disrupt the cell monolayer. Drain excess buffer from the culture vessel by inverting over a paper towel.
4. For isolation of cellular fractions, add 3 mL of homogenization buffer (total) to each treatment group (approx 0.5 mL per 60-mm dish). Using a cell scraper or rubber policeman, scrape the cells into the homogenization buffer. Transfer cells in buffer to a corresponding tube, taking precautions not to leave behind any residual cellular material. This process is repeated for each sample group and tubes are kept on ice until harvest is complete.
5. For isolation of RNA, add 1 or 2 mL of TRIzol (or equivalent reagent) to each well of a six-well plate or 60-mm dish, respectively, and scrape cells with a cell scraper. Pipet the sample up and down several times until the sample is dissolved completely (this step might take longer with samples overlaid with extracellular matrix). Transfer samples to the corresponding RNase-free tube, seal tube tightly, and store on ice. Repeat process for each sample until harvest is complete.
6. Store all samples at –80°C (in screw-cap or snap-cap tubes) or process immediately to prepare cellular fractions.

3.6. Overlay With Extracellular Matrix (Optional)

Extracellular matrix composition and configuration have been proposed to play a key role in the maintenance of hepatocyte structure and function in vitro *(25–28)*. Many different matrix conditions have been tested and found to be appropriate given that the proper cell density is maintained (*see* **Fig. 5**). An overlay with extracellular matrix such as Matrigel or collagen is recommended in most cases to avoid variability in monolayer quality and to restore normal cell polarity and cytoskeletal distribution (*see* **Fig. 6**). In addition, the addition of an overlay of ECM can be more "forgiving" of misjudgments on the part of inexperienced scientists or unforeseen differences in cell attachment efficiency.

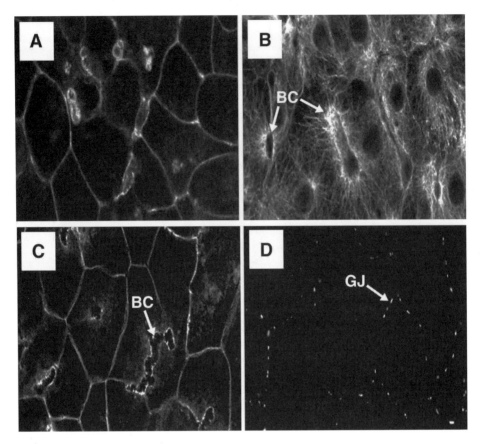

Fig. 6. Immunolabeling of primary cultures of human hepatocytes maintained for 3 d in a sandwich configuration showing the normal distribution of (A) actin microfilaments, (B) microtubules, (C) E-cadherin, and (D) gap junctions (Cx-32). BC: bile canaliculus; GP: gap junction.

3.6.1. Collagen Sandwich

1. Prepare the required amount of gelled collagen as described in **Table 2**. All solutions must be kept on ice and must be handled with cold glass pipets. The final concentration of gelled collagen will be approx 1.5 mg/mL. (Note that volumes only apply if using rat-tail collagen, type I, from Collaborative Research, BD Biosciences.)
2. In the order shown in **Table 2**, add the components listed into a tube on ice and gently mix.
3. After cells have attached (from **Subheading 3.5.1.**), aspirate the medium. Swirl dishes well prior to removal of medium to ensure all unattached cells and debris are removed.

Table 2
Preparation of Collagen Solutions for Overlaying Hepatocyte Monolayers

Final volume	5 mL	10 mL	15 mL	20 mL
Collagen	2 mL	4 mL	6 mL	8 mL
Sterile water	2 mL	4 mL	6 mL	8 mL
10X DMEM	0.5 mL	1 mL	1.5 mL	2 mL
0.2 N NaOH	0.5 mL	1 mL	1.5 mL	2 mL

4. Tilt dishes at an approx 45° angle against a tray and let them stand for a few seconds to allow excess medium to collect at the edge of the dish; then aspirate it.
5. Gently add 5–10 μL of diluted collagen per cm^2 culture area (i.e., 200 μL per 60-mm dish) (*see* **Note 15**). Use a cold 1-mL pipet and place the drops in the center of the dish. Only handle a maximum of five dishes at any one time, to prevent gelling of the collagen prematurely.
6. Gently tilt and rotate the dishes to spread the collagen evenly over the surface of the monolayers and place them back in the incubator. Leave for 45–60 min to allow the collagen to gel. Place any remaining collagen in the incubator; this provides a way of checking the gelling process.
7. Carefully add back appropriate volume of warm medium according to **Table 1** to the center of the dish or well (*see* **Note 16**).

3.6.2. Overlay With Matrigel

Both dilute (5 mg/mL) and concentrated (10–13 mg/mL) Matrigel stocks can be used for the overlay. Dilute Matrigel stocks provide the advantage of being easier to work with and are less likely to gel when handled.

1. Calculate the amount of Matrigel required to give a final concentration of 0.25 mg Matrigel/mL of the desired medium (*see* **Table 1** and **Note 17**).
2. Slowly thaw out Matrigel stock by placing in slushy ice. It will take at least 2–3 h for the Matrigel to be fully thawed (*see* **Note 18**).
3. Place refrigerated culture medium on ice, and using an ice-cold glass pipet, add the required volume of Matrigel to the culture medium and mix well by swirling.
4. Rinse the pipet out with the cold medium after transferring Matrigel to ensure that none is left behind in the pipet. Ensure that Matrigel is well mixed in the medium. In the event that an entire vial or tube of stock Matrigel is required, rinse the vial or tube out with cold medium to remove any residual matrix material from the bottom and sides of the container.
5. Remove medium from the cultures, ensuring that all of the unattached cells and debris is removed by swirling the culture vessel.
6. Add appropriate volume of Matrigel-containing medium per dish or well and then return cells to the incubator (*see* **Note 13** and **Table 1**).
7. Leave cultures undisturbed for 24 h, after which the medium should be replaced with Matrigel-free medium for subsequent experiments and treatments.

4. Notes

1. The Liver Tissue Procurement and Distribution System (LTPADS) is a National Institutes of Health (NIH) service contract to provide human liver from regional centers for distribution to scientific investigators throughout the United States. LTPADS provide liver tissue and isolated hepatocytes from "normal" human liver to NIH investigators. NIH investigators are always given preference for tissue requests. Supporting letters for NIH new or renewal grant requests can be provided. Direct inquires can be made to Harvey L. Sharp, M.D. or Sandy K. Dewing, LTPADS Coordinator, University of Minnesota, Minneapolis, MN (http://www.peds.umn.edu/Centers/ltpads).

2. Instructions for materials, setup, and use of basic perfusion equipment are described by David et al. *(17)*. Recommended pump system and tubing are the Masterflex® L/S digital economy drive pump with an easy-load #2 pump-head (model 77200-52) and Masterflex® silicone tubing (#96420) (size 14–16). A self-contained, temperature-regulated, HEPA-filtered, stainless-steel organ perfusion unit has been specially designed and is available through Blue Collar Scientific, Inc. (BCS, Pittsboro, NC) for the perfusion of liver tissue as either whole organs or resected remnants. Although excellent results have been obtained using a variety of standard laboratory components, the BCS unit is designed specifically to provide a more efficient, standardized, and reproducible isolation of primary hepatocytes from liver tissue, thereby greatly expanding the range of personnel capable of successfully isolating primary hepatocytes.

3. Most liver perfusions are done with collagenase preparations that are partially purified. Different companies indicate the degree of purification with a company-specific nomenclature and one must read the company's literature to learn the details of the nomenclature and its implications for the extract or purified factor(s) being sold. Generally, the liver perfusions are done with a preparation that is intermediate in purity (e.g., type IV in Sigma's series, CLS2 in Worthington's series, or Type B or C in the Boehringer Mannheim series), because both collagenase and one or more proteases are required for optimal liver digestion. Moreover, it has been learned only relatively recently that the most effective liver perfusions are achieved with a mixture of purified collagenase and purified elastase at precise ratios *(29,30)*. An additional commercially available mixture of digestive enzymes for perfusion of a number of tissues is called Liberase from Roche Applied Science (Indianapolis, IN). However, its use has been limited because of its high cost. With any preparation of collagenase, it is essential to prescreen individual lots or batches to determine the optimum concentration and perfusion times. Optimal collagenase digestion conditions are a function of temperature, time, and concentration. Every batch of collagenase will inflict damage and be potentially lethal to cells; therefore, one must determine the balance between achieving the highest yields and minimizing cell damage and death. In general, prolonged perfusion times (>30 min) are detrimental to cells, especially from tissues that have been in cold storage for long periods, and should be avoided. It is preferred to increase the collagenase concentration while minimizing the perfusion times.

4. Glucocorticoids (e.g., dexamethasone or hydrocortisone) can have significant effects on the basal expression of many genes in vitro, such as albumin and the cytochromes P450 *(6,31)*.

5. As with any human-derived tissue or cells, universal biohazard precautions should be taken at all times when handling liver tissue samples.

6. The supernatant contains nonparenchymal and progenitor cells, which can be isolated separately according to a number of published methods.

7. The use of a Percoll gradient generally improves the quality of monolayer formation with all preparations of cells by removing cell debris and most dead cells. If the viability of the hepatocyte preparation is <75% after the initial centrifugation step, then the following Percoll separation step should be utilized: Mix 22.5 mL PBS with 7.5 mL of Percoll (final Percoll concentration = 25%) in a 50-mL tube. Hepatocytes in suspension medium are gently layered on top of the isotonic Percoll solution. (Maximum density should be 100×10^6 cells per 50-mL tube.)

8. If the liver has a high fat content, then the buoyant density of the hepatocytes is altered considerably. In this case, the Percoll concentration must be adjusted accordingly by reducing the volume of Percoll used; for example, 39.5 mL of cells in DMEM + 10.5 mL of 90% isotonic Percoll per 50-mL tube. The centrifugation time can be increased to 10 min to better resolve the distinct layers of cells.

9. Store 9–10 million cells for d 0 biochemical assessment. Centrifuge 5 min at 75g, resuspend pellet in 3 mL of appropriate buffer, such as homogenization buffer, TRIzol, RNAeasy, and store at –80°C.

10. Of all of the issues discussed thus far relative to the cultivation of human hepatocytes in vitro, proper seeding density ranks first, by far, in terms of importance for restoring the optimal induction response to treatment with drugs and other xenobiotics. Several studies have shown that this is related to the restoration and maintenance of proper cell–cell interactions *(23,32,33)*. Plating densities in the range of 125,000–150,000 cells/cm^2 appear to be optimal for the formation of confluent monolayers *(1,6,23)*. Notably, higher seeding densities can be used for Matrigel-coated dishes and plates; however, densities that are too high on any type of substratum will interfere with cell attachment and cause less than optimal monolayer formation.

11. Example calculation:

$$\text{Volume of cell stock required} = \frac{\text{Volume of cell suspension needed} \times \text{Seeding density}}{\text{Stock cell density}}$$

Need to seed 15 × 60-mm dishes. Require 3 mL/dish. So total cell suspension needed = 45 mL.

Make 50 mL of cell suspension:

Seeding density = 1.33×10^6 viable cells/mL. Stock cell density = 1×10^7 viable cells/mL

$$\text{Volume of cell stock needed} = \frac{50 \text{ mL} \times 1.33 \times 10^6 \text{ cells/mL}}{1 \times 10^7 \text{cells/mL}} = 6.65 \text{ mL}$$

Take 6.65 mL of stock cell suspension and dilute to 50 mL in DMEM.

12. Generally do not pipet more than one stack of dishes (15 mL per five 60-mm dishes) or one plate (12 mL/plate) at a time to minimize settling of cells during plating.
13. In our experience, serum-free medium formulations, such as modified Chee's medium (MCM), Williams' E medium (WEM), or Hepatocyte Maintenance Medium (HMM) (Biowhittaker, CC-3197), supplemented with insulin (4–6 µg/mL), transferrin (4–6 µg/mL), selenium (5–6 ng/mL), and BSA/linoleic acid (1 mg/mL) are adequate for performing CYP450 induction studies and maintaining monolayer integrity and hepatocyte morphology for at least 1 wk. However, experiments requiring longer culture periods (>2 wk) might require more specialized medium formulations and additives *(12,18,19)*.
14. Experimental evidence suggests that primary human hepatocytes are refractory to modulating agents until normal cell–cell contacts are restored *(1,6)*.
15. For example, 200 µL/60-mm dish, 100 µL/well of 6-well plate, 50 µL/well of 12-well plate, and so forth.
16. When adding medium back to the culture vessels, the medium should form droplets that dance across the gel. This is a good sign that the collagen has gelled sufficiently. Appropriate volumes are shown in **Table 1**.
17. Example calculation for Matrigel dilution:
    - Have 10 multiwell plates, 12 mL total medium per plate; therefore, require 120 mL of medium.
    - The amount of overlay per dish must be 0.25 mg/mL.
    - 0.25 × 120 = 30 mg of Matrigel are required in total.
    - Stock solution is 10 mg/mL, 30 ÷ 10 = 3.
    - Therefore, must add 3 mL of the stock 10 mg/mL Matrigel to 117 mL of medium.
18. Do not try to speed up the thawing process by placing Matrigel at room temperature or warming, as this will cause Matrigel to gel. Allow enough time for Matrigel to thaw (2–3 h on ice), so that it is ready to use once the media has to be changed after attachment.

Acknowledgments

Funding for these studies derives in part from grants from ECVAM (19471-2002-05-F1 ED ISP FR) and the Food and Drug Administration (CDER). The authors would like to thank Darryl Gilbert and Lynn Johnson for technical assistance (USA). We also acknowledge the invaluable contributions of Dr. Benjamin Calvo, Dr. Kevin Behrns, and Dr. David Gerber (USA), and the staff of Dr. Daniel Jaeck, Dr. Georges Mantion, and Dr. Bruno Heyd (France) for assistance with the procurement of human liver tissue in support of this project.

References

1. Maurel, P. (1996) The use of adult human hepatocytes in primary culture and other *in vitro* systems to investigate drug metabolism in man. *Adv. Drug Deliv. Rev.* **22,** 105–132.
2. Guillouzo, A. (1998) Liver cell models in *in vitro* toxicology. *Environ. Health Perspect.* **106(Suppl. 2),** 511–532.
3. Komai, T., Shigehara, E., Tokui, T., et al. (1992) Carrier-mediated uptake of pravastatin by rat hepatocytes in primary culture. *Biochem. Pharmacol.* **43(4),** 667–670.
4. Goll, V., Alexandre, E., Viollon-Abadie, C., Nicod, L., Jaeck, D., and Richert, L. (1999) Comparison of the effect of various peroxisome proliferators on peroxisomal enzyme activities, cell proliferation and apoptosis in rat and human hepatocyte cultures. *Toxicol. Appl. Pharmacol.* **160,** 21–32.
5. LeCluyse, E. L., Madan, A., Hamilton, G., Carroll, K., Dehaan, R., and Parkinson, A. (2000) Expression and regulation of cytochrome P450 enzymes in primary cultures of human hepatocytes. *J. Biochem. Mol. Toxicol.* **14,** 177–188.
6. LeCluyse, E. L. (2001) Human hepatocyte culture systems for the *in vitro* evaluation of cytochrome P450 expression and regulation. *Eur. J. Pharm. Sci.* **13(4),** 343–368.
7. Parkinson, A. (1996) An overview of current cytochrome P450 technology for assessing the safety and efficacy of new materials. *Toxicol. Pathol.* **24,** 45–57.
8. Parkinson, A. (2001) Biotransformation of xenobiotics, in *Casarett and Doull's Toxicology. The Basic Science of Poisons*, 6th ed. (Klaassen, C. D., ed.), McGraw-Hill, New York, pp. 133–224.
9. Richert, L., Lamboley, C., Viollon-Abadie, C., et al. (2003) Effects of clofibric acid on mRNA expression profiles in primary cultures of rat, mouse and human hepatocytes. *Toxicol. Appl. Pharmacol.* **191,** 130–146.
10. Donato, M. T., Castell, J. V., and Gomez-Lechon, M. J. (1995) Effect of model inducers on cytochrome P450 activities of human hepatocytes in primary culture. *Drug Metab. Dispos.* **23,** 553–558.
11. Strom, S. C., Pisarov, L. A., Dorko, K., Thompson, M. T., Schuetz, J. D., and Schuetz, E. G. (1996) Use of human hepatocytes to study P450 gene induction. *Methods Enzymol.* **272,** 388–401.
12. Ferrini, J. B., Pichard, L., Domergue, J., and Maurel, P. (1997) Long-term primary cultures of adult human hepatocytes. *Chem.–Biol. Interact.* **107,** 31–45.
13. Kern, A., Bader, A., Pichlmayr, R., and Sewing, K. F. (1997) Drug metabolism in hepatocyte sandwich cultures of rats and humans. *Biochem. Pharmacol.* **54(7),** 761–772.
14. Li, A. P., Colburn, S. M., and Beck, D. J. (1992) A simplified method for the culturing of primary adult rat and human hepatocytes as muticellular spheroids. *In Vitro Cell. Dev. Biol.* **28A,** 673–677.
15. Li, A. P., Roque, M. A., Beck, D. J., and Kaminski, D. L. (1992) Isolation and culturing of hepatocytes from human livers. *J. Tissue Culture Methods* **14,** 139–146.

16. Silva, J. M., Morin, P. E., Day, S. H., et al. (1998) Refinement of an *in vitro* cell model for cytochrome P450 induction. *Drug Metab. Dispos.* **26**, 490–496.

17. David, P., Viollon, C., Alexandre, E., et al. (1998) Metabolic capacities in cultured human hepatocytes obtained by a new isolating procedure from non-wedge small liver biopsies. *Hum. Exp. Toxicol.* **17**, 544–553.

18. Runge, D., Runge, D. M., Jager, D., et al. (2000) Serum-free, long-term cultures of human hepatocytes: maintenance of cell morphology, transcription factors, and liver-specific functions. *Biochem. Biophys. Res. Commun.* **269(1)**, 46–53.

19. Runge, D., Kohler, C., Kostrubsky, V. E., et al. (2000) Induction of cytochrome P450 (CYP)1A1, CYP1A2, and CYP3A4 but not of CYP2C9, CYP2C19, multidrug resistance (MDR-1) and multidrug resistance associated protein (MRP-1) by prototypical inducers in human hepatocytes. *Biochem. Biophys. Res. Commun.* **273**, 333–341.

20. Macdonald, J., Xu, A., Kubota, H., et al. (2001) Protocols for isolation and ex vivo maintenance of hepatic stem cells and their normal and transformed derivatives. *PNAS* **24**, 12,132–12,137.

21. Richert, L., Alexandre, E., Lloyd, T., et al. (2004) Tissue collection, transport and isolation procedures required to optimize human hepatocyte isolation from waste liver surgical resections. A multi-laboratory study. *Liver Int.,* in press.

22. Alexandre, E., Cahn, M., Abadie-Viollon, C., et al. (2002) Influence of pre-, intra- and post-operative parameters of donor liver on the outcome of isolated human hepatocytes. *Cell Tissue Bank.* **3**, 223–233.

23. Hamilton, G. A., Jolley, S. L., Gilbert, D., Coon, D. J., Barros, S., and LeCluyse, E. L. (2001) Regulation of cell morphology and cytochrome P450 expression in human hepatocytes by extracellular matrix and cell–cell interactions. *Cell Tissue Res.* **306(1)**, 85–99.

24. Kostrubsky, V. E., Lewis, L. D., Strom, S. C., et al. (1998) Induction of cytochrome P4503A by taxol in primary cultures of human hepatocytes. *Archiv. Biochem. Biophys.* **355**, 131–136.

25. Bissell, D. M., Arenson, D. M., Maher, J. J., and Roll, F. J. (1987) Support of cultured hepatocytes by a laminin-rich gel. Evidence for a functionally significant subendothelial matrix in normal rat liver. *J. Clin. Invest.* **79(3)**, 801–812.

26. Ben-Ze'ev, A., Robinson, G. S., Bucher, N. L., and Farmer, S. R. (1988) Cell–cell and cell–matrix interactions differentially regulate the expression of hepatic and cytoskeletal genes in primary cultures of rat hepatocytes. *Proc. Natl. Acad. Sci. USA* **85**, 2161–2165.

27. Brill, S., Zvibel, I., Halpern, Z., and Oren, R. (2002) The role of fetal and adult hepatocyte extracellular matrix in the regulation of tissue-specific gene expression in fetal and adult hepatocytes. *Eur. J. Cell Biol.* **81**, 43–50.

28. Richert, L., Binda, D., Hamilton, G., et al. (2002) Evaluation of the effect of culture configuration on morphology, survival time, antioxidant status and metabolic capacities of cultured rat hepatocytes. *Toxicol. In Vitro* **16**, 89–99.

29. Gill, J. F., Chambers, L. L., Baurley, J. L., et al. (1995) Safety testing of Liberase, a purified enzyme blend for human islet isolation. *Transplant. Proc.* **27(6)**, 3276–3277.

30. Olack, B. J., Swanson, C. J., Howard, T. K., and Mohanakumar, T. (1999) Improved method for the isolation and purification of human islets of Langerhans using Liberase enzyme blend. *Hum. Immunol.* **60(12),** 1303–1309.
31. Pascussi, J. M., Drocount, L., Fabre, J. M., Maurel, P., and Vilarem, M. J. (2000) Dexamethasone induces pregnane X receptor and retinoid X receptor-alpha expression in human hepatocytes: synergistic increase of CYP3A4 induction by pregnane X receptor activators. *Mol. Pharmacol.* **58,** 361–372.
32. Greuet, J., Pichard, L., Ourlin, J. C., et al. (1997) Effect of cell density and epidermal growth factor on the inducible expression of CYP3A and CYP1A genes in human hepatocytes in primary culture. *Hepatology* **25(5),** 1166–1175.
33. Ferrini, J. B., Pichard, L., Domergue, J., and Maurel, P. (1997) Long-term primary cultures of adult human hepatocytes. *Chem. Biol. Interact.* **107(1–2),** 31–45.

16

Primary Kidney Proximal Tubule Cells

Mary Taub

Summary

Primary rabbit kidney epithelial cell cultures can be obtained that express renal proximal tubule functions. Toward these ends, renal proximal tubules are purified from the rabbit kidney by the method of Brendel and Meezan. To summarize, each kidney is perfused with iron oxide, which becomes associated with glomeruli. The renal cortex is sliced and homogenized to liberate nephron segments. Renal proximal tubules and glomeruli are purified by sieving. The glomeruli, covered with iron oxide, are removed using a magnet. After a brief collagenase treatment (to disrupt basement membrane), the tubules are plated in hormonally defined serum-free medium supplemented with 5 µg/mL bovine insulin, 5 µg/mL human transferrin, and 5×10^{-8} M hydrocortisone. After 5–6 d of incubation, confluent monolayers are obtained that possess multicellular domes, indicative of their capacity for transepithelial solute transport.

Key Words: Primary culture; kidney; renal proximal tubule; serum-free medium; epithelial cell growth.

1. Introduction

An important application of hormonally defined serum-free media is for preparing primary cultures of differentiated cells that lack fibroblast overgrowth. Of particular interest in this regard is the serum-free culture of primary kidney tubule epithelial cells *(1–4)*. Investigations with primary kidney cell cultures are particularly advantageous for several reasons. First, kidney cells can be grown in vitro from the animal of choice. Thus, the results of tissue culture studies can be more closely correlated with animal studies. Second, new tissue culture systems can be developed that more closely resemble the kidney cells in vivo than presently available established kidney cell lines. Third, the use of serum-free medium permits a precise analysis of the mechanisms by which hormones and growth factors regulate growth and expression of differentiated function.

From: *Methods in Molecular Biology, vol. 290: Basic Cell Culture Protocols, Third Edition*
Edited by: C. D. Helgason and C. L. Miller © Humana Press Inc., Totowa, NJ

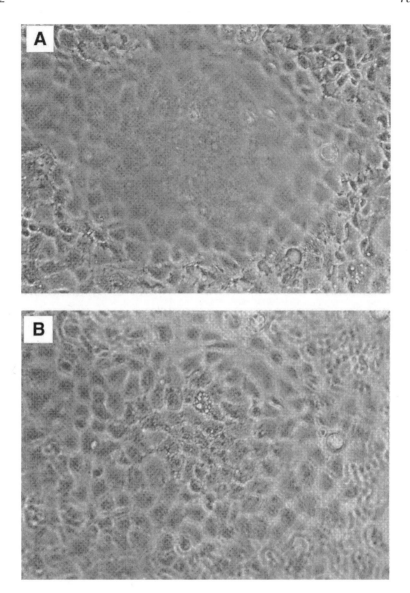

Fig. 1. Dome formation by rabbit kidney proximal tubule cell cultures. A photo-micrograph was taken of a confluent monolayer under an inverted microscope at ×100 magnification. **Panel A** focuses on the cells in the monolayer, whereas **panel B** focuses on the cells in the dome.

Of particular interest to this chapter is the use of serum-free medium to grow primary cultures of rabbit kidney epithelial cells that express renal proximal tubule functions. Rabbit kidney proximal tubules are first purified from the

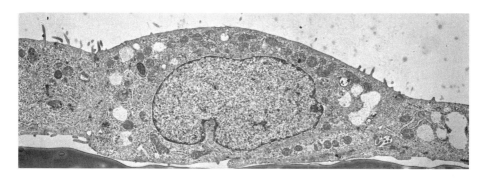

Fig. 2. Transmission electron microscopy (TEM) of primary renal proximal tubule cells. Primary rabbit kidney proximal tubule cells were cultured on a plastic substratum in serum-free medium (as in **Fig. 1**), processed for TEM *(9)*, and photographed at ×4000 magnification.

renal cortex by a modification *(4–6)* of the Brendel and Meezan method *(7,8)* and then placed into tissue culture dishes containing serum-free medium supplemented with three growth supplements: insulin, transferrin, and hydrocortisone. Within the first day of culture, the tubules attach to the culture dish. Subsequently, epithelial cells grow out from the tubule explants. After 1 wk, confluent monolayers are obtained, which possess multicellular domes (*see* **Fig. 1**), indicative of their capacity for transepithelial solute transport from the cells' apical surface (facing the culture medium) through the basolateral surface (facing the culture dish). An examination of the primary kidney epithelial cells by transmission electron microscopy (TEM) (*see* **Fig. 2**) indicates that the cells do indeed possess such a polarized morphology and are interconnected by tight junctions. An examination of the apical surface by scanning electron microscopy (SEM) (*see* **Fig. 3**) also shows the presence of dense clusters of microvilli, typical of the renal proximal tubule. In addition to retaining a polarized morphology, the primary cultures express a number of renal proximal tubule functions, including γ-glutamyl transpeptidase, phosphoenolpyruvate carboxykinase, a sodium–glucose cotransport system, and a *p*-aminohippurate transport system, which are distinctive of renal proximal tubule cells (*see* **Table 1**).

These monolayer cultures can be used for a large number of purposes, ranging from infection with viruses, and subsequent viral production, to transfection with plasmid DNAs containing oncogenes for subsequent cell immortalization. The confluent monolayers are amenable to biochemical studies when cultured on plastic, as well as electrophysiologic studies when cultured on permeable supports. Now that many of the differentiated functions of the renal proximal tubule cells have been defined and appropriate genes have been cloned, the

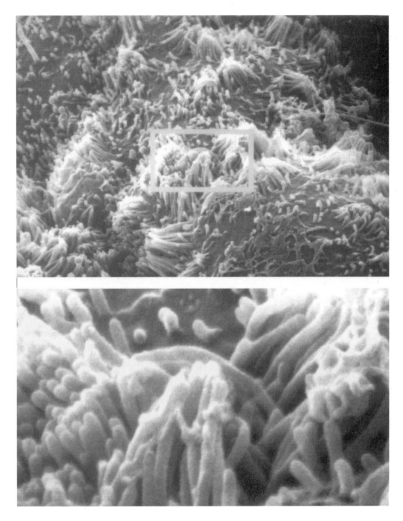

Fig. 3. Scanning electron microscopy (SEM) of primary renal proximal tubule cells. A monolayer of primary rabbit kidney proximal tubule cells was examined by SEM at ×6000 magnification (insert is ×30,000).

primary cultures are amenable for molecular biology studies concerning the control of the expression of differentiated function.

2. Materials

One of the most critical steps involved in the initial setup prior to culturing renal proximal tubule cells includes the preparation of the basal medium (*see* **Note 1**). As the cells are grown serum-free, additional medium supplements

Table 1
Properties of Primary Rabbit Kidney Proximal Tubule Cell Cultures

1. Morphology *(4,10,11)*
   Domes
   Form polarized monolayers
   Adjacent cells form tight junctions
   Brush border (although not as elaborated as in vivo)
2. Transport properties
   Sodium–glucose cotransport system *(4,10)*
   Sodium–phosphate cotransport system *(6)*
   p-Aminohippurate transport system *(12)*
   Amiloride-sensitive sodium transport *(13)*
3. Responses to hormones and other effector molecules
   Parathyroid hormone-sensitive cyclic AMP production *(4)*
   Regulation of phosphate transport by glucocorticoids and estrogens *(14,15)*
   Regulation of sodium transport by angiotensin II *(16)*
4. Enzymes; Metabolic properties
   Leucine aminopeptidase *(4)*
   Alkaline phosphatase *(4)*
   Gamma glutamyl transpeptidase *(4)*
   Glutathione *(17,18)*
   Glutathione-*S*-transferase *(17,19)*
   Angiotensin-converting enzyme *(11)*
   Phosphoenolpyruvate carboxykinase *(20)*
   Aerobic metabolism *(10)*
   Hexose monophosphate shunt *(10)*
   Gluconeogenesis *(21)*
5. Growth properties
   Cell growth in serum-free medium *(4)*
   Growth in response to insulin, transferrin and hydrocortisone *(4,20)*
   Growth in response to estrogens *(15,22)*
   Growth in response to laminin, collagen, and fibronectin
   Growth improved on colloidal silica *(9)*
   Growth in glucose-free medium *(21)*
   Can undergo two passages *(9)*
6. Responsiveness to toxicants
   Mercury toxicity *(17,18)*
   Ifosfamide toxicity *(23,24)*

are required, including insulin, transferrin, and hydrocortisone. In addition, antibiotics can be added to the medium as a preventative against bacterial growth (*see* **Note 2**). Finally, implements required for the dissection of the kidney and purification of renal proximal tubules must be sterilized prior to use.

1. Milli-Q reagent-grade water.
2. Dulbecco's modified Eagle's medium (DMEM)/Ham's F12 (F12) (50:50 mixture). Mix 10 L DMEM (with 4.5 g/L D-glucose and L-glutamine, and without either sodium pyruvate or sodium bicarbonate) and 10 L of Ham's F12 medium to obtain a 50:50 mixture (DMEM/F12).
3. Peristaltic pump.
4. Kanamycin. Sterile aliquots of 1000X concentrated kanamycin (100 mg/mL) are prepared by dissolving 5.0 g of kanamycin in 50 mL Milli-Q water containing 0.425 g NaCl. The solution is sterilized through a 0.22-μm syringe filter, aliquoted into 5-mL sterile polystyrene tubes, and stored at −20°C. Aliquots in use are stored at 4°C.
5. Penicillin. Sterile aliquots of 1000X concentrated penicillin (92,000 IU/mL) are prepared by dissolving 3.05 g penicillin in 50 mL Milli-Q water containing 0.425 g NaCl/50 mL. The solution is sterilized through a syringe filter (0.22 μm), aliquoted into 5-mL sterile polystyrene tubes, and stored at −20°C. Aliquots being used are stored at 4°C.
6. Streptomycin. Sterile aliquots of 1000X concentrated streptomycin (200 μg/mL) are prepared by dissolving 10 g streptomycin in 50 mL Milli-Q water. The solution is sterilized through a syringe filter (0.22 μm) and aliquoted into 5-mL sterile polystyrene tubes. The aliquots are stored at −20°C for up to 1 yr. Aliquots in use are stored at 4°C.
7. Na^+ bicarbonate.
8. HEPES.
9. Basal media. The basal medium consists of DMEM/F12 supplemented with 15 mM HEPES buffer (pH 7.4) and 20 mM sodium bicarbonate (DMEM/F12). The medium is sterilized using a Millipak filter unit with a 0.22-μm pore size. A Millipore peristaltic pump is utilized to pump the medium from a reservoir (a 20-L carboy), through the Millipak filter, into sterile medium bottles. Individual bottles of medium are stored frozen at −20°C, until ready for use. After thawing, the medium is kept at 4°C for up to 2 wk.
10. Antibiotic-supplemented basal media. This medium is used throughout the tubule isolation protocol. It is prepared by adding 92 IU penicillin and 200 μg/mL streptomycin to the basal medium (*see* **Note 2**).
11. Complete growth media. The basal medium is supplemented with 5 μg/mL bovine insulin, 5 μg/mL human transferrin, and 5×10^{-8} M hydrocortisone on the day of use. The sterile stock solutions of 5 mg/mL bovine insulin, 5 mg/mL human transferrin, and 10^{-3} M hydrocortisone are prepared for this purpose. In addition, the growth medium also contains 92 IU penicillin and 0.1 mg/mL kanamycin (*see* **Note 2**).
12. Bovine insulin: Bovine insulin (Sigma, cat. no. I5500) is solubilized in 0.01 N HCl at a concentration of 5 mg/mL, sterilized by passage through a 0.22-μ filter, and distributed into sterile 12, 75-mm polystyrene tubes using 1-mL aliquots. Insulin is kept at 4°C and can be used for up to 1 yr as long as sterility is maintained.

13. Human transferrin. Human apo-transferrin (iron poor, > 97% pure; Sigma, cat. no. T2252) is prepared at a concentration of 5 mg/mL in water, sterilized using a 0.22-μm filter unit, and distributed into sterile polystyrene tubes. Individual aliquots are kept at –20°C. Aliquots of transferrin can be frozen and thawed up to four times.

14. Hydrocortisone: Hydrocortisone (Sigma, cat. no. H4001) is solubilized in 100% ethanol at 10^{-3} M, aliquoted into sterile 5-mL polypropylene tubes, and stored at 4°C for 3–4 m.

15. Phosphate-buffered saline (PBS).

16. Sodium hydroxide (NaOH).

17. Ferrous sulfate.

18. Potassium nitrate.

19. Oxygen-saturated water.

20. Iron oxide solution (0.5% w/v). While perfusing the kidney, a 0.5% iron oxide solution (w/v), prepared as described by Cook and Pickering *(25)* is required for removal of the contaminating glomeruli. The solution is prepared by dissolving 2.6 g sodium hydroxide and 20 g potassium nitrate in 100 mL of oxygen-saturated water. Ferrous sulfate (9 g) is dissolved in 100-mL aliquots of oxygen-saturated water. These two 100-mL solutions are mixed in a flask and boiled for 20 min. The resulting black precipitate (iron oxide) is washed 5–10 times with Milli-Q water. During each wash, the precipitate is first resuspended in water and is then brought down to the bottom of the flask using a strong magnet. The wash water is then decanted away. After the final wash, the iron oxide is resuspended in 1 L of 0.9% NaCl, distributed into 250-mL bottles, and autoclaved. Immediately prior to use, a portion of the iron oxide solution is diluted fourfold in PBS.

21. NaCl.

22. EDTA–trypsin in PBS (Invitrogen Corp.). A trypsin solution is prepared for subculturing. Sterile 0.25% trypsin–1 mM (ethylenedinitrilo)tetraacetic acid (EDTA) in PBS (Invitrogen Corp.), used for the trypsinization of renal proximal tubule cell cultures, is prepared by diluting a 10X concentrate (Invitrogen Corp.) into PBS. Trypsin solutions are filter-sterilized, aliquoted, and frozen at –20°C. Aliquots for immediate use are maintained at 4°C.

23. Soybean trypsin inhibitor (Invitrogen Corp.). A soybean trypsin inhibitor solution is prepared to inhibit proteases in collagenase. In addition, soybean trypsin inhibitor is used to inhibit trypsin action after trypsinization has come to an end (a necessity when using serum-free medium, unlike the case with medium containing serum, which itself inhibits trypsin action). A 0.1% soybean trypsin inhibitor solution is prepared in PBS. Soybean trypsin inhibitor solutions are filter-sterilized, aliquoted, and frozen at –20°C. Aliquots for immediate use are maintained at 4°C.

24. Collagenase Class 4 (162 U/mg; Worthington, cat. no. 4188): A collagenase preparation that permits the outgrowth of cells from nephron segments is prepared in basal medium at 10 mg/mL and filter-sterilized. The collagenase solution is prepared on the day of use (*see* **Note 3**).

25. New Zealand white rabbits, 4–5 lb.
26. 100% CO_2
27. Curved-nose scissors.
28. 100-mm-Diameter glass Petri dishes (2).
29. Heavy suture thread.
30. Kelly hemostat (5-1/2 in. straight end).
31. 20-Gage metal needle (blunt ended with a file).
32. 50-mL Glass syringe.
33. 15-mL Dounce homogenizer (Type A, loose pestle).
34. Metal spatula (9 in.).
35. 2-in. Magnetic stir bar.
36. 1000-mL Beaker.
37. Nylon nitrex screening fabric, both 253 µm and 85 µm (TETCO, Inc., Depew, NY).
38. 4-in. Plastic embroidery hoops (two sets).
39. Sieves are prepared for the purification of renal proximal tubules and glomeruli from a suspension of nephron segments obtained from disrupted renal cortical tissue in the procedure described in **Subheading 3.1.5.** The sieving procedure involves the use of Nylon nitrex screening fabric (both 253 µm and 85 µm) cut to a size permitting them to be held in place in a 4-in.-diameter plastic embroidery hoop.
40. Disposable tubes: 50 mL polypropylene; 5 mL polystyrene; 5 mL polypropylene.
41. Tissue culture dishes (35 mm in diameter).
42. Transwells (Corning, cat. no. 3450-Clear), 24 mm in diameter, 0.4 µm pore size.
43. Filtration apparatuses (0.22 µm).

3. Methods

The methods described outline (1) the purification of renal proximal tubule cells and (2) the culturing of the primary cells.

3.1. Purification of Renal Proximal Tubule Cells

Immediately prior to the cell culture procedure, implements must be sterilized. A sterile collagenase solution must be prepared and growth factors must be added to the basal medium. Subsequently, renal proximal tubules are purified as outlined here for immediate use for primary cell cultures.

3.1.1. Preparation of Medium, Reagents, and Implements Prior to Primary Culture

1. Implements to be used for primary renal proximal tubule cell culture are wrapped appropriately in aluminum foil and sterilized in an autoclave (1 h at 135°C, 35 psi in a Castle autoclave). Included among the implements are two sets of forceps, curved-nose scissors, a hemostat, the blunt-ended needle (18 gage), a metal spatula, suture, two magnetic stir bars (2 in. long), a 1000-mL beaker, two 100-mm-diameter glass Petri dishes, a 15-mL Dounce tissue homogenizer with a loose pestle (type A), and a 50-mL glass syringe (*see* **Note 4**).

2. The Nylon mesh placed within embroidery hoops is sterilized by prior soaking (overnight) in a 2-L polypropylene beaker containing 95% ethanol. One 253-μm mesh and one 85-μm mesh are used per kidney. If both kidneys are utilized for culturing, then two 253-μm meshes and two 85-μm meshes should be prepared (*see* **Note 4**).
3. Prepare the sterile collagenase solution (*see* Materials, item 24).
4. Prepare the antibiotic-supplemented basal media and the complete growth medium for isolation and subsequent culture, respectively, of the tubules.

3.1.2. Initial Dissection of Kidneys

1. A rabbit is sacrificed in a container filled with 100% CO_2 (*see* **Note 4**).
2. Prior to the removal of each kidney, the ureter is removed. The renal artery and vein are separated with forceps, so as to facilitate the insertion of the needle into the renal artery, for perfusion of the kidney (as described in **Subheading 3.1.3.**).
3. Each kidney is then removed (with the renal artery and vein intact) using sterile scissors and placed in a sterile 50-mL conical tube containing ice-cold antibiotic-supplemented basal medium. The tubes containing the left and right arteries are labeled, as the length of the artery varies with the kidney (the left kidney has the longer artery, which can be somewhat easier to insert the needle for perfusion, as described in **Subheading 3.1.3.**).
4. The kidneys are kept ice cold prior to perfusion, by placing the 50-mL tubes in ice until the kidneys are used for culturing.

3.1.3. Perfusion of Kidneys

1. Each kidney is placed in a 100-mm-diameter glass Petri dish and washed with ice-cold antibiotic-supplemented basal medium. A sterile, blunt-ended needle (18 gage) is inserted into the renal artery, and the artery is sutured. A hemostat is also used to keep the needle in place in the renal artery.
2. A 50-mL syringe is then connected to the needle, and the kidney is perfused with 30–40 mL of ice-cold PBS (to remove the blood).
3. After the blood is removed (the kidney becomes blanche in color), the kidney is perfused with 30–40 mL of PBS containing iron oxide (such that the kidney becomes gray–black in color). For this purpose, the iron oxide is diluted fourfold in PBS.

3.1.4. Removal of Renal Cortex and Homogenization

1. To remove the renal capsule, the kidney is grasped with two pairs of sterile forceps (4-1/2-in. straight, toothed-end). One of the forceps is used to pierce the renal capsule and gradually peel it off the kidney. Using the remaining forceps, the kidney is immediately transferred into another sterile 100-mm glass Petri dish containing 1–2 mL of ice-cold antibiotic-supplemented basal medium. At this point, great care must be taken with regard to sterility.
2. Slices of the renal cortex are removed from the kidney using a sterile, curve-nosed scissors (*see* **Note 5**).

3. The slices of the renal cortex are transferred into a sterile 15-mL Dounce homogenizer containing antibiotic-supplemented basal medium. Although ice-cold
medium is preferred at this point, successful cultures can still be obtained after
carrying out the procedure at room temperature. The tissue is disrupted with four
to five strokes of a loose pestle (type A). After tissue disruption, the homogenizer
can be covered with the lid of a sterile tissue culture dish, to maintain sterility
while setting up the sieves for the next purification step (*see* **Note 6**).

3.1.5. Purification of Renal Proximal Tubules and Glomeruli by Sieving

1. The nephron segments are separated using the Nylon mesh sieves. Toward this
end, a 85-μm sieve is placed over a sterile 1000-mL beaker, sitting directly on the
mouth of the beaker. Then, the wider 253-μm sieve is placed over the 85-μm sieve.
2. The sieves are washed with 100–200 mL of antibiotic-supplemented basal
medium (DMEM/F12) to remove the ethanol while maintaining sterility.
3. The suspension of disrupted tissue in the Dounce homogenizer (which contains
tubule segments and glomeruli) is poured over the top sieve. Subsequently,
700–800 mL of antibiotic-supplemented basal medium is slowly poured over the
sieves. During this process, care is taken so as to wash the tubules and glomeruli
through appropriate sieves. Undisrupted material remains on the top of the first
sieve 253 μm. Proximal tubules and glomeruli collect on the top of the second
sieve 85 μm, because of their large diameter. Narrower tubule segments and
debris pass into the beaker.
4. The tubules and glomeruli are removed from the top of the 85-μm sieve with a
sterile metal spatula (preferably with a rounded end) and are transferred into a
sterile 50-mL plastic conical tube containing 40 mL antibiotic-supplemented
basal medium.

3.1.6. Removal of Glomeruli and Disruption of the Basement Membrane

1. The glomeruli in the tubule suspension are removed by placing a sterile magnetic
stir bar in the 50-mL conical tube (*see* **Note 7**). The glomeruli (which are covered
with iron oxide) are attracted to the stir bar. Then, the tubule suspension is carefully poured from the conical tube (without the stir bar) into another 50-mL conical tube.
2. In order to disrupt the basement membrane, soybean trypsin inhibitor (0.1 mL of
a 0.1% solution) followed by collagenase (0.2 mL of a 10 mg/mL solution) is
added to the tubule suspension (to obtain final concentrations of 0.0025 and
0.050 mg/mL, respectively). The tubules are incubated with collagenase for
2 min at 23°C (*see* **Note 8**).
3. The collagenase treatment is stopped by centrifugation of the tube containing the
tubules in a desktop centrifuge for 5 min at 500 rpm ($21g$).
4. The pellet containing the renal proximal tubules is resuspended in 40 mL antibiotic-supplemented basal medium and centrifuged once again for 5 min at
500 rpm ($21g$).

3.2. Primary Renal Proximal Tubule Cell Cultures

As many as 60 confluent monolayers in 35-mm culture dishes can be obtained from the purified renal proximal tubules, following the procedure outlined here (*see* **Note 9**).

3.2.1. Plating of Renal Proximal Tubules for Monolayer Cell Culture

1. After centrifugation, the renal proximal tubules are suspended in 30 mL of complete growth medium in order to obtain 60 confluent monolayers in 35-mm culture dishes.
2. Prior to adding the purified renal proximal tubules to the 35-mm culture dishes, 1.5 mL of complete growth medium is added to each dish.
3. The tubule suspension is inoculated into the media-containing 35-mm diameter tissue culture dishes at 0.5 mL/dish using a 5-mL pipet (*see* **Note 10**).
4. The culture dishes are placed in a 5% CO_2/95% air humidified environment at 37°C.
5. The culture medium is changed the day after plating the tubules (to remove debris and unattached nephron segments). The medium is changed routinely every 2 d thereafter. Initially, cells grow out from the nephron segments that attach to the culture dish surface. Subsequently, monolayers form and can become confluent following 6–7 d in culture (*see* **Notes 11–14**).

3.2.2. Plating of Proximal Tubules Into Transwells

As an alternative to preparing monolayer cultures on plastic dishes, primary cultures can also be prepared on transwells or other semipermeable supports. Attachment of tubules and growth to confluence can be obtained on 3450-Clear transwells (24 mm in diameter; 0.4 μm pore size), which also permit visualization of the monolayers through an inverted microscope. Growth of tubules on transwells permits studies requiring accessibility to either the basolateral or the apical membrane.

1. After centrifugation, the tubules are suspended in 24 mL complete growth medium
2. Prior to the addition of the tubules to the culture dishes, 2.6 mL of complete growth medium is added to the bottom of each of the six wells present within a plate containing transwells.
3. Tubules are added to each well of the transwells in a total volume of 1.5 mL, taking care to have an equal distribution of material in each transwell (*see* **Note 15**). After 2 d in culture, medium is changed every other day.

3.2.3. Passaging of Primary Cultures

Primary rabbit kidney proximal tubule cells on plastic dishes can be subcultured using EDTA–trypsin. Confluent first-passage cultures can be obtained if care is taken to minimize the trypsinization period. In some cases, proximal tubule monolayers can even be obtained following a second passage into plastic dishes.

1. In order to obtain first-passage cells, the culture medium is removed by aspiration and the cells are washed with PBS.
2. A solution of EDTA–trypsin is then added. The majority of the EDTA–trypsin solution is immediately removed, so that only a film of trypsin covers the cells, providing a gentler trypsinization.
3. The cells are transferred to a 37°C incubator for a short time (as short as 1 min) and examined under an inverted microscope at ×100 magnification to determine whether the cells have detached from the dish surface. Trypsin action is stopped by the addition of an equimolar concentration of soybean trypsin inhibitor (0.5 mL/dish). Basal medium is added to bring the volume to 5 mL (*see* **Note 16**).
4. The cells are removed from the dish into a 12-mL plastic centrifuge tube, followed by centrifugation for 5 min at 500 rpm (21*g*). The cells are resuspended in complete growth medium and inoculated into plastic tissue culture dishes. Confluent monolayers may be obtained following a 1:1, a 1:2, or a 1:4 passage (*see* **Notes 16** and **17**).

4. Notes

1. Water to be used for the preparation of sterile medium and other sterile reagents is purified with a Milli-Q reagent-grade water system as previously described *(4,5)*. The feed water for the Milli-Q reagent-grade water system is obtained from a Millipore reverse osmosis system. A separate set of glassware, bottles, and stir bars is utilized for the preparation of medium, as is the case with all other tissue culture solutions. Glassware and other implements are to be washed using a phosphate-free detergent such as 7X (ICN Biomedical, Costa Mesa, CA).
2. The medium used during the purification of the renal proximal tubules contains both penicillin and streptomycin, to kill micro-organisms associated with the kidney preparation. However, the medium used for the growth and maintenance of primary renal proximal tubule cells should not be supplemented with streptomycin, which is a nephrotoxin. Streptomycin apparently has no deleterious effects during the 1- to 2-h period during which the proximal tubules are being purified and two antibiotics in combination are more effective than only one to prevent the growth of micro-organisms. However, our culture results indicate that over more extended incubation periods at 37°C nephrotoxic effects of streptomycin are indeed elicited. Streptomycin not only impedes the initial attachment of nephron segments to the culture dish but also affects the initial outgrowth of cells from the nephron segments. If penicillin is to be added, a final concentration of 0.92×10^5 IU/L is to be employed. We have successfully cultured primary proximal tubule cells in medium that is completely antibiotic-free. As a second antibiotic, kanamycin can be added at a final concentration of 100 μg/mL.

3. Because of lot-to-lot differences, each lot of collagenase should be tested to evaluate whether a particular lot promotes the outgrowth of cells from renal proximal tubules or has deleterious effects.

4. If proximal tubule cells are to be isolated from a different animal than the rabbit, then several different implements would be required. If a mouse or a rat kidney is to be used, then the kidneys of these animal would be perfused *in situ*, using a sterile Terumo Surflo winged infusion set (19G × 3/4 in. needle and 12-in. tubing). In addition, meshes that differ in size from that utilized with the rabbit kidney would need to be identified. The use of iron oxide to perfuse was originally described by Meezan et al. *(8)* for the isolation of glomeruli from rat kidneys.

5. The medulla can be distinguished from the cortex after perfusion with iron oxide, as only the cortex becomes gray–black in color following this procedure. Care should be taken not to remove slices of the medulla, as during subsequent processing, the medullary slices cause difficulty in the homogenization step and also result in the presence of a large quantity of debris in the final cell cultures (which might be prohibitive to cell growth).

6. During homogenization of the renal cortex, renal tubule fragments are released. Because the renal proximal tubules and glomeruli are wider in diameter than distal tubules and loops of Henle, the proximal tubules and glomeruli do not pass through the second sieve (85-μm) used during the sieving process. Thus, a purified preparation of renal proximal tubules and glomeruli is obtained after sieving.

7. Iron oxide is used to remove glomeruli from the preparation of purified proximal tubules and glomeruli. After perfusion of the iron oxide through the renal artery, the iron particles become associated with glomeruli, but are too large to pass through the glomeruli and enter the lumen of the nephron. After homogenization of the kidney, a suspension of purified nephron segments is obtained. The suspension is further enriched with renal proximal tubules and glomeruli following the sieving procedure (as described in **Subheading 3.1.5.**). The glomeruli in this suspension are selectively attracted to a magnetic stir bar. The glomeruli are then removed with the magnet, leaving a suspension of purified proximal tubules. If a substantial number of glomeruli with iron oxide remain after removing the magnetic stir bar, a second sterile magnetic stir bar can be added to the suspension of renal proximal tubules and, once again, the second magnet can be removed. However, care must be taken not to lose tubules during this process.

8. Empirically, we have observed that the use of collagenase is required if the outgrowth of epithelial cells from the tubules is to occur in vitro. Soybean trypsin inhibitor is used to inhibit proteases in collagenase. In addition, soybean trypsin inhibitor inhibits trypsin action after cells in treated cultures have detached from the culture dish. This is a necessity when using serum-free medium. In contrast, serum-supplemented medium contains serum components, which inhibit trypsin action.

9. During the process of pipetting the proximal tubules into culture dishes, care should be taken to keep the tubules in suspension, as they are denser than isolated cells in suspension and could rapidly settle even while pipetting. Thus, the tubule suspension should be shaken to resuspend after each pipetting. Alternatively, to maintain uniformity, the tubule suspension can be placed in a sterile bottle with a sterile stir bar and continuously stirred at a medium speed (a setting of 3–4) using a stirring motor. These steps should be carried out in a tissue culture hood.

    Tubules can be plated in larger-diameter dishes, including 60-mm-diameter dishes (30 dishes/kidney preparation), as well as 100-mm-diameter dishes (20 dishes/kidney preparation). Cultures prepared in this manner are particularly useful for such applications as Northern analysis and enzyme activity measurements. For growth studies (where low plating densities are desired), up to 200 mL of medium can be used to suspend the renal proximal tubules obtained from a single rabbit kidney and, thus, many more cultures in 35-mm dishes can be obtained. Ultimately, confluent monolayers can even be obtained from such cultures plated at lower densities.

10. If primary rabbit kidney proximal tubule cell cultures do not grow to confluence, a number of problems might have occurred. First, enough tubules must be added to the culture dishes in order to obtain confluent monolayers (If desired the protein content, approx 0.5 mg protein/mL tubule suspension) can be determined by the Bradford method *(26)* immediately after obtaining the suspension of purified renal proximal tubules). The tubules can easily be lost if they are not carefully harvested at the end of the sieving procedure.

11. Second, cell cultures obtained from kidneys of young adults (as opposed to older animals) are the most successful. A third point of concern is the tissue culture medium. The purity of the water is critical in defined medium studies. Loss of purity because of contamination (from a dirty pH probe, for example) could result in medium that does not support cell growth. Our laboratory determines pH using samples of the medium rather than placing the probe in the medium to be used for tissue culture studies. In addition, a set of glassware is used in medium preparation that is specifically designated for that purpose.

12. Another point of caution is the hormone supplements. Improper preparation or storage of the growth supplements might be deleterious to cell growth. The growth stimulatory effect of insulin might be lost if the stock solution is frozen. Furthermore, the medium supplements should be added to the medium immediately prior to use for tissue culture, as these supplements are not necessarily stable in the tissue culture medium.

13. Care should be taken that the incubator be maintained at a constant temperature of 37°C and in a constant 5% CO_2/95% air environment. Animal cell growth in the absence of serum is more sensitive to shifts in temperature and to changes in the medium pH than in the presence of serum. The addition of HEPES buffer to the medium alleviates this latter problem to some extent.

14. Finally, the primary rabbit kidney proximal tubule cells are less adherent to plastic dishes than many other cell types. Thus, the cells might detach during their manipulation for cell growth studies or for transport studies (for example). The problem of adhesion might be alleviated by growing the cultures on tissue culture dishes coated with basement membrane components such as laminin, or Type IV collagen, with Matrigel, or with such biomaterials as silica (**9**). Renal proximal tubule cell cultures on plastic cell culture dishes can readily be passaged once. Although one or two additional passages can be obtained, confluence is obtained with difficulty. Proximal tubule cell cultures grow more rapidly and achieve a higher saturation density and a higher passage number on laminin-coated dishes.

15. Care must be taken at this step to evenly distribute the tubules throughout the surface of each transwell (otherwise confluent monolayers that completely cover the transwell will not be obtained). Chopstick electrodes (Millipore Corp.), or another similar apparatus, are used to test the transepithelial resistance when confluence is indeed achieved, so as to determine if an electrically tight monolayer has been obtained.

16. Trypsinization can also be conducted at room temperature. The cells are loosely attached to the bottom of the culture dish and are readily detached by an incubation with trypsin as short as 5 min. Trypsinization at room temperature is gentler to the cells and does not result in the type of cell damage and death that can occur with these delicate cultures at 37°C.

17. When passaging primary cultures on a 1:4 basis, we have been able to reproducibly obtain confluent first-passage cultures within 1 wk. Similarly, second-passage cultures are readily obtained within 1 wk when passaging on a 1:4 basis. However, following the third passage, cultures do not readily undergo many cell divisions. Thus, in order to obtain confluent monolayers after more than two passages, either more cells must be used originating from larger dishes (e.g., passaging 1:4 from a 100-mm dish into a 35-mm dish) and/or lower passage ratios can be utilized (i.e., 1:1 or 1:2). Other options that might permit a greater success during passages include the addition of growth factors (such as epidermal growth factor, 10 ng/mL, or fibroblast growth factor, 50 ng/mL) to the culture medium, as well as coating culture dishes with such matrix components as laminin or collagen IV.

Acknowledgments

Dr. Thaddeus Szczesny of the State University of New York at Buffalo is thanked for his preparation of transmission electron micrographs and Dr. Peter Bush of the State University of New York at Buffalo is thanked for his assistance in scanning electron microscopy. This work was supported by NHLBI grant no. 1RO1HL69676-01 to MT.

References

1. Taub, M., Chuman, L., Saier, M. H., Jr., and Sato, G. (1979) Growth of Madin Darby Canine Kidney epithelial cell (MDCK) line in hormone-supplemented serum-free medium. *Proc. Natl. Acad. Sci. USA* **76,** 3338–3342.
2. Chuman, L., Fine, L. G., Cohen, A. I., and Saier, M. H., Jr. (1982) Continuous growth of proximal tubular epithelial cells in hormone-supplemented serum-free medium. *J. Cell Biol.* **94,** 506–510.
3. Taub, M. and Sato, G. (1979) Growth of functional primary cultures of kidney epithelial cells in defined medium. *J. Cell. Physiol.* **105,** 369–378.
4. Chung, S. D., Alavi, N., Livingston, D., Hiller, S., and Taub, M. (1982) Characterization of primary rabbit kidney cultures that express proximal tubule functions in a hormonally defined medium. *J. Cell Biol.* **95,** 118–126.
5. Taub, M. (1985) Primary culture of proximal tubule cells in defined medium. *J. Tissue Culture Methods* **9,** 67–72.
6. Waqar, M. A., Seto, J., Chung, S. D., Hiller-Grohol, S., and Taub, M. (1985) Phosphate uptake by primary renal proximal tubule cells grown in hormonally defined medium. *J. Cell. Physiol.* **124,** 411–423.
7. Brendel, K. and Meezan, E. (1975) Isolation and properties of a pure preparation of proximal kidney tubules obtained without collagenase treatment. *Fed. Proc.* **34,** 803.
8. Meezan, E. K., Brendel, J., Ulreich, J., and Carlson, E. C. (1973) Properties of a pure metabolically active glomerular preparation from rat kidneys. I. Isolation. *J. Pharm. Exp. Ther.* **187,** 332–341.
9. Taub, M., Axelson, E., and Park, J. H. (1998) Improved method for primary cell cultures: use of tissue culture dishes with a colloidal silica surface. *Biotechniques* **25,** 990–994.
10. Sakhrani, L. M., Badie-Dezfooly, B., Trizna, W., et al. (1984) Transport and metabolism of glucose by renal proximal tubular cells in primary culture. *Am. J. Physiol.* **246,** F757–F764.
11. Matsuo, S., Fukatsu, A., Taub, M. L., Caldwell, P. R. B., Brentjens, J. R., and Andres, G. (1987) Nephrotoxic glomerulonephritis induced in the rabbit by antiendothelial antibodies. *J. Clin. Invest.* **79,** 1798–1811.
12. Yang, I. S., Goldinger, J. M., Hong, S. K., and Taub, M. (1987) The preparation of basolateral membranes that transport *p*-aminohippurate from primary cultures of rabbit kidney proximal tubule cells. *J. Cell. Physiol.* **135,** 481–487.
13. Fine, L. G. and Sakhrani, L. M. (1986) Proximal tubular cells in primary culture. *Miner. Electrolyte Metab.* **12,** 51–57.
14. Park, S. H., Taub, M., and Han, H. J. (2001) Regulation of phosphate uptake in primary cultured rabbit renal proximal tubule cells by glucocorticoids: evidence for nongenomic as well as genomic mechanisms. *Endocrinology* **142,** 710–720.
15. Han, H. J., Lee, Y. H., Park, S. H., and Taub, M. (2002) Estradiol-17β stimulates phosphate uptake and is mitogenic for primary rabbit renal proximal tubule cells. *Exp. Nephrol.* **10,** 355–364.

16. Han, H. J., Park, S. H., Koh, H. J., and Taub, M. (2000) Mechanism of regulation of Na⁺ transport by angiotensin II in primary renal cells. *Kidney Int.* **57,** 2457–2467.

17. Aleo, M. D., Taub, M. L., Olson, J. R., and Kostyniak, P. J. (1990) Primary cultures of rabbit renal proximal tubule cells: II. Selected phase I and phase II metabolic capacities. *Toxicol. In Vitro* **4,** 727–733.

18. Aleo, M. D., Taub, M. L., and Kostyniak, P. J. (1992) Primary cultures of rabbit renal proximal tubule cells: III. Comparative cytotoxicity of inorganic and organic mercury. *J. Toxicol. Appl. Pharm.* **112,** 310–317.

19. Aleo, M. D., Taub, M. L., Olson, J. R., Nickerson, P. A., and Kostyniak, P. J. (1987) Primary cultures of rabbit renal proximal tubule cells as an in vitro model of nephrotoxicity: effects of 2 mercurials, in *In vitro Toxicology: Approaches to Validation* (Goldberg, A. M., ed.), Mary Ann Liebert, Inc., New York, Vol. 5, pp. 211–225.

20. Wang, Y. and Taub, M. (1991) Insulin and other regulatory factors modulate the growth and the phosphoenolpyruvate carboxykinase (PEPCK) activity of primary rabbit kidney proximal tubule cells in serum free medium. *J. Cell. Physiol.* **147,** 374–382.

21. Jung, J. C., Lee, S. M., Kadakia, N., and Taub, M. (1992) Growth and function of primary rabbit kidney proximal tubule cells in glucose-free serum-free medium. *J. Cell. Physiol.* **150,** 243–250.

22. Han, H. J., Jung, J. C., and Taub, M. (1999) Response of primary rabbit kidney proximal tubule cells to estrogens. *J. Cell. Physiol.* **178,** 35–43.

23. Springate, J., Davies, S., Chen, K., and Taub, M. (1999) Toxicity of ifosfamide and its metabolite chloroacetaldehyde in cultured renal tubule cells. *In Vitro Cell Dev. Biol.* **35,** 314–317.

24. Zaki, E. L., Springate, J. E., and Taub, M. (2003) Comparative toxicity of ifosfamide metabolites and protective effect of mesna and amifostine in cultured renal tubule cells. *Toxicol. In Vitro* **17,** 397–402.

25. Cook, W. F. and Pickering, G. W. (1958) A rapid method for separating glomeruli from rabbit kidney. *Nature* **182,** 1103–1104.

26. Bradford, M. M. (1976) A rapid and sensitive method for the quantitation of microgram quantities of protein utilizing the principle of protein-dye binding. *Anal. Biochem.* **72,** 248–254.

17

Enzymatic Dissociation and Culture of Normal Human Mammary Tissue to Detect Progenitor Activity

John Stingl, Joanne T. Emerman, and Connie J. Eaves

Summary

Normal human mammary tissue is composed of a glandular epithelium embedded within a fibrous and fatty stroma. Collagenase and hyaluronidase digestion of normal reduction mammoplasty specimens followed by differential centrifugation yields a suspension of single cells and cell aggregates that contain elements of the terminal ductal lobular units and stromal components of the mammary gland. The terminal ductal lobular units (TDLU) can be further dissociated to complete viable single-cell suspensions by treatment with trypsin, dispase II, and deoxyribonuclease I. These suspensions are suitable for cell separation and analysis in culture. Such studies indicate the existence of biologically distinct subpopulations of luminal-restricted, myoepithelial-restricted, and bipotent mammary epithelial progenitors detected by their ability to generate colonies of the corresponding progeny types in serum-free cultures. This review summarizes the methodology of the techniques required to generate and characterize the colonies obtained in vitro from these progenitors, as well as the special considerations and potential pitfalls associated with performing these protocols.

Key Words: Human mammary epithelial cells; cell culture; colony assays; tissue dissociation; breast cancer.

1. Introduction

The human mammary gland is a compound tubulo-alveolar gland composed of a series of branched ducts that drain saclike alveoli. These ducts and alveoli are composed of two general lineages of epithelial cells: the cells that line the lumen of the ducts and alveoli and an underlying smooth-muscle-like myoepithelial cell population. Recent reports have demonstrated the presence of phenotypically distinct progenitor and nonprogenitor cell subpopulations within

From: *Methods in Molecular Biology, vol. 290: Basic Cell Culture Protocols, Third Edition*
Edited by: C. D. Helgason and C. L. Miller © Humana Press Inc., Totowa, NJ

the mammary epithelium *(1–6)*. The progenitor types that can be isolated include the luminal-restricted progenitor, the myoepithelial-restricted progenitor, and the bipotent progenitor, with the latter generating mixed colonies of luminal and myoepithelial cells in vitro. These progenitor populations can be isolated from normal human mammary mammoplasties by an initial collagenase and hyaluronidase digestion to yield intact terminal ductal lobular units (TDLUs). Further enzymatic digestion yields single-cell suspensions suitable for cell separation and colony assays. The growth and differentiation properties of the populations thus obtained can then be evaluated using both clonal and nonclonal culture systems. This chapter describes the methods to (1) obtain single-cell suspensions from surgically excised normal human mammary tissue, (2) propagate mammary epithelial cells at limiting as well as nonlimiting cell densities, and (3) phenotypically characterize cultured normal human mammary epithelial cells and their progenitors.

2. Materials

1. Human reduction mammoplasty specimens obtained from normal donors with informed consent.
2. Dulbecco's modified Eagle's medium (1000 mg glucose/L)/Nutrient Mixture F12 Ham (DMEM/F12); 1:1 (v:v) (StemCell Technologies, Vancouver, BC, Canada) supplemented with 10 mM HEPES [H (Sigma Chemical Co., St. Louis, MO)] and 5% fetal bovine serum (FBS) (Gibco Laboratories, Grand Island, NY).
3. Sterile glass Petri dishes.
4. Scalpels.
5. Tissue dissociation flasks (StemCell Technologies; Bellco Glass, Inc., Vineland, NJ).
6. EpiCult-B™ human mammary epithelial cell culture medium (StemCell Technologies). Stable for at least 1 yr when stored according to manufacturer's instructions. Stable for 2 wk at 4°C once reconstituted.
7. Collagenase/hyaluronidase enzyme dissociation mixture (10X concentrated stock; StemCell Technologies). Stable for at least 1 yr when stored at –20°C. Avoid repeated freezing and thawing.
8. Dimethyl sulfoxide (DMSO) (Sigma).
9. Hank's balanced salt solution (HBSS) modified (StemCell Technologies) supplemented with 2% FBS.
10. 0.25% Porcine trypsin and 1.0 mM ethylenediaminetetraacetic acid (EDTA)·4Na in HBSS (Ca^{2+}- and Mg^{2+}-free) (StemCell Technologies). Stable for 1 yr when stored at –20°C. Avoid repeated freezing and thawing.
11. 5 mg/mL Dispase in Hank's balanced salt solution modified (StemCell Technologies). Stable for at least 1 yr when stored at –20°. Avoid repeated freezing and thawing.
12. 1.0 mg/mL Deoxyribonuclease I (DNAse I) (StemCell Technologies; Sigma). Stable for at least 1 yr when stored at –20°C. Avoid repeated freezing and thawing.

13. 40-μm Cell strainer (StemCell Technologies; Becton Dickinson Labware, Franklin Lakes, NJ).
14. 0.8% Ammonium chloride cell lysis solution (StemCell Technologies). Stable for 1 yr at −20°C or 2 mo at 4°C.
15. NIH 3T3 mouse embryonic fibroblasts (American Type Tissue Culture Collection [ATCC], Manassas, VA).
16. Dulbecco's modified Eagle's medium (4500 mg glucose/L; StemCell Technologies) supplemented with 5% FBS.
17. Acetone:methanol (1:1).
18. Wright's Giemsa stain (Fisher Scientific, Vancouver, BC, Canada).
19. Ovine prolactin (Sigma). Store according to manufacturer's instructions.
20. Matrigel (Becton Dickinson Biosciences Discovery Labware). Store according to manufacturer's instructions.
21. Murine collagen IV (Becton Dickinson Biosciences Discovery Labware, Bedford, MA). Store according to manufacturer's instructions.
22. Rat tail collagen type I (Becton Dickinson Biosciences Discovery Labware). Store according to manufacturer's instructions.

3. Methods

The methods described outline (1) a sequential enzymatic dissociation procedure for obtaining single-cell suspensions from surgically excised normal human mammary tissue, (2) a two-dimensional liquid culture procedure for the selective propagation of mammary epithelial cells at limiting as well as nonlimiting cell densities, and (3) the phenotypic characterization of cultured normal human mammary epithelial cells and their progenitors. Although the methods described pertain specifically to human tissue obtained from reduction mammoplasties, many are also relevant to tumor biopsy and mastectomy samples. However, these do not favor the isolation of malignant populations when these are present and, in fact, typically select for the outgrowth of residual normal cells in culture (*see* **Note 1**).

3.1. Dissociation of Human Mammary Tissue

We have found that the best yield of human mammary progenitors is obtained when a two-step enzymatic dissociation of human mammary tissue is adopted. The first involves incubating the tissue in collagenase and hyaluronidase, which allows epithelial organoids (TDLUs) and stromal cells to be liberated from the tissue samples. The second step is to incubate the mammary organoids thus obtained in enzymes that allow their digestion into a single-cell suspension.

3.1.1. Collagenase and Hyaluronidase Digestion
of Human Mammary Tissue

1. Normal tissue from reduction mammoplasties is best transported from the operating room on ice in sterile specimen cups in DMEM/F12/H supplemented with

DAY 1: Dissociate tissue in dissociation flask for 16 hours

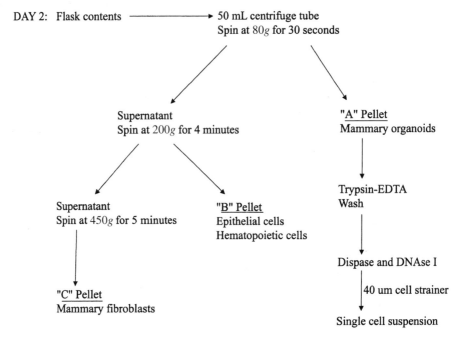

DAY 2: Flask contents ─────────→ 50 mL centrifuge tube
                                  Spin at 80*g* for 30 seconds

Supernatant
Spin at 200*g* for 4 minutes

"A" Pellet
Mammary organoids

Supernatant
Spin at 450*g* for 5 minutes

"B" Pellet
Epithelial cells
Hematopoietic cells

Trypsin-EDTA
Wash

"C" Pellet
Mammary fibroblasts

Dispase and DNAse I

40 um cell strainer

Single cell suspension

Fig. 1. Flow diagram demonstrating the differential centrifugation steps to isolate human mammary organoids and human mammary fibroblasts.

5% FBS. The size of the specimen container will depend on the amount of tissue. Upon delivery to the laboratory, the tissue is transferred in a vertical laminar airflow hood to sterile Petri dishes and minced with scalpels. Large lobes of fat can be trimmed at this time; however, it is not necessary to remove all the fat because this will liquefy during the dissociation process. In fact, it is better to be conservative in trimming away fat to minimize potential losses of parenchymal tissue.

2. After mincing, the tissue is suspended with EpiCult-B supplemented with 10% collagenase and hyaluronidase enzyme mixture and placed in sterile dissociation flasks. The final concentration of collagenase and hyaluronidase is 300 U/mL and 100 U/mL, respectively. The total volume of the suspended tissue in the dissociation flask should not exceed the widest portion of the flask. Multiple flasks should be used, as dictated by the amount of tissue to be dissociated. The dissociation flasks are then sealed with sterile aluminum foil and placed on a rotary shaker in a 37°C incubator. If the rotary shaker is not in a 5% CO_2 equilibrated incubator, the flask should also be sealed with Parafilm® to prevent the dissociation mixture from becoming too alkaline.

3. The tissue should be kept on the rotary shaker for approx 16–18 h (overnight); however longer dissociation times might be required for tough fibrous samples. Tissue dissociation is complete when the bulk of the cell suspension can be drawn through the bore of a 10-mL plastic serological pipet. When handling human mammary epithelial cells, glass pipets should be avoided unless it is siliconized glass. Fragments of tissue that have not undergone complete digestion should be discarded, or, if sufficiently numerous, these can be allowed to settle and collected for a second round of digestion with collagenase and hyaluronidase.

4. The dissociated organoids and stromal cell-enriched preparations are then enriched by differential centrifugation (*see* **Fig. 1**). Briefly, the dissociated tissue should be transferred to 50-mL centrifuge tubes and centrifuged for 30 s at 80*g*. After removal of the overlying liquefied fat layer, the pellet (the "A" pellet) is highly enriched for epithelial organoids. If the supernatant is transferred to a new 50-mL centrifuge tube and centrifuged at 200*g* for 4 min, a second pellet (the "B" pellet) is obtained that contains variable numbers of epithelial cells, stromal cells, and red blood cells. The supernatant from this second centrifugation is particularly enriched for human mammary fibroblasts (and their precursors). These latter cells can be collected by transferring the supernatant to a third 50-mL centrifuge tube and harvesting the pellet obtained after centrifugation at 450*g* for 5 min.

5. These different fractions of cells can then be cryopreserved with high-viability postthaw. We recommend cryopreserving the cells in EpiCult-B supplemented with 50% FBS and 6% DMSO (*see* **Note 2**).

3.1.2. Generation of Single-Cell Suspensions From Human Mammary Organoids

Mammary organoids ("A" pellets) are the best source of human mammary epithelial cell progenitors. These can also be obtained from "B" pellets, but success rates are more variable because of the variable epithelial cell content of this fraction. Mammary organoids should first be further dissociated into a near-single-cell suspension, otherwise the cells will not adhere well to the tissue culture flask. Typical yields of cells from dissociated human mammary tissue "A" pellets using the protocols outlined here is approx $(5–20) \times 10^6$ cells/g tissue.

1. Cryovials containing the mammary organoids should be rapidly thawed at 37°C and the contents transferred to a 50-mL centrifuge tube. The suspension should then be slowly diluted with cold Hank's balanced salt solution modified supplemented with 2% FBS (now referred to as HF). Use 10 mL of HF for every 1 mL of cryopreserved material thawed. The suspension should then be centrifuged at 450*g* for 5 min and the supernatant discarded.

2. Add 1–5 mL of prewarmed trypsin–EDTA to the organoid pellet such that the organoids are well suspended. Gently pipet with a P1000 for 2–5 min. The sample should become very stringy as a result of lysis of dead cells and the release of

DNA. Add 10 mL of cold HF and spin at 450g for 5 min. Remove as much of the supernatant as possible, but do not decant because the pellet will become dislodged and will be eluted with the supernatant. The pellet could be a large viscous mass at the bottom of the tube.

3. Several milliliters (i.e., 2–5 mL) of prewarmed (37°C) dispase and 200 μL of 1 mg/mL DNAse I is then added to fully resuspend the pellet. The sample is then triturated for 1 min using a P1000 pipet. This should cause the sample to become cloudy, but not stringy. If still stringy, more DNAse I should be added.

4. To generate a single-cell suspension, dilute the cells with 10 mL of cold HF and filter through a 40-μm cell strainer into a new 50-mL centrifuge tube and spin at 450g for 5 min. If the sample is heavily contaminated with red blood cells, the pellet should be resuspended in a 1:4 mixture of cold HF and ammonium chloride cell lysis solution and centrifuged again at 450g for 5 min. The cells in the final pellet can then be counted and are suitable for immunomagnetic separation, flow cytometry, or clonal cell cultures (*see* **Note 3**).

3.2. Culture of Human Mammary Epithelial Cells

3.2.1. Clonal Mammary Epithelial Cell Cultures

Seeding cells at clonal densities offers the opportunity to quantitate progenitor frequencies in the input cell population. Clonal cultures also offer the opportunity to investigate the ability of different factors to alter the initiation or differentiation behavior of specific subtypes of mammary epithelial cell progenitors. Optimal growth of human mammary epithelial progenitors seeded at clonal densities (<800 cells/cm^2) requires the presence of a suitable feeder layer, most readily provided by using a pre-established layer of irradiated, but viable NIH 3T3 cells (*see* **Note 4**).

1. Harvest some NIH 3T3 cells from *subconfluent* (<60%) cultures. The NIH 3T3 cells can be cultured in DMEM supplemented with 5% FBS. Resuspend the harvested cells at 10^6 cells/mL in HF and irradiate with 5 × 10^3 cGy. Seed the irradiated cells into the desired tissue culture vessels (e.g., 60-mm culture dishes) at 1 × 10^4 cells/cm^2 in DMEM supplemented with 5% FBS.

2. Following attachment of the feeder cells, fresh (or freshly thawed) nonsorted mammary epithelial cells should then be added to the culture dishes containing the feeders after removal of the overlying DMEM (*see* **Notes 5** and **6**). The nonsorted mammary cells should be seeded at 2 × 10^3 cells per 60-mm dish in 4 mL of EpiCult-B medium supplemented with 5% FBS (*see* **Notes 7** and **8**). Failure to include serum in the medium during initiation of mammary epithelial cell cultures results in poor adherence of the epithelial progenitors to the plastic (*see* **Note 9**).

3. Two days later, replace the medium with the same volume (4 mL per 60-mm dish) of EpiCult-B but without any serum supplementation. Failure to remove the serum can result in overgrowth of the culture with contaminating stromal cells.

4. Cultures are then incubated at 37°C, 5% CO_2, and 5–7 d later the medium should then be removed from each dish and 4 mL of acetone:methanol (1:1) added for 5 s. The fixative should then also be removed and the cultures air-dried. Five to seven days is a suitable length of time to permit the salient features of the pure luminal, pure myoepithelial, and mixed colonies to become evident.
5. After the plates have air-dried, rinse gently once with tap water, and stain for 30 s with Wright's Giemsa stain. Rinse plates with water and air-dry one last time.
6. Colonies can then be readily scored microscopically at a low magnification. Refer to **Subheading 3.3.** for descriptions and photos.

3.2.2. Bulk Cultures

1. Dissociated mammary cells can also be seeded into tissue culture flasks at higher densities ([1–5] × 10^4 cells/cm^2) in EpiCult-B + 5% FBS.
2. Twenty-four to forty-eight hours later, the culture medium should be changed to EpiCult-B without serum supplementation. Thereafter, this medium should be replaced with fresh medium 1–3 times per week, depending on the confluency of the cells.
3. Mammary epithelial cells can be subcultured by first washing the adherent cells with HBSS modified followed by incubation with prewarmed trypsin–EDTA. Once the cells have detached from the culture vessel, an equal volume of cold HBSS + 5% FBS should be added and the cell suspension centrifuged at 450g. Collected cells can then be split 1:3 and reseeded into tissue culture as initially described.

3.2.3. Culture of Human Mammary Fibroblasts

Human mammary fibroblast cultures can be initiated using the stromal-cell-enriched fraction from the differential centrifugation following collagenase and hyaluronidase digestion.

1. Seed cells at (1–5) × 10^4 cells/cm^2 in DMEM/F12/H supplemented with 5% FBS and maintain with weekly media changes.
2. Stromal cell cultures can be subcultured as per mammary epithelial cell cultures (*see* **Subheading 3.2.2., step 3**).

3.3. Characterization of Cultured Human Mammary Epithelial Cells

Three types of mammary epithelial progenitor have been identified and differentially isolated from human mammary tissue *(1,4,5)*. These are luminal-restricted progenitors, myoepithelial-restricted progenitors, and bipotent progenitors (*see* **Note 10**). Examples of the types of colony they generate after 6–9 d in vitro are illustrated in **Fig. 2**. Pure luminal cell colonies are characterized by the tight arrangement of the cells they contain with indistinct cell borders. In the smaller colonies, the cells form tightly arranged clumps that give the appearance of rounded spherical structures (*see* **Fig. 2A**). In the larger of

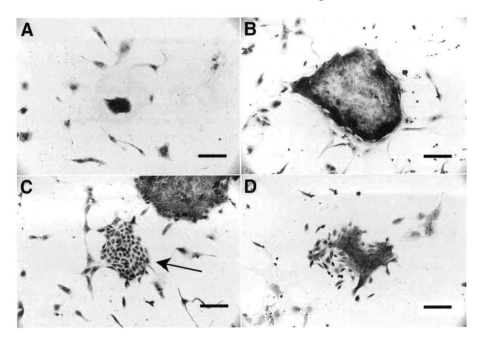

Fig. 2. Pure luminal cell (**panels A** and **B**), pure myoepithelial cell (arrow in **panel C**), and mixed (**panel D**) colonies after 1 wk in vitro in serum-free culture in the presence of NIH 3T3 feeders. Note the close cellular arrangement of the luminal epithelial cells and the dispersed arrangement of the myoepithelial cells. Bar = 200 µm.

the pure luminal cell colonies (*see* **Fig. 2B**), the cells at the periphery appear more tightly arranged than those at the center. Pure luminal cell colonies are also characterized by their smooth colony boundaries. The majority of the cells within such pure luminal cell colonies express the luminal markers MUC1, keratins 8/18, epithelial cell adhesion molecule (EpCAM), and keratin 19. They do not express the basal-cell-specific markers keratin 14 and histoblood group antigen H type 2 (BGA2) *(5,7)*. Cells having a luminal phenotype can be induced to further differentiate into casein-producing cells by the addition of fresh culture medium supplemented with of 1 µg/mL ovine prolactin and 50% Matrigel *(6)*.

Colonies composed solely of myoepithelial-like basal cells (*see* **Fig. 2C**) are rare in primary cultures and cells of such colonies (as well as those generated in the mixed colonies described below) do not exhibit a fully differentiated myoepithelial phenotype, in that they fail to express smooth muscle actin, a feature inversely associated with the growth state of these cells *(8)*. Very few of the cells within pure myoepithelial cell colonies have any further prolifera-

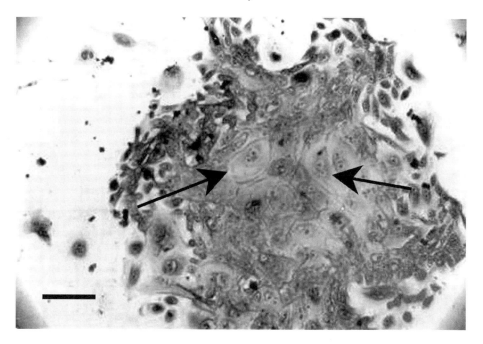

Fig. 3. Squamous metaplastic differentiation within a mixed colony. Note the presence of the large flat cells (arrows) that are often multinucleated. Bar = 200 µm.

tive capacity under the culture conditions examined and do not form new colonies upon replating *(5)*.

Bipotent progenitors generate colonies that contain cells expressing luminal-specific markers as well as cells containing myoepithelial-associated markers. The single cell origin of these colonies has been established *(5)*. The typical arrangement of the cells within these colonies is a central core of cells expressing luminal cellular characteristics (close cell arrangement, expression of luminal specific epitopes such as MUC1, EpCAM, and keratin 19, and lack of expression of keratin 14 and BGA2) surrounded by a halo of highly refractile dispersed teardrop-shaped cells (*see* **Figs. 2D** and **3**). The latter cells express the basal cell markers keratin 14 and BGA2, but not the luminal-associated markers MUC1, EpCAM, and keratin 19. Although keratin 18 is a marker of luminal cells in vivo *(9)*, expression of keratin 18 is not a reliable marker of luminal cells in vitro because a substantial portion of these myoepithelial cells will express this protein *(5,7)*. The centrally located cells in these mixed colonies occasionally exhibit a squamous phenotype (*see* **Fig. 3**), a phenomenon associated with the presence of cholera toxin, cell proliferation, and the menstrual cycle status of the patient from whom the sample was obtained *(10,11)*.

Tissue specimens from women who are in the late stages of the menstrual cycle are more susceptible to undergo squamous metaplasia than those in the early stages of the cycle. The centrally located cells, particularly when they have a squamous phenotype, can also occasionally be observed to be multinucleated (*see* **Fig. 3** and **Note 11**). It should be noted that all of the colonies described here and illustrated in the **Figs. 2** and **3** have been chosen as the most distinct examples of the different types of colony generated. However, in reality, this distinction is somewhat arbitrary, as varying distributions of both cell types are often seen.

Human mammary epithelial progenitor cells will typically divide 10–20 times prior to senescing *(5,12,13)*. As senescence (or more the accurate term "selection" *[12]*) becomes imminent, the cells become large, flattened, and vacuolated. It should be noted that cultures that have been maintained for extended periods of time under the conditions described here, the proportion of cells exhibiting a myoepithelial phenotype generally increases concomitant with a loss of cells exhibiting a luminal phenotype (*see* **Note 12**).

4. Notes

1. Malignant human breast epithelial cells isolated from biopsy, mastectomy, and pleural fluids have proven notoriously difficult to grow in vitro and, indeed, their definitive distinction from contaminating nonmalignant human breast epithelial cells require markers of malignancy that are often difficult to apply *(14–16)*. At first, this might appear paradoxical, because malignant cells are thought to be proliferating at a faster rate in vivo than their normal counterparts. However, it must be remembered that most cell culture media have been designed to maximize the proliferation of normal mammary epithelial cells, under which conditions they might display a proliferative advantage over transformed cells. Indeed, analysis of normal and malignant cell growth rates in vitro also support the concept that tumor cells have slower doubling times ($\geq$120 h) compared to normal luminal (approx 48 h) or myoepithelial (approx 24 h) cells maintained under the same conditions *(17)*. Therefore, caution should be used when interpreting results obtained from cultures initiated with tissue specimens containing malignant cells. It is likely that the culture conditions will not reflect the tumor heterogeneity present in vivo and that the in vitro environment might favor the outgrowth of normal cells or at least select for a subset of tumor cells with distinct growth factor requirements *(16)*. Several novel systems have been reported to promote the selective growth of malignant mammary epithelial cells. These include the use of a reconstituted basement membrane *(18)*, culture conditions that simulate the microenvironment of breast tumors *(19,20)*, the use of irradiated NIH 3T3 feeders *(21)*, and optimization of other culture parameters *(15,16,22)*.
2. Dimethyl sulfoxide is toxic; therefore, the cells should not be left for extended periods of time at room temperature in DMSO-containing media.

3. It is essential that these suspensions be kept cold at all times to prevent reaggregation and clumping of the mammary epithelial cells. Usually, treatment with DNAse I and gentle pipetting can resolve such problems, but, occasionally, a second round of cell filtration might be necessary. If the mammary epithelial cells are to be cultured at nonclonal densities, the cell filtration step can be omitted to maximize cell recoveries.

4. We recommend the use of NIH 3T3 mouse embryonic fibroblasts as feeders for clonal cultures. Although human mammary fibroblasts can also be used as feeders, our experience has been that such feeders are less effective in supporting the maintenance of mammary cells exhibiting a luminal cell phenotype when compared to NIH 3T3 cells *(3,4)*. In addition, the use of a cell line as a feeder reduces inevitable variability between different sources of human mammary fibroblasts.

5. Although human mammary epithelial cells grow well on tissue culture plastic, enhanced colony formation can be achieved by precoating the culture dishes with murine collagen IV at 5 $\mu g/cm^2$. This does not increase the cloning efficiency of mammary epithelial progenitors, but it does dramatically increase the size of the colonies obtained (our own unpublished observations). Coating of tissue culture plates with type I collagen has also been reported to enhance the growth of mammary epithelial cells *(23,24)*.

6. Immunocytochemical analysis using nonfluorescent methods can be performed directly on the adherent cultured cells. However, if fluorescence microscopy is to be performed, sterile collagen-coated glass cover slips should be placed in the culture vessels prior to seeding the cells so that the cover slips can be removed and stained separately to avoid the background fluorescence of many tissue culture plastics.

7. To streamline the colony assay process, NIH 3T3 cells obtained from subconfluent cultures can be stored as frozen irradiated aliquots and added to the mammary epithelial cell suspension at the same time the latter are seeded into the assay cultures. Feeder cells should then be added directly to the EpiCult-B + 5% FBS seeding solution.

8. Typical cloning efficiencies of human mammary epithelial progenitors obtained from freshly dissociated mammary organoids range from 1% to 5%. Accordingly, seeding 6×10^3 cells into 3 culture dishes should yield a total of 60–300 colonies. If the initially isolated cells are first cultured for 1–6 d in EpiCult-B, the cloning efficiency can be increased twofold. If additional procedures are used to enrich for specific progenitor subtypes (*see* **Note 10**) cloning efficiencies of 10–20% can be reproducibly achieved. Under such circumstances, the number of cells seeded per culture needs to be correspondingly decreased to prevent overplating of the cells.

9. Recent reports have demonstrated that human mammary epithelial cells maintained in serum-free suspension cultures under conditions that prevent cell attachment supports the generation of three-dimensional colonies that have been called "mammospheres" *(6,25,26)* because of perceived features suggesting their origin from mammary stem cells, as originally described in the neural system for

the generation of "neurospheres" *(27–29)*. These mammospheres display some self-renewal ability upon disaggregation and are enriched for multipotent epithelial progenitors. At this point, it is not clear whether the mammosphere-initiating cells are human mammary stem cells, but if so, this would represent a new and exciting avenue for mammary gland research.

10. Luminal-restricted and bipotent progenitors can be enriched by flow cytometry from 3-d-old cultures of dissociated human mammary organoids on the basis of differential expression of the MUC1 glycoprotein, EpCAM, and CD49f (α-6 integrin) *(5)*. Luminal-restricted progenitors have a MUC1$^+$/EpCAM$^+$/CD49f$^+$ phenotype, whereas the bipotent progenitors have a MUC1$^-$/EpCAM$^+$/CD49f$^+$ phenotype. A nonclonogenic fraction of mammary epithelial cells can also be enriched within the EpCAM$^+$/CD49f$^-$ cell fraction.

11. Normal human mammary epithelial cells typically undergo 10–20 population doublings prior to reaching a growth plateau. However, a rare subset of human mammary epithelial cells present at a frequency of 10^{-5}–10^{-4} can bypass this growth plateau when the cultures are maintained for several wk after reaching this plateau *(12,30)*. These cells are characterized by methylation of the p16^{INK4A} promoter, eroding telomeric sequences, multiple nuclei, and a susceptibility to acquire further genomic abnormalities *(12,13)*.

12. Robust growth of both luminal and myoepithelial cells can be obtained using EpiCult-B and similar media; however, repeated passaging of human mammary epithelial cells in serum-free medium does promote the generation of cells expressing a myoepithelial phenotype *(2,5,15,31)*. This can be inhibited by preconditioning the EpiCult-B medium by a 3-d exposure to irradiated NIH 3T3 feeder (our own unpublished observations). Similarly, it has been reported that the very complex chemically defined medium CDM6 promotes the generation of cells exhibiting a luminal phenotype while inhibiting the generation of cells with a myoepithelial phenotype *(2)*. Survival and expansion of cells expressing luminal cell characteristics can also be obtained by including FBS within the culture medium. However it is not used on a routine basis because FBS also promotes the proliferation of mammary fibroblasts that eventually dominate the cultures *(15,24,31)*. A strategy that permits the use of serum within the culture medium and avoids the overgrowth of stromal cells is to remove the stromal cells by an appropriate immunoselective procedure (e.g., positive selection of MUC1$^+$ cells) *(5,32)*. Alternatively, confluent lawns of irradiated NIH 3T3 cells can be preseeded to inhibit the growth of contaminating endogenous stromal cells *(3)*.

Acknowledgments

The authors would like to thank Darcy Wilkinson, Gabriela Dontu, and Afshin Raouf for insightful discussions on the culture of human mammary epithelial cells. We also thank Dr. Patty Clugston, Dr. Jane Sproul, Dr. Peter Lennox, and Dr. Richard Warren for supplying the surgical specimens. Grants from the British Columbia Health Research Foundation, the Canadian Breast

Cancer Research Initiative of the National Cancer Institute of Canada, the Canadian Breast Cancer Foundation (BC/Yukon Chapter), and the Natural Sciences and Engineering Research Council of Canada supported this work.

References

1. Stingl, J., Eaves, C. J., Kuusk, U., and Emerman, J. T. (1998) Phenotypic and functional characterization in vitro of a multipotent epithelial cell present in the normal adult human breast. *Differentiation* **63,** 201–213.
2. Pechoux, C., Gudjonsson, T., Ronnov-Jessen, L., Bissell, M. J., and Petersen, O. W. (1999) Human mammary luminal cells contain progenitors to myoepithelial cells. *Dev. Biol.* **206,** 88–99.
3. Matouskova, E., Dudorkinova, D., Krasna, L., and Vesely, P. (2000) Temporal *in vitro* expansion of the luminal lineage of human mammary epithelial cells achieved with the 3T3 feeder layer technique. *Breast Cancer Res. Treat.* **60,** 241–249.
4. Stingl, J., Eaves, C. J., and Emerman, J. T. (2000) Characterization of normal human breast epithelial cell subpopulations isolated by fluorescence-activated cell sorting and their clonogenic growth in vitro, in *Methods in Mammary Gland Biology and Breast Cancer Research* (Ip, M. and Asch, B. B., eds.), Kluwer Academic/Plenum, New York, pp. 177–193.
5. Stingl, J., Zandieh, I., Eaves, C. J., and Emerman, J. T. (2001) Characterization of bipotent mammary epithelial progenitor cells in normal adult human tissue. *Breast Cancer Res. Treat.* **67,** 93–109.
6. Dontu, G., Abdallah, W. M., Foley, J. M., et al. (2003) In vitro propagation and transcriptional profiling of human mammary stem/progenitor cells. *Genes Dev.* **17,** 1253–1270.
7. Karsten, U., Papsdorf, F., Pauly, A., et al. (1993) Subtypes of non-transformed human mammary epithelial cells cultured in vitro: Histo-blood group antigen H type 2 defines basal cell-derived cells. *Differentiation* **54,** 55–66.
8. Petersen, O. W. and van Deurs, B. (1988) Growth factor control of myoepithelial-cell differentiation in cultures of human mammary gland. *Differentiation* **39,** 197–215.
9. Taylor-Papadimitriou, J. and Lane, E. B. (1987) Keratin expression in the mammary gland, in *The Mammary Gland: Development, Regulation and Function* (Neville, M. C. and Daniel, C. W., eds.), Plenum, New York, pp. 181–215.
10. Schaefer, F. V., Custer, R. P., and Sorof, S. (1983) Squamous metaplasia in human breast culture: induction by cyclic adenine nucleotide and prostaglandins, and influence of menstrual cycle. *Cancer Res.* **43,** 279–286.
11. Rudland, P. S., Ollerhead, G. E., and Platt-Higgins, A. M. (1991) Morphogenetic behaviour of simian virus 40-transformed human mammary epithelial stem cell lines on collagen gels. *In Vitro Cell. Dev. Biol.* **27A,** 103–112.
12. Romanov, S., Kozakiewicz, K. B., Holst, C. R., Stampfer, M. R., Haupt, L. M., and Tlsty, T. D. (2001) Normal human mammary epithelial cells spontaneously escape senescence and acquire genomic changes. *Nature* **409,** 633–637.

13. Tlsty, T. D., Romanov, S. R., Kozakiewics, K. B., Holst, C. R., Haupt, L. M., and Crawford, Y. G. (2001) Loss of chromosome integrity in human mammary epithelial cells subsequent to escape from senescence. *J. Mammary Gland Biol. Neoplasia* **6**, 235–243.

14. Cailleau, R., Young, R., Olive, M., and Reeves, W. J. J. (1974) Breast tumor cell lines from pleural effusions. *J. Natl. Cancer Inst.* **53**, 661–666.

15. Ethier, S. P., Mahacek, M. L., Gullick, W. J., Frank, T. J., and Weber, B. L. (1993) Differential isolation of normal luminal mammary epithelial cells and breast cancer cells from primary and metastatic sites using selective media. *Cancer Res.* **53**, 627–635.

16. Ethier, S. P. (1996) Human breast cancer cell lines as models of growth regulation and disease progression. *J. Mammary Gland Biol. Neoplasia* **1**, 111–121.

17. Petersen, O. W. and van Deurs, B. (1987) Preservation of defined phenotypic traits in short-term cultured human breast carcinoma derived epithelial cells. *Cancer Res.* **47**, 856–866.

18. Petersen, O. W., Ronnov-Jessen, L., Howlett, A. R., and Bissell, M. J. (1992) Interaction with basement membrane serves to rapidly distinguish growth and differentiation pattern of normal and malignant human breast epithelial cells. *Proc. Natl. Acad. Sci. USA* **89**, 9064–9068.

19. Bergstraesser, L. M. and Weitzman, S. A. (1993) Culture of normal and malignant primary human mammary epithelial cells in a physiological manner simulates in vivo growth patterns and allows discrimination of cell type. *Cancer Res.* **53**, 2644–2654.

20. Dairkee, S. H., Deng, G., Stampfer, M. R., Waldman, F. M., and Smith, H. S. (1995) Selective cell culture of primary breast carcinoma. *Cancer Res.* **55**, 2516–2519.

21. Krasna, L., Dudorkinova, D., Vedralova, J., et al. (2002) Large expansion of morphologically heterogeneous mammary epithelial cells, including the luminal phenotype, from human breast tumours. *Breast Cancer Res. Treat.* **71**, 219–235.

22. Pandis, N., Heim, S., Bardi, G., Limon, J., Mandahl, N., and Mitelman, F. (1992) Improved technique for short-term culture and cytogenetic analysis of human breast cancer. *Genes Chromosomes Cancer* **5**, 14–20.

23. Ethier, S. P. (1985) Primary culture and serial passage of normal and carcinogen-treated rat mammary epithelial cells in vitro. *J. Natl. Cancer Inst.* **74**, 1307–1318.

24. Emerman, J. T. and Wilkinson, D. A. (1990) Routine culturing of normal, dysplastic and malignant human mammary epithelial cells from small tissue samples. *In Vitro Cell. Dev. Biol.* **26**, 1186–1194.

25. Dontu, G., Abdallah, W. M., Chen, Q., and Wicha, M. S. (2002) Partial purification of human mammary stem cells. *Proc. Am. Assoc. Cancer Res.* **43**, 1051.

26. Dontu, G., Jackson, K. W., Abdallah, W. M., Foley, J. M., Kawamura, M. J., and Wicha, M. S. (2003) Role of LIF in cell fate determination of normal human mammary epithelial cells. *Proc. Am. Assoc. Cancer Res.* **44**, 855.

27. Reynolds, B. A. and Weiss, S. (1992) Generation of neurons and astrocytes from isolated cells of the adult mammalian central nervous system. *Science* **255**, 1707–1710.

28. Reynolds, B. A. and Weiss, S. (1996) Clonal and population analyses demonstrate that an EGF-responsive mammalian embryonic CNS precursor is a stem cell. *Dev. Biol.* **175,** 1–13.

29. Rietze, R. L., Valcanis, H., Brooker, G. F., Thomas, T., Voss, A. K., and Bartlett, P. F. (2001) Purification of a pluripotent neural stem cell from the adult mouse brain. *Nature* **412,** 736–739.

30. Holst, C. R., Nuovo, G. J., Esteller, M., et al. (2003) Methylation of p16 (INK4a) promoter occurs in vivo in histologically normal human mammary epithelia. *Cancer Res.* **63,** 1596–1601.

31. Gomm, J. J., Coope, R. C., Browne, P. J., and Coombes, R. C. (1997) Separated human breast epithelial and myoepithelial cells have different growth requirements in vitro but can reconstitute normal breast lobuloalveolar structure. *J. Cell. Physiol.* **171,** 11–19.

32. Stingl, J., Eaves, C. J., Emerman, J. T., Choi, D., Woodside, S. M., and Peters, C. E. (2003) Column-free FACS-compatible immunomagnetic enrichment of EpCAM-positive and CD49f-positive human mammary epithelial progenitor cells. *Proc. Am. Assoc. Cancer Res.* **44,** 855.

18

Generation and Differentiation of Neurospheres From Murine Embryonic Day 14 Central Nervous System Tissue

Sharon A. Louis and Brent A. Reynolds

Summary

Murine embryonic day 14 or E14 neural stem cells (NSCs), first isolated and characterized as a stem cell in culture, are a unique population of cells capable of self-renewal. In addition, they produce a large number of progeny capable of differentiating into the three primary phenotypes—neurons, astrocytes, and oligodendrocytes—found in the adult mammalian central nervous system (CNS). A defined serum-free medium supplemented with epidermal growth factor (EGF) is used to maintain the NSCs in an undifferentiated state in the form of clusters of cells, called neurospheres, for several culture passages. When EGF is removed and serum added to the medium, the intact or dissociated neurospheres differentiate into the three primary CNS phenotypes. This chapter outlines the simple NSC culture methodology and provides some of the more important details of the assay to achieve reproducible cultures.

Key Words: Murine; embryonic neural stem cells; neurospheres; differentiation; CNS; culture; stem cells.

1. Introduction

Culture systems for the isolation, expansion and differentiation of central nervous system (CNS) stem cells provides a unique and powerful in vitro model system for studying and elucidating the molecular and cellular properties of development, plasticity, and regeneration. Neural stem cells (NSCs) have been isolated from the mammalian CNS and can be maintained in vitro for extended periods of time without loss of their proliferative or differentiation potential *(1)*. Murine embryonic day 14 or E14 NSCs are a unique population of cells that exhibit stem cell functions, including self-renewal and production of a large number of progeny capable of differentiation into the three primary phenotypes found in the adult mammalian CNS *(1)*. Stem cells isolated from the

From: *Methods in Molecular Biology, vol. 290: Basic Cell Culture Protocols, Third Edition*
Edited by: C. D. Helgason and C. L. Miller © Humana Press Inc., Totowa, NJ

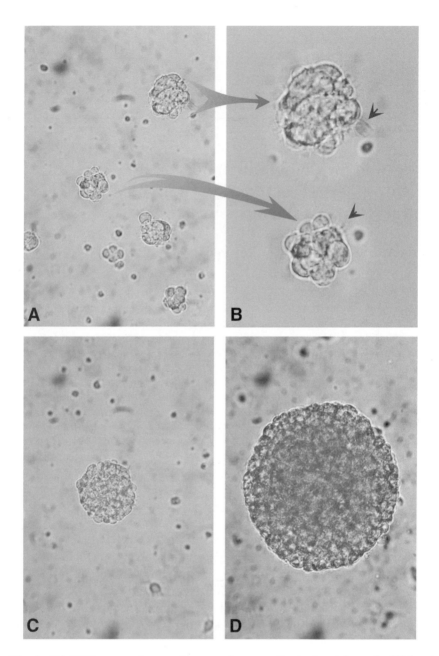

Fig. 1. **(A)** EGF-responsive murine neural stem cells, isolated from the E14 striatum, were grown for 7 d in culture and then passaged. Two days after passaging, small clusters of cells can be identified. **(B)** Two spheres from **(A)** are enlarged showing the appearance of microspikes (arrow heads) that are commonly seen on young healthy neurospheres. **(C)** By 4 d in vitro (DIV) neurospheres have grown in size, detached from the substrate and float in suspension. **(D)** A floating 7-DIV neurosphere. Magnification: **A, C, D** = ×200; **B** = ×400.

murine E14 CNS can be maintained in an undifferentiated proliferative state in a defined serum-free medium supplemented with epidermal growth factor (EGF). After approx 7 d in this growth medium, the proliferating EGF-responsive NSCs form spheres, called neurospheres, measuring 100–200 µm in diameter and composed of approx 10,000 cells (*see* **Fig. 1**). At this stage, the neurospheres can be passaged. This procedure can be repeated weekly resulting in an arithmetic increase in total cell numbers. Embryonic-murine-derived neurospheres treated in this manner have been passaged for 10 wk with no loss in their proliferative ability, resulting in a 10^7-fold increase in cell number. When removed from the growth medium and plated on an adhesive substrate, either as intact clusters or dissociated cells in a low-serum-containing medium, the stem cell progeny can be differentiated into the three primary CNS phenotypes—neurons, astrocytes, and oligodendrocytes (*see* **Fig. 2**). This culture system provides a robust and reliable in vitro assay for studying developmental processes and elucidating the role of genetic and epigenetic factors on the potential of CNS stem cells and the determination of CNS phenotypes. Once established, this culture methodology is simple to apply. However, relatively strict adherence to the procedures is required in order to achieve reliable and consistent results. The original protocols and medium formulations are based on the work of Reynolds and Weiss *(1,2)*. Versions of these protocols have been published elsewhere *(3,4)*. This chapter describes the protocols for the culture of NSCs isolated from various regions of the E14 murine embryonic brain in an attempt to provide standardized, accurate, and reproducible assays for defining NSCs. These protocols assume a basic knowledge of murine embryonic brain anatomy. The reader is referred to **ref. 3** for reference on this topic, the procedures of which is essential for culturing murine NSCs.

2. Materials
2.1. Dissection Equipment

1. Large scissors (1).
2. Small fine scissors (1).
3. Extrafine spring microscissors (1) (cat. no. 15396-01, Fine Science Tools).
4. Small forceps (1) (cat. no. 11050-10, Fine Science Tools).
5. Small fine forceps (1) (cat. no. 11272-30, Fine Science Tools).
6. Ultrafine curved forceps (1) (cat. no., 11251-35, Fine Science Tools).
7. Phosphate-buffered saline (PBS) (e.g., cat. no. 37350, StemCell Technologies Inc.) containing 2% glucose, cold, sterile.
8. Plastic Petri dishes, 100 mm, sterile (four to five plates to hold uteri, embryos, heads, and brain).
9. Plastic Petri dishes, 35 mm, sterile (four to five plates to hold dissected brain regions).
10. Tubes, 17 × 100-mm polystyrene test tubes, sterile (e.g., Falcon, cat. no. 2057).
11. Isopropanol or 70% ethanol.

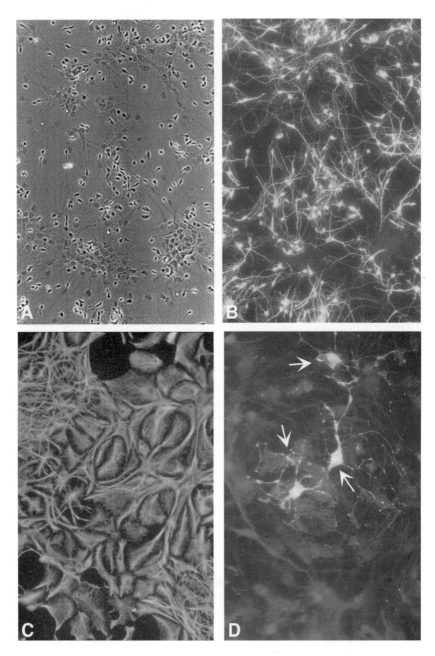

Fig. 2. (**A**) Neurospheres are differentiated as detailed in **Subheading 3.2.** Phase-contrast micrograph shows cells with processes (mostly neurons and oligodendrocytes) sitting on top of a layer of astrocytes. (**B**) Neurons were identified with a fluorescent-labeled antibody raised against β-tubulin (a neuron-specific antigen found in cell

12. Bead Sterilizer (cat. no. 250, Fine Science Tools).
13. Dissecting microscope (Zeiss Stemi 2000, 1:7 zoom).

2.2. General Equipment

1. Biological safety cabinet (e.g., Canadian Cabinets) certified for Level II.
2. Low-speed centrifuge (e.g., Beckman TJ-6) equipped with biohazard containers.
3. 37°C Incubator with humidity and gas control to maintain >95% humidity and an atmosphere of 5% CO_2 in air (e.g., Forma 3326).
4. Vortex (e.g., Vortex Genie).
5. Pipet-aid (e.g., Drummond Scientific).
6. Hemacytometer (e.g., Brightline).
7. Trypan blue (e.g., cat. no. 07050, StemCell Technologies Inc.).
8. Routine light microscope for hemacytometer cell counts.
9. Inverted microscope with flatfield objectives and eyepieces to give object magnification of approx ×20–×30, ×80, and ×125 (e.g., Nikon Diaphot TMD).
10. Fire-polished glass pipets, sterile.

2.3. Tissue Culture Equipment

1. 25-cm^2 Flask (cat. no. 156367, Nunc; or cat. no. 5056, Corning).
2. T-162 cm^2 Flask (cat. no. 3151, Corning).
3. Tubes, 17 × 100-mm polystyrene test tubes, sterile (e.g., cat. no. 2057, Falcon).
4. Tubes, 50, polypropylene, sterile (e.g., cat. no. 2057, Falcon).
5. 24-Well culture dishes (e.g., cat. no. 3526, Corning).
6. Round glass cover slips, sterile.
7. Precoated eight-well culture chamber slides; poly-D-lysine/laminin (cat. no. 35-4688) or human fibronectin (cat. no. 35-4631) (BioCoat Becton Dickinson).
8. Fire-polished glass pipets, sterile.

2.4. Media and Supplements

1. 30% Glucose (Sigma, cat. no. G-7021). Mix 30 g of glucose in 100 mL of distilled water. Filter-sterilize and store at 4°C.
2. 7.5% Sodium bicarbonate ($NaHCO_3$) (Sigma, cat. no. S-5761). Mix 7.5 g of $NaHCO_3$ in 100 mL of distilled water. Filter-sterilize and store at 4°C.
3. 1 M HEPES solution (Sigma, cat. no. H-0887).
4. 10X Stock solution of DMEM/F12. Mix five 1-L packages each of DMEM powder (high glucose with L-glutamine, minus sodium pyruvate, minus sodium bicarbonate [Gibco–Invitrogen Corp., cat. no. 12100-046]) and F12 powder

---

[Fig. 2 caption continued] bodies and processes). A large number of positively labeled cells with a neuronal morphology can be identified. (**C**) Both protoplasmic and stellate astrocytes are identified with a fluorescent-tagged antibody against the astrocyte specific protein GFAP. (**D**) Three oligodendrocytes (arrows) labeled with an antibody against myelin basic protein (MBP). Magnification: **A–C** = ×200; **D** = ×400.

[contains L-glutamine, no sodium bicarbonate [Gibco–Invitrogen Corp., cat. no. 21700-075]) in 1 L water. Filter-sterilize and store at 4°C.

5. 3 mM Sodium selenite (Na_2SeO_3) (Sigma, cat. no. S-9133). Mix 1 mg of Na_2SeO_3 with 1.93 mL of distilled water. Filter-sterilize and store at –20°C.

6. 2 mM Progesterone (Sigma, cat. no. P-6149). Add 1.59 mL of 95% ethanol to a 1-mg stock of progesterone and mix. Aliquot into sterile tubes and store at –20°C.

7. Apo transferrin. Bovine Transferrin Iron Poor (APO) (Serologicals, cat. no. 820056-1). Dissolve 400 mg of Apo transferrin directly to the 10X hormone mix.

8. Insulin: Dissolve 100 mg insulin in 4 mL of sterile 0.1 N HCl. Mix in 36 mL of distilled water and add entire volume directly to the 10X hormone mix.

9. Putrescine (Sigma, cat. no. P-7505). Dissolve 38.6 mg putrescine in 40 mL of distilled water and add entire volume directly to the 10X hormone mix.

10. 200 mM L-Glutamine (e.g., cat. no. 07100, StemCell Technologies Inc.).

11. Basal medium. To prepare 450 mL of basal medium, the individual components are added in the following order: 375 mL of ultrapure distilled water, 50 mL of 10X DMEM/F12, 10 mL of 30% glucose, 7.5 mL of 7.5% sodium bicarbonate, 2.5 mL of 1 M HEPES, and 5 mL of 20 mM L-glutamine. Mix components well and filter-sterilize (*see* **Note 1**).

12. 10X Hormone mix. To prepare 10X hormone mix, the individual components are added in the following order: 300 mL of ultrapure distilled water, 40 mL of 10X DMEM/F12, 8 mL of 30% glucose, 6 mL of 7.5% sodium bicarbonate, and 2 mL of 1 M HEPES. Mix components well at this point. The following components are then added to the above mixture in the order listed: 400 mg of Apo transferrin, 40 mL of 2.5 mg/mL insulin stock, 40 mL of 10 mg/mL putrescine stock, 40 µL of 3 mM sodium selenite, and 40 µL of 2 mM progesterone. Mix all components well and filter-sterilize. Aliquot into 10- or 50-mL volumes in sterile tubes and store at –20°C (*see* **Note 1**).

13. Hormone-supplemented neural culture media. This media is prepared as follows: Thaw an aliquot of the 10X hormone mix from **item 12**. Add 50 mL of the 10X hormone mix to 450 mL of basal medium from **item 11** to give a 1:10 dilution. The hormone-supplemented neural culture media should be stored at 4°C and used within 1 wk (*see* **Note 1**).

14. Human recombinant epidermal growth factor (rhEGF) (cat. no. 02633, StemCell Technologies Inc.). A stock solution of 10 µg/mL of rhEGF is made up in the hormone-supplemented neural culture media from **item 13** and stored as 1-mL aliquots at –20°C until required for use (*see* **Note 2**).

15. Human recombinant basic fibroblast growth factor (rhFGF) (cat. no. 02634, StemCell Technologies Inc.). A stock solution of 10 µg/mL of rhFGF is made up in the hormone-supplemented neural culture media from **item 13** and stored as 1-mL aliquots at –20°C until required for use (*see* **Note 2**).

16. "Complete" NSC medium. Add 2 µL of rhEGF to every 1 mL of the hormone-supplemented neural culture medium from **item 13** to give a final concentration of 20 ng/mL of rhEGF (*see* **Note 2**).

17. "Complete" hormone and serum-supplemented NSC differentiation media. Thaw an aliquot of the 10X hormone mix from **item 12**. Add 50 mL of the 10X hormone mix to 450 mL of basal medium to give a 1:10 dilution. Add 5 mL of fetal bovine serum (FBS) (cat. no. 06550, StemCell Technologies Inc.) to the 500 mL of hormone-supplemented neural culture medium. The media henceforth called complete NSC differentiation medium is now ready for use (*see* **Note 1**). The complete NSC differentiation medium should be stored at 4°C and used within 1 wk.

18. Poly-L-ornithine (cat. no. P3655, Sigma). Dissolve 0.15 mg in 10 mL of PBS to yield a 15-µg/mL stock solution.

19. Poly-L-ornitine-coated glass cover slips. Round glass cover slips are soaked in EtOH and individually hand cleaned with a Kimwipe. Cover slips are sterilized by autoclaving. Using a sterile forceps, transfer a single glass cover slip per well into a 24-well plate. Dispense 1 mL of a 15-µg/mL poly-L-ornithine solution into each well and incubate glass cover slips for a minimum of 3 h at 37°C. After the incubation, remove the poly-L-ornithine solution from each well by aspiration and rinse each well three times for 15 min with sterile PBS. The poly-L-ornitine-coated glass cover slips should be used on the day of preparation.

20. 4% Para-formaldehyde (in PBS, pH 7.2). Dissolve 4 g of paraformaldehyde powder in 100 mL of PBS (cat. no. 37350, StemCell Technologies Inc.) in a fume hood. Stir the solution overnight (gentle heating can help to dissolve the powder). The next day, filter the solution with Whatman filter paper to remove any debris. Store at room temperature for up to 1 mo.

21. Phosphate-buffered saline (PBS) (cat. no. 37350, StemCell Technologies Inc.).

22. PBS containing 0.3% Triton X-100.

23. Appropriate normal serum: goat, rat, mouse (various suppliers; e.g., goat serum, cat. no. G6767, Sigma) (*see* **Note 3**).

24. Fluorosave™ reagent (cat. no. 345789, Calbiochem).

25. Primary antibodies (*see* **Note 4**). The suggested primary antibodies for detection of the various types of neural lineages are indicated in **Table 1**.

26. Secondary antibodies (*see* **Note 4**). The suggested secondary antibodies for fluorescent detection of the various types of neural lineages are indicated in **Table 2**.

3. Methods

3.1. Establishment and Subculture of Primary Neurospheres

All culture procedures including dissections of CNS regions should be performed in Level II Biosafety cabinets using aseptic technique and universal safety precautions.

3.1.1. Dissection of Different CNS Regions

1. Mice (e.g., CD1 albino) are mated overnight (*see* **Note 5**). The next morning, female and male mice are separated and the female mice are checked for the presence of a gestational plug. If a plug is present, this day is counted as d 0 (E0). At E14, the pregnant female is sacrificed in accordance with rules dictated by the animal ethics committee and the embryos are collected for dissection.

Table 1
Suggested Primary Antibodies and Targeted Antigens for the Different Neural Lineages

Targeted antigen	Primary antibody			
	Clone	Isotype	Working dilution	Catalog no.[a]
Neurons				
Neuronal Class III β-Tubulin	TUJ1	Mouse IgG_{2a}	1:1000	01409
Microtubule-associated protein-2 (MAP2)	AP20	Mouse IgG_1	1:200	01410
Neurotransmitters				
GABA	—	Rabbit polyclonal	1:200	01411
Tyrosine hydroxylase-2	TH2	Mouse IgG_1	1:400	01412
Astrocytes				
Glial fibrillary acidic protein (GFAP)	—	Rabbit polyclonal	1:100	01415
Oligodendrocyte				
Oligodendrocyte marker	O4	Mouse IgM	1:50	01416
Myelin basic protein (MBP)	—	Rabbit polyclonal	1:200	01417
Undifferentiated cells				
Nestin	Rat 401	Mouse IgG_1	1:50	01418

[a]StemCell Technologies Inc.

Table 2
Suggested Secondary Antibodies Conjugated
to Different Fluorophores to Detect Primary Antibodies Listed in Table 1

Secondary antibody	Catalog no.[a]
Affini-Pure sheep anti-mouse IgG (H+L) FITC-conjugated	10210
Affini-Pure goat anti-mouse IgM, μ-chain-specific FITC-conjugated	10211
Affini-Pure goat anti-rabbit IgG (H+L) FITC-conjugated	10212
Affini-Pure goat anti-mouse IgG (H+L) Texas red dye-conjugated	10213
Affini-Pure goat anti-rabbit IgG (H+L) AMCA-conjugated	10214

[a]StemCell Technologies Inc.

2. The intact brains are removed from the embryos and transferred to a 35-mm plate containing PBS plus 2% glucose for further dissection procedures.
3. Dissect out striata, ventral mesencephalon, cortex, or other desired brain regions and place in PBS containing 2% glucose, on ice (*see* **Note 6**).
4. When dissections are complete, allow tissues to settle and pipet off supernatant.

3.1.2 Primary Cultures

1. Resuspend tissues in 2 mL of complete NSC medium.
2. Using a fire-polished glass pipet, triturate the tissue approx ten times until a fine single-cell suspension is achieved. If undissociated tissue remains, allow the suspension to settle for 1–2 min and then pipet off the supernatant containing single cells into a fresh tube (*see* **Note 7**).
3. Add more complete NSC medium to the undissociated cells for a total volume of 2 mL. Continue to triturate, transfer, and pool the supernatant containing single cells. Repeat trituration if necessary (*see* **Note 8**).
4. Centrifuge the cells at 800 rpm (110g) for 5 min. Remove supernatant and resuspend the cells with a brief trituration in 2 mL of complete NSC medium.
5. Measure the precise volume and count cell numbers using a dilution in trypan blue (1/5 or 1/10 dilution) and hemacytometer.
6. For primary cultures, seed cells at a density of 2×10^6 cells per 10 mL (T-25 cm^2 flask) or 8×10^6 cells in 40 mL media (T-162 cm^2 flask), in complete NSC medium.

3.1.3. Subculturing

Cells should proliferate to form spheroids, called neurospheres, which, in general, detach from the tissue culture plastic and float in suspension (*see* **Fig. 1**). The neurospheres should be ready for subculture 6–8 d after plating (*see* **Fig. 3** and **Note 9**).

1. Observe the neurosphere cultures under a microscope to determine if the NSCs are ready for passaging (*see* **Fig. 3**). If neurospheres are attached to the culture substrate, tapping the culture flask against the benchtop should detach them.

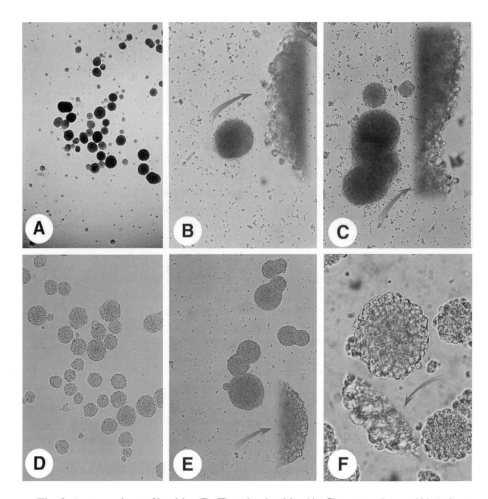

Fig. 3. A comparison of healthy (**D–F**) and unhealthy (**A–C**) neurospheres. (**A**) At least two different types of neurosphere can be identified in this micrograph: dark, dense spheres and light, translucent spheres. The dark spheres are unhealthy and are composed of more dead cells than the lighter colored spheres. (**B**) An unhealthy sphere. The inset on the upper right-hand corner (arrow) is a higher magnification of the left side of the sphere. Note the irregular surface and a high proportion of dead cells on the sphere's periphery. (**C**) Often when the spheres are past their prime, they begin to clump together, as can be seen here. The right-side inset is a magnification of a small section of the clumped spheres. Note the large number of unhealthy or dead cells on the outer edge of the sphere. (**D**) The neurospheres here are healthy and ready to be passaged. If left in culture for an additional 2–3 d, the majority of the spheres would become unhealthy. (**E**) Although these spheres are relatively healthy, they are beginning to reach the end of their prime; they should have been passed 1 d ago. The largest sphere (at the bottom) is becoming dark in the center, a sign that a number of cells are dying.

2. Remove medium with suspended cells and place in an appropriately sized sterile tissue culture tube. If some cells remain attached to the substrate, detach them by shooting a stream of media across the attached cells. Spin at 400 rpm ($75g$) for 5 min.

3. Remove the supernatant and resuspend cells in a maximum of 2–3 mL of complete NSC medium (this volume allows for the most efficient trituration manipulations). If more than one tube was used to harvest cultures, resuspend each pellet in a small volume (0.5 mL) of complete NSC medium and pool all cell suspensions. With a fire-polished Pasteur pipet, triturate the neurosphere until single-cell suspension is achieved (*see* **Notes 7** and **8**). If undissociated tissue remains, follow the instructions in the procedure for establishing primary cultures in **Subheading 3.2.**

4. Centrifuge pooled single-cell suspension at 800 rpm ($110g$) for 5 min. Remove the supernatant and resuspend cells by trituration in an appropriate (approx 1–2 mL) volume of complete NSC medium.

5. Measure the precise volume and count cell numbers using a dilution in trypan blue (1/5 or 1/10 dilution) and hemacytometer.

6. Set up the cells for the next culture passage in complete NSC medium at 5×10^5 cells per 10 mL. For example, in a T-162 cm^2 flask, set up 2×10^6 cells in 40 mL complete NSC medium.

3.2. Differentiation of Neural Stem Cells

In the presence of EGF, NSCs and their progeny are in a relatively undifferentiated state. Upon removal of the growth factor and the addition of a small amount of serum, differentiation of the NSC progeny into neurons, astrocytes, and oligodendrocytes is induced (*see* **Fig. 2**). Neurospheres can be differentiated in at least two ways: as whole spheres cultured at low density or as dissociated cells at high density. The techniques for both methods are provided.

3.2.1. Differentiation of Whole Neurospheres

1. After 7–8 d of culture, remove the medium with suspended cells and place in an appropriately sized sterile tissue culture tube. If some cells remain attached to the substrate, detach them by shooting a stream of media across the attached cells. Spin at 400 rpm ($75g$) for 5 min.

2. The EGF-containing supernatant is pipetted off and discarded.

(Fig. 3 caption continued) The inset (arrow) represents a higher magnification of the right side of this sphere. Although the center of the sphere is becoming dark, the outer portion of the sphere is still light and translucent with light, round healthy cells on the outer edge. **(F)** The sphere in the center is light in color and translucent—a sign that it is composed primarily of live healthy cells. A higher magnification of the lower left corner of this sphere (arrow) reveals the absence of dead cells and abundance of large, healthy round cells.

3. Resuspend the neurospheres in 5 mL of complete NSC differentiation medium.
4. Transfer the suspended neurospheres to a 60-mm dish (or any appropriate vessel to allow isolation of single neurospheres with a disposable plastic pipet tip).
5. Between 1 and 10 neurospheres are isolated with a pipet and deposited on poly-L-ornithine-coated (15 μg/mL) glass cover slips in individual wells of 24-well culture dishes containing complete NSC differentiation medium (1.0 mL/well).
6. Observe cultures daily for 6–8 d with an inverted light microscope and determine if cells have differentiated and are viable (*see* **Note 10**).
7. Cover slips containing differentiated neural cells can be removed and processed immediately for indirect immunofluorescence.
8. If precoated chamber slides were used, proceed to the immunofluorescence procedures in **Subheading 3.3.** (*see* **Note 11**).

3.2.2. Differentiation of Dissociated Cells

1. After 7–8 d of culture, remove the medium with suspended cells and place in an appropriately sized sterile tissue culture tube. If some cells remain attached to the substrate, detach them by shooting a stream of media across the attached cells. Spin at 400 rpm (75g) for 5 min.
2. The EGF-containing supernatant is pipetted off and discarded.
3. Resuspend the neurospheres in 2 mL of complete NSC differentiation medium.
4. With a fire-polished Pasteur pipet, triturate the neurosphere suspension until a single-cell suspension is achieved. If undissociated tissue remains, follow the instructions in the procedure for subculturing primary cultures (*see* **Subheading 3.1.3.**). Measure the precise volume and count cell numbers using a dilution in trypan blue (1/5 or 1/10 dilution) and hemacytometer.
5. Using the complete NSC differentiation medium, prepare an appropriate volume for plating the number of desired wells with a cell density of 5×10^5 cells/mL.
6. Observe cultures after 6–8 d with an inverted light microscope and determine if cells have differentiated and are viable (*see* **Note 10**).
7. Cover slips containing differentiated neural cells can be removed and processed immediately for indirect immunofluorescence.
8. If precoated chamber slides were used, proceed to the immunofluorescence procedures in **Subheading 3.3.**

3.3. Immunostaining of Differentiated Cells

3.3.1. Fixation

1. Add 1 mL of 4% para-formaldehyde (in PBS, pH 7.2) to the required number of wells in a 24-well plate.
2. Transfer cover slips containing differentiating neurosphere cells into the para-formaldehyde solution. Fix cells in 4% para-formaldehyde by incubation at room temperature for 30 min (*see* **Note 11**).
3. For ease, remove the para-formaldehyde solution using an aspiration system connecting to a vacuum pump.

Table 3
Appropriate Blocking Serum Use for Different Secondary Conjugated Antibodies

Secondary antibody	Serum
Affini-Pure sheep anti-mouse IgG (H+L) FITC-conjugated	Sheep
Affini-Pure goat anti-mouse IgM, μ-chain-specific FITC-conjugated	Goat
Affini-Pure goat anti-rabbit IgG (H+L) FITC-conjugated	Goat
Affini-Pure goat anti-mouse IgG (H+L) Texas red dye-conjugated	Goat
Affini-Pure goat anti-rabbit IgG (H+L) AMCA-conjugated	Goat

4. Add PBS (pH 7.2) to the samples and incubate for 5 min. Aspirate PBS using a vacuum pump and repeat this washing procedure two more times for a total of three wash steps.

3.3.2. Permeabilization

1. Permeabilize cells by adding 1 mL of PBS containing 0.3% Triton X-100 to each well and incubate for 5 min at room temperature.
2. Remove PBS/Triton-X 100 by aspiration. Perform two times for 5 min each in PBS as in **Subheading 3.3.1., step 4**.

3.3.3. Blocking and Labeling With Primary Antibodies

1. Make up a solution of PBS with 10% serum (*see* **Note 3** regarding the choice of the appropriate serum to be used). This will be used as the diluent for the primary antibody. **Table 3** provides a list of the appropriate serum for use with the various secondary antibodies.
2. Dilute the primary antibody in the appropriate serum-containing diluent according to **Table 1** to give the right working dilution for labeling (*see* **Note 12**). Add diluted antibodies to the 24-well plate in a minimum volume of 250 μL or alternately place a small volume of antibody (approx 50 μL) directly on the cover slip containing the differentiating cells and place a clean second cover slip directly on top. Place in a hydrating chamber.
3. Incubate for 2 h at 37°C or overnight at 4°C.
4. Wash off primary antibody with three 5-min washes using PBS.

3.3.4. Secondary Antibody Staining

1. Prepare a 1:100 dilution of the secondary antibodies in PBS+ 2% serum (*see* **Note 12**; the serum used here is the same as in the diluent for the primary antibody).
2. Add the secondary antibody to the 24-well plate in a minimum volume of 250 μL.
3. Incubate secondary antibodies for 30 min at 37°C (*see* **Note 13**).
4. Wash off secondary antibody with three 5-min washes using PBS.
5. After the last wash, add distilled water to each well.

Table 4
**Peak Wavelengths of Absorption and Emission
for Different Fluorophore-Conjugated Secondary Antibodies**

Fluorophore	Absorption peak (nm)	Emission peak (nm)
Aminomethylcoumarin, AMCA	350	450
Fluorescein, FITC	492	520
Rhodamine red-X, RRX	570	590
Texas red, TR	596	620

3.3.5. Mounting

3.3.5.1. MOUNTING OF PRECOATED SLIDES

1. If precoated chamber slides are used, follow manufacturer's protocol for removal of the chambers from the glass slides. Rinse slides in distilled water in a Coplin jar.
2. Add about 5 µL of mounting medium in each chamber slot and cover with a 75-mm cover slip, avoiding trapping any air bubbles.
3. Visualize immunostaining under a fluorescent microscope using the appropriate filters for each fluorophore (*see* **Table 4**).

3.3.5.2. MOUNTING OF GLASS COVERSLIPS

1. On a clean glass cover slip, add 10 µL of Fluorosave reagent.
2. Remove stained cover slip from the 24-well plate and gently tap corner of the cover slip to remove excess water.
3. Place cover slip sample side down onto the mounting medium, avoiding any air bubbles.
4. Visualize immunostaining under a fluorescent microscope using the appropriate filters for each fluorophore (*see* **Table 4**).

4. Notes

1. The generation of neurospheres is critically dependent on optimized media and cell density plating conditions. If media is being made in the laboratory, use only tissue-culture-grade components. Optimized reagents for the culture and differentiation of neurospheres are available from StemCell Technologies Inc. (www.stemcell.com). The reagents available include basal medium (NeuroCult™ NSC basal media, cat. no. 05700); 10X hormone mix (NeuroCult™ NSC proliferation supplements, cat. no. 05701), and NeuroCult™ NSC Differentiation supplements (cat. no. 05703, StemCell Tech. Inc.). Follow instructions provided in the manual (StemCell Technologies Inc., NeuroCult™ Technical Manual 28704 or visit www.stemcell.com/stemcell/html/Product_Pages/literature/F_product_literature.html) for preparation of the complete media required using StemCell Technologies Inc. products.
2. Both EGF and bFGF have been shown to be mitogens for CNS stem cells. In general, the number of neurospheres generated and the rate of expansion is enhanced

when the two mitogens are used simultaneously. Each growth factor can act on different populations of stem cells *see* **ref. 6**.

3. The type of serum used depends on the host in which the secondary antibody was generated. For example, if the secondary antibody is a sheep anti-mouse IgG (H+L), use 10% sheep serum in PBS as the diluent. If the secondary antibody is goat anti-mouse, then 10% goat serum in PBS will be the diluent. For immunostaining procedures, it is important to determine the appropriate blocking conditions.

4. Primary and secondary antibodies can be purchased from various suppliers (e.g., StemCell Technologies Inc., Chemicon, coVance, SIGMA, Jackson Laboratories). Because the efficiency of each antibody varies in immunostaining procedures, the working dilutions for each primary or secondary antibody must be worked out for the individual application.

5. Any strain of mice can be used to obtain primary CNS tissues; however, the CD1 albino strain was the strain used in the initial experiments to culture neurospheres *(5)*.

6. Embryonic NSCs have been isolated from nearly all regions of the CNS, including the striatum, cortex, ventral mesencephalon, septum, spinal cord, and thamalus. For reference regarding brain anatomy relevant to dissection of the various regions, *see* **ref. 3**.

7. The mechanical dissociation of cells by trituration with a fire-polished pipet is not a particularly gentle procedure on the cells and is known to cause cell death. However, some precautionary steps can be performed during trituration to diminish the negative effects. For example, avoid forcing air bubbles into the cell suspensions. Also, it is important to wet the glass pipet with a small amount of media before sucking the cells into the pipet to reduce the number of cells sticking to the glass surface.

8. Trituration must be repeated until cell clumps and intact neurospheres are dissociated. Because clumps of cells are heavier than single cells, these will settle to the bottom of the tube when left standing for about 5 min. If all of the collected cells are not required for subculturing, the clumps can be allowed to settle, then the single-cell suspension can be removed to a fresh sterile tube and used for subsequent cultures, leaving the undissociated clusters at the bottom of the tube.

9. The cell density for plating primary striatum, cortex, ventral mesencephalon, or other regions of the E14 murine brain is higher than that for subculturing conditions. Initially, single cells should proliferate to form small clusters of cells that might lightly adhere to the culture vessel. These will lift off from the substratum as the density of the sphere increases. Viable neurospheres will, for the most part, be semitransparent, with many of the cells on the outer surface displaying microspikes (*see* **Figs. 1** and **3**). Cells should be passaged earlier rather than later and before the neurospheres grow too large (>150 μm in diameter). If the neurospheres are allowed to grow too large, the cells within the inside of the neurospheres lack appropriate gas and nutrient/waste exchange and die (*see* **Fig. 3**). Larger neurospheres are also more difficult to dissociate. Therefore, it is important that cultures be monitored each day to determine the conditions of the neurospheres (round bright phase spheres with a smooth periphery) and media.

10. In the differentiation assays, depending on the number of cells plated, the medium might not have to be changed during the differentiation procedure. Plates should be checked daily. If the medium becomes acidic, it should be changed by removing approx 50% of the medium and replacing with fresh complete NSC differentiation medium.

11. If precoated chamber slides are used, remove the culture medium from each chamber containing differentiating cells (taking care not to remove all of the medium and expose the unfixed cells to air) and add 1 mL of the 4% para-formaldehyde solution directly into the chamber. Incubate for 30 min at room temperature. After this step, proceed on to the steps described in **Subheadings 3.3.1. (step 3)–3.3.5.**

12. Primary and secondary antibodies are diluted fresh before each immunostaining applications. Diluted antibodies should then only be used within the same day.

13. Secondary antibody is sensitive to light, and, therefore, whenever possible, keep samples in the dark to prevent bleaching.

Acknowledgments

The authors would like to thank Ravenska Wagey for her assistance in the preparation of this chapter.

References

1. Reynolds, B. A. and Weiss, S. (1996) Clonal and population analyses demonstrate that an EGF-responsive mammalian embryonic CNS precursor is a stem cell. *Dev. Biol.* **175,** 1–13.
2. Reynolds, B. and Weiss, S. (1992) Generation of neurons and astrocytes from isolated cells of the adult mammalian central nervous system. *Science* **255,** 1701–1710.
3. O'Connor, T. J., Vescovi, A. L., and Reynolds, B. A. (1998) Isolation and propagation of stem cells from various regions of the embryonic mammalian central nervous system, in *Cell Biology: A Laboratory Handbook* (Celis, J. E., eds.), Vol. 1, Academic Press, London, pp. 149–153.
4. Gritti, A., Galli, R., and Vescovi, A. L. (2001) Cultures of stem cells of the central nervous system, in *In Protocols for Neural Cell Culture* (Federoff, S. and Richardson, A., eds.), Humana, Totowa, New Jersey, pp. 173–197.
5. Reynolds, B. A., Tetzlaff, W., and Weiss, S. (1992) A multipotent EGF-responsive striatal embryonic progenitor cell produces neurons and astrocytes. *J. Neurosci.* **12,** 4565–4574.
6. Tropepe, V., Sibilia, M., Ciruna, B. G., Rossant, J., Wagner, E. F., and van der Kooy, D. (1999) Distinct neural stem cells proliferate in response to EGF and FFG in the developing mouse telencephalon. *Dev. Biol.* **208,** 166–188.

19

Isolation and Culture of Skeletal Muscle Myofibers as a Means to Analyze Satellite Cells

Gabi Shefer and Zipora Yablonka-Reuveni

Summary

Myofibers are the functional contractile units of skeletal muscle. Mononuclear satellite cells located between the basal lamina and the plasmalemma of the myofiber are the primary source of myogenic precursor cells in postnatal muscle. This chapter describes protocols used in our laboratory for isolation, culturing, and immunostaining of single myofibers from mouse skeletal muscle. The isolated myofibers are intact and retain their associated satellite cells underneath the basal lamina. The first protocol discusses myofiber isolation from the flexor digitorum brevis (FDB) muscle. Myofibers are cultured in dishes coated with Vitrogen collagen, and satellite cells remain associated with the myofibers undergoing proliferation and differentiation on the myofiber surface. The second protocol discusses the isolation of longer myofibers from the extensor digitorum longus (EDL). Different from the FDB myofibers, the longer EDL myofibers tend to tangle and break when cultured together; therefore, EDL myofibers are cultured individually. These myofibers are cultured in dishes coated with Matrigel. The satellite cells initially remain associated with the myofiber and later migrate away to its vicinity, resulting in extensive cell proliferation and differentiation. These protocols allow studies on the interplay between the myofiber and its associated satellite cells.

Key Words: Satellite cells; skeletal muscle; myofiber isolation; single myofiber culture; flexor digitorum brevis; extensor digitorum longus; mouse; Vitrogen collagen; Matrigel.

1. Introduction

Myofibers are the functional contractile units of skeletal muscle. Although they are established during embryogenesis by fusion of myoblasts into myotubes, processes involved in their growth and repair continue throughout life. The development of myofibers and their regenerative potential depends on the availability of myogenic precursor cells. Mononuclear satellite cells

From: *Methods in Molecular Biology, vol. 290: Basic Cell Culture Protocols, Third Edition*
Edited by: C. D. Helgason and C. L. Miller © Humana Press Inc., Totowa, NJ

located between the basal lamina and the plasmalemma of the myofiber are classically considered to be the myogenic precursors in postnatal muscle *(1,2)*. Although in a growing muscle, at least some of the satellite cells are proliferating and adding myonuclei to the enlarging muscle fibers, in a normal adult muscle, most satellite cells are quiescent. However, in response to a variety of conditions, ranging from increased muscle utilization to muscle injury, quiescent satellite cells can enter the cell cycle, replicate, and fuse into existing myofibers or form new myofibers (reviewed in **ref. 2**). The cascade of cellular and molecular events controlling satellite cell myogenesis is therefore of interest for understanding the mechanisms of muscle maintenance during the lifespan, as well as for developing strategies to enhance muscle repair after severe trauma or during myopathic diseases.

Two main in vitro strategies have been employed in the study of satellite cells: (i) myogenic cultures prepared from mononucleated cells dissociated from the whole muscle (i.e., primary myogenic cultures) and (ii) cultures of isolated myofibers where the satellite cells remain in their *in situ* position underneath the myofiber basal lamina. Protocols for obtaining primary myogenic cultures aim at releasing as many satellite cells as possible from the entire muscle. Steps of mincing, enzymatic digestion, and repetitive trituration of the muscle are required for breaking down the connective tissue network and myofibers in order to release the satellite cells from the muscle bulk. These steps are followed by procedures aimed at removing tissue debris and reducing the contribution of nonmyogenic cells typically present in primary isolates of myogenic cells *(3–5)*. In contrast, approaches for isolating myofibers aim at releasing intact myofibers that retain the satellite cells in their native position underneath the basal lamina *(4)*. The first method, based on breakage of the myofibers and release of satellite cells, provides a means for studying parameters affecting the progeny of satellite cells as they proliferate, differentiate, and fuse into myotubes. The second method, based on isolating intact myofibers, allows studying satellite cells in their *in situ* position as well as studying their progeny after migrating from the myofibers.

This chapter describes the two approaches used in our laboratory for isolation and culture of single myofibers from mouse skeletal muscle. One approach, first introduced by Bekoff and Betz *(6)* and further developed by Bischoff *(7,8)*, has been adopted by us for studies of satellite cells in isolated myofibers from both rat *(4,9)* and mouse *(10)*. In this case, myofibers are isolated from the flexor digitorum brevis (FDB) muscles of the hind feet, and multiple myofibers are typically cultured together. The FDB has been used as donor muscle because it consists of short myofibers that do not tangle (and consequently break) when cultured together. A second approach, introduced by Rosenblatt and colleagues *(11,12)*, is suitable for the isolation of longer myofibers from a variety of limb muscles (e.g., extensor digitorum longus [EDL], tibialis anterior [TA]) *(11,13,14)*. Different from the FDB myofibers, the longer myofibers tend to tangle and break if cultured together. Hence, typically when working with muscles such as EDL or TA, the isolated myofibers are cultured individually. In both approaches, the culture dishes are coated with commercially available matrixes that facilitate rapid and firm adherence of the myofibers to the dish surface.

Table 1 compares in brief the two approaches for myofiber isolation and the specific use of each procedure. Both protocols for myofiber isolation can produce high yields of intact myofibers retaining their satellite cells underneath the basal lamina. However, delicate handling of the donor muscle only at the tendons throughout harvesting and processing, the type and specific source of the digesting enzyme, the length of the enzymatic digestion period, and the degree of trituration of the digested muscle are all important factors that should be well controlled during the isolation procedure. Myofibers that are damaged in the course of the isolation procedure will not survive and can be easily distinguished from the intact myofibers because they typically hypercontract.

Protocols for immunocytochemical analysis of satellite cells and their progeny in cultures of FDB and EDL myofibers are also included in the chapter.

Representative micrographs of FDB and EDL myofiber cultures are shown in **Figs. 1** and (FDB, panels **A–D**) and **Fig. 2** (EDL, panels **A–D**).

2. Materials

2.1. General Comments

1. As a general rule, only sterile materials and supplies are to be used. All solutions, unless otherwise noted, are sterilized by filtering through 0.22-μm filters, all glassware and dissection tools are sterilized by autoclaving, and all cell-culturing steps are performed using sterile techniques.
2. The cultures are maintained at 37.5°C and 5% CO_2 in a humidified tissue culture incubator.
3. All culture media are stored at 4°C and used within 3 wk of preparation.
4. Before starting isolation, the tissue culture medium is prewarmed to 37°C and then held at room temperature throughout the procedures (do not leave medium at 37°C for an extended period of time). Before transferring solutions/media into the tissue culture hood, spray the glass/plastic containers with 70% ethanol.
5. The quantities of glassware, media, and reagents as well as the time intervals for enzymatic digestion described in this chapter are appropriate for the isolation of myofibers from one adult mouse of the age and strain detailed in **Subheading 2.4.** Adjustments are needed when isolating myofibers from younger/older mice, other mouse strains, mutant mice, or other laboratory rodents such as rats.
6. Muscles used for preparing isolated myofibers are harvested from the hind limbs.

Table 1
Characteristics of Myofiber Cultures From FDB and EDL Muscles of Adult Mice

Donor muscle	Flexor digitorum brevis (FDB)	Extensor digitorum longus (EDL)
Relative myofiber length	Short	Long
Number of fibers per culture dish	~50–100	1
Typical tissue culture dish	35-mm Dish	24-Well multiwell dish
Dish coating	Thick, gel-like layer of native collagen type I prepared from bovine dermal collagen [Vitrogen, Cohesion Technologies (*9,10,15*)] (*see* **Note 1**).	Thin coating of diluted, growth-factor-reduced Matrigel. Matrigel is a basement membrane preparation isolated from a mouse tumor (BD Biosciences) (*11*) (*see* **Note 2**).
Medium	Dulbecco's modified Eagle's medium (DMEM)-based, mitogen-depleted serum; specific exogenous growth factors are added to study their effect on satellite cell activation, proliferation, and differentiation (*9,10,15*).	DMEM-based, serum rich/mitogen rich; medium can be modified to a serum poor/mitogen-poor one to allow analysis of satellite cell activation (*11,12,16*).
Satellite cell profile after culturing	Satellite cells remain at the surface of the parent myofiber as they proliferate and differentiate. Satellite cells undergo a limited number of proliferative cycles and rapidly differentiate without fusing with the parent myofiber.	Satellite cells emigrate from the parent myofiber and undergo multiple rounds of proliferation, giving rise to an elaborate network of myotubes, resembling regular primary cultures of cells dissociated from whole muscle.
Summary	Cultures can model in vivo behavior of satellite cells in intact fibers during growth and routine muscle utilization. Cultures typically have been maintained short-term and employed for studies on recruitment of satellite cells into the cell cycle. Steps of proliferation and differentiation are highly synchronous (*9,10*). Cultures can be further used to study cells emigrating from the myofibers as described for the EDL fiber cultures.	Cultures can model cellular events after muscle trauma where new myofibers are formed. Cultures typically have been maintained long term and employed in studies of myogenic cells, progeny of satellite cells that emigrate from the myofiber to the myofiber surrounding (*11*). Cultures can also be used for analysis of molecular and cellular events associated with the first round of satellite cell proliferation, as in FDB cultures (*16*).

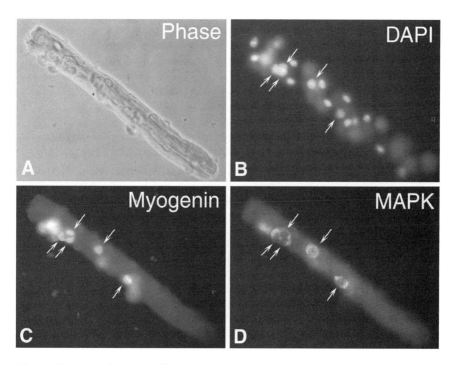

Fig. 1. Phase and immunofluorescent micrographs of an isolated FDB myofiber with associated satellite cells undergoing myogenesis. Myofibers were isolated from a 3-mo-old mouse and cultured in 35-mm tissue culture dishes coated with isotonic Vitrogen collagen. Cultures were maintained for 4 d in basal medium containing fibroblast growth factor 2 (FGF2, 2 ng/mL) and fixed with methanol as described in **Subheading 3.3.1.1.** The culture shown in this figure was reacted via double immuno-fluorescence with a monoclonal antibody against myogenin that stains the nuclei of myogenic cells that have entered the differentiated step of myogenesis (**panel C**) and a polyclonal antibody against ERK1/ERK2 mitogen-activated protein kinases (MAPKs), which stains the cytoplasm of all fiber-associated cells (**panel D**). Reactivity with the monoclonal and polyclonal antibodies was traced with a fluorescein- and rhodamine-labeled secondary antibody, respectively. Parallel phase image (**panel A**) and DAPI staining image (**panel D**; both myofiber nuclei and satellite cell nuclei are stained) are shown as well. Arrows in parallel panels point to the location of the same cell. Additional immunopositive cells present on the myofiber are not shown, as not all positive nuclei or cells on the fibers are in the same focal plane. All micrographs were taken with a ×40 objective. Additional details regarding the source of the antibodies and the rationale of using these antibodies are provided in our previous publications (*10,15*).

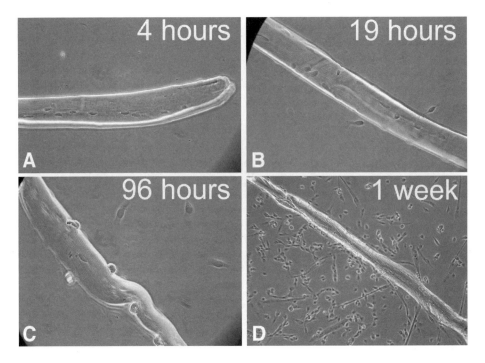

Fig. 2. Phase micrographs of EDL myofibers depicting the temporal development of myogenic cultures from cells emanating from individual myofibers. Myofibers were isolated from 3-mo-old mice and cultured individually in 24-well multiwell tissue culture dishes coated with Matrigel. Cultures were maintained in serum-rich/mitogen-rich growth medium and fixed with paraformaldehyde, as described in **Subheading 3.3.1.2.** Satellite cells begin to emigrate from the myofiber within the first day in culture and continue to emigrate during subsequent days. Satellite cells that have emigrated from the myofibers proliferate, differentiate, and fuse into myotubes, establishing a dense myogenic culture. Satellite cells remained attached to the muscle fiber during the first hours after culturing (**panel A**). Nineteen hours after culturing, two to three cells detached from the fiber but remained in close proximity to the fiber (**panel B**). Four days following culturing, more cells are seen in the vicinity of the myofibers (only four cells shown in **panel C**). By d 7, progeny of satellite cells that emigrated from the myofiber have established a culture containing mostly proliferating myoblasts and some myotubes (**panel D**). Micrographs in **panels A–C** were taken with a ×40 objective to show details of the few cells that emigrated from the myofiber, whereas the micrograph in **panel D** was taken with a ×10 objective to show the establishment of a dense myogenic culture.

2.2. General Equipment

The following facilities are required for the cultures described in this chapter:

1. Standard humidified tissue culture incubator (37.5°C, 5% CO_2 in air).
2. Tissue culture hood.
3. Phase-contrast microscope.
4. Stereo dissecting microscope with transmitted light base (microscope is placed inside a tissue culture hood).
5. Low-speed agitator placed in the tissue culture incubator (Labline Instruments, Inc., model no. 1304). The agitator is used for gently agitating the muscle during enzymatic digestion; a shaking water bath set at 37°C can be used instead of the low-speed agitator.
6. Bunsen or alcohol burner inside the tissue culture hood.
7. Water bath (37°C).
8. Hair trimmer (optional, for shaving hair from the hind limbs prior to muscle dissection).

2.3. Surgical Tools

1. Straight operating scissors: V. Mueller, fine-tipped, Sharp/Sharp stainless steel, 165 (6.5-in.) (VWR Scientific Inc., cat. no. 25601-142), for delicate cutting and fine incisions.
2. Dissecting scissors: stainless steel, 140-mm (5.5-in.) length. Both blades blunt (VWR Scientific Inc., cat. no. 25877-103), protects the surrounding tissue from any unwanted nicks.
3. Dressing forceps: V. Mueller, serrated, stainless steel, rounded points, 140-mm (5.5-in.) length (VWR Scientific Inc., cat. no. 25601-072).
4. Two, very fine-point forceps: extrafine tips, smooth spring action, stainless steel. Straight, 110 mm (4.5 in.; VWR Scientific Inc., cat. no. 25607-856).
5. Microscissors, Vannas scissors: 8 cm long, straight 5-mm blades, 0.1-mm tips (World Precision Instruments, cat. no. 14003).
6. Scalpel handle: size 3 for blades 10–15 (Bard-Parker, cat. no. 371030) and sterile blade (no. 10; Bard-Parker, cat. no. 371110).
7. Placement instrument (VWR Scientific Inc., cat. no. 1790-034)
8. Two straight, 5-in. hemostatic forceps (VWR Scientific Inc., cat. no. 25607-302).
9. Dissecting board.

2.4. Animals

C57BL/6 mice, 2–5 mo-old, maintained according to institutional animal care regulations. Various other mouse strains have been used in our studies following the myofiber isolation procedures described in this chapter.

2.5. Muscles

The information in this subsection is provided to assist in the identification and isolation of the FDB and EDL muscles.

2.5.1. Flexor Digitorum Brevis

The FDB is a superficial, multipennate, broad, and thin muscle of the foot and paw *(8,17)*; it arises from the tendon of the plantaris as three slender muscles converging into long tendons. At the base of the first phalanx, it divides into two, passes around the tendon of the flexor hallucis longus obliquely across the dorsum of the foot, and ends as the tendons insert into the second phalanx of the second through the fifth digits. As the FDB contracts, digits 2–5 are flexed. For additional details about the anatomy of the FDB muscle, *see* **Note 3**.

2.5.2. Extensor Digitorum Longus

The EDL muscle is situated at the lateral part of the hind limb running from the knee to the ankle, extending to the second to fifth digits *(17)*. The EDL actually consists of four combined muscle bellies and their tendons; the bellies arise from the lateral condyle of the tibia and the front edge of the fibula (two tendons at the origin of the muscle). The tendons lie close to each other and appear as one glistening white tendon that continues down to the surface of the ankle. At the ankle joint, it separates to four tendons, each attached to one of the second to fifth digits. As the EDL contracts, the four digits are extended. For additional details about the anatomy of the EDL muscle, *see* **Note 3**.

2.6. Plastic and Glassware for Myofiber Isolation and Culture

2.6.1. FDB Myofiber Isolation and Culture

1. Standard 9-in. Pasteur pipets (VWR Scientific Products, cat. no. 0035904).
2. Standard 5-in. sterile glass Pasteur pipets (VWR Scientific Products, cat. no. 0035901).
3. Wide-mouth pipets prepared from the standard 5-in. Pasteur pipets. Cut the tip of a pipet about 3-in. from its narrow end using a file or a diamond knife. Shake the pipet to remove any glass fragments. Use flame in hood to fire-polish the distal ends of all Pasteur pipets listed in **items 1–3** to smoothen sharp edges that can damage myofibers.
4. Syringe filters, 0.2 and 0.45 µm (Millex-GS, Millipore, cat. no. SLGS0250S and SLHA0250S, respectively), and 10-cm^3 syringes.
5. Sterile conical tubes, 15 and 50 mL (BD Biosciences/Falcon, cat. no. 352098 and 302097, respectively).
6. Three glass Corex tubes, 15 mL (Sorvall centrifuge tubes; or alternatively 15-mL bicarbonate Sorvall tubes).
7. Wide-bore 100-µL micropipet tips. Trim 100-µL tips 3 mm from the end to minimize myofiber shearing when transferring or dispensing FDB myofibers.
8. Tissue culture dishes, 35 mm (Corning Incorporated, cat. no. 430165).
9. Two L-shaped bent pipet spreaders prepared from standard 9-in. Pasteur pipets. Use flame to first seal the distal end, then flame about ¾ in. from the sealed end

until the pipet starts to bend. The bent pipets are used to spread the coating solution on the tissue culture dishes. Spreaders should be prepared in advance and allowed to cool before use.

2.6.2. EDL Myofiber Isolation and Culture

1. Standard 9-in. and 5-in. sterile Pasteur pipets, syringe filters and conical tubes listed and treated as described in **items 1–6** in **Subheading 2.6.1.**
2. Three gradually narrower-bore pipets prepared from standard 5-in. Pasteur pipets. Use a file or a diamond knife to prepare a set of pipets with bore diameter of approx 2.5, 2, and 1 mm. Shake the pipet to remove any glass fragments and fire-polish the sharp ends. These pipets are used to triturate the digested muscle in order to release single myofibers.
3. Six plastic Petri dishes, 60 × 15 mm (Becton Dickinson Biosciences, Falcon, cat. no. 351007).
4. Twenty four-well Falcon multiwell tissue culture dish (Becton Dickinson Biosciences, cat. no. 353047) (*see* **Note 4**).

2.7. Media and Cell Culture Reagents

2.7.1. FDB Myofiber Isolation and Culture

1. DMEM (Dulbeco's modified Eagle's medium; high glucose, with L-glutamine, with 110 mg/L sodium pyruvate, with piridoxine hydrochloride (Gibco–Invitrogen Life Technologies, cat. no. 11995065) supplemented with 50 U/mL penicillin and 50 mg/mL streptomycin (Gibco–Invitrogen, cat. no. 15140-122).
2. Horse serum (HS); standard, not heat inactivated (HyClone, cat. no. SH30074.03); stored at –20°C (*see* **Note 5**).
3. HS, 20 mL, freshly filtered (on the day of use) through a 0.45-μm filter.
4. DMEM, 100 mL, containing 10% filtered HS. All Pasteur pipets and micropipet tips are preflushed with DMEM containing 10% HS to prevent sticking of myofibers during manipulation.
5. Controlled Process Serum Replacement (CPSR) (Sigma–Aldrich, stored at –20°C). Alternative serum replacement products (e.g., Sigma–Aldrich, cat. no. S9388) *(18)* can also be used depending on experimental requirements (*see* **Note 6**).
6. FDB myofiber culture medium is made up of DMEM (supplemented with antibiotics), 20% CPSR and 1% HS.
7. Vitrogen collagen in solution (Cohesion Technologies, cat. no. FXP-019) for coating 35-mm tissue culture dishes. Vitrogen collagen in solution is the recommended product and the use of collagen from other companies would require prescreening to ensure compatibility (*see* **Note 1**).
8. 7X DMEM made from powder DMEM (1-L package; Sigma–Aldrich, cat. no. D3656); used to prepare isotonic Vitrogen collagen (*see* **Note 1**).
9. Collagenase (type I, Sigma–Aldrich, cat. no. C-0130) used for muscle digestion as described in **Subheading 3.**

2.7.2. EDL Myofiber Isolation and Culture

1. DMEM and HS as listed and prepared in **items 1–4** in **Subheading 2.7.1.**
2. Fetal bovine serum (FBS; standard, not heat inactivated) (Sigma–Aldrich, cat. no. F-2442; stored at –20°C) (*see* **Note 7**).
3. Chicken embryo extract (CEE) (Gibco–Invitrogen, cat. no. 16460024), stored at –20°C; or, as in our studies, prepared by the investigator (*see* **Notes 8 and 9**).
4. EDL myofiber culture medium made up of DMEM (supplemented with antibiotics), 20% FBS, 10% HS, and 1% CEE.
5. Matrigel (*see* **Note 2**) for coating 24-well multiwell dishes. Matrigel can be purchased in its standard format (BD Biosciences, cat. no. 354234) or in its growth-factor-reduced format (BD Biosciences, cat. no. 354230). In our studies, we use the growth-factor-reduced format.
6. Collagenase, as listed in **item 9** in **Subheading 2.7.1.**

2.8. Reagents and Solutions
for Fixing and Immunostaining Myofiber Cultures

2.8.1. FDB Myofiber Cultures

1. Prefixation rinse solution: DMEM as in **item 1** in **Subheading 2.7.1.**
2. Fixative: ice-cold 100% methanol (*see* **Note 10**).
3. Rinse solution: Tris-buffered saline (TBS); 0.05 M Tris-HCl, 0.15 M NaCl, pH 7.4 (*see* **Note 11**).
4. Detergent: Tween-20 (Sigma, cat. no. P1379).
5. Detergent solution: TBS containing 0.05% Tween-20 (TBS-TW20).
6. Blocking reagent: normal goat serum (Sigma–Aldrich, cat. no. G9023).
7. Blocking solution: TBS containing 1% normal goat serum (TBS-NGS).
8. Mounting medium: Vectashield (Vector Laboratories, Inc., Burlingham; cat. no. H-1000).
9. Cover glass, 22 mm^2 (Corning Labware and Equipment, cat. no. 48371-045).

2.8.2. EDL Myofiber Cultures

1. Fixative: 4% paraformaldehyde containing 0.03 M sucrose (*see* **Notes 12 and 13**).
2. Rinse solution: TBS as in **item 3** in **Subheading 2.8.1.**
3. Detergents: Triton X-100 (Sigma, cat. no. T6878); Tween-20 as in **item 4** in **Subheading 2.8.1.**
4. Detergent solution: TBS containing 0.5% Triton X-100 (TBS-TRX100); TBS-TW20 as in **item 5** in **Subheading 2.8.1.**
5. Blocking reagent and solution: same as **items 6** and **7** in **Subheading 2.8.1.**
6. Mounting medium: same as **item 8** in **Subheading 2.8.1.**
7. Microcover glass (VWR Scientific Inc., cat. no. 12CIR-1).

3. Methods

3.1. Isolation of Single Myofibers From the Flexor Digitorum Brevis Muscle

3.1.1. Initial Steps Prior to Harvesting the Muscle and Preparation of Digestive Enzyme

1. Add 3 mL of DMEM to six 35-mm tissue culture dishes and place the dishes in the tissue culture incubator until muscle dissection begins.
2. Add 3 mL of DMEM containing 10% HS to three 35-mm tissue culture dishes and place them in the tissue culture incubator until needed for the isolated single myofibers.
3. Add 6 mg of collagenase type I to 3 mL of DMEM in order to prepare 0.2% (w/v) collagenase type I solution. Use a 0.22-µm filter attached to a 10-cm^3 syringe to filter the collagenase solution into a 35-mm tissue culture dish (*see* **Note 14**).

3.1.2. Dissection of FDB Muscle

1. Euthanize one mouse according to institute regulations.
2. Shave the hind limbs and spray them lightly with 70% ethanol.
3. Secure the mouse, lying on its back, to the dissecting board by pinning down the forelimb diagonally across from the limb being dissected.
4. Use a scalpel to cut through the skin all around and just above the ankle (after this initial circular cut, the skin below resembles a sock).
5. Cut the skin in a straight line along the center of the ventral part of the foot almost all the way to the digits (the cut as viewed from the front of the foot should resemble a "T" shape).
6. Clamp a hemostatic forceps to one of the upper corners of the cut tissue (at the junction of the circular and longitudinal cuts), shifting the skin away from the foot.
7. Hold the scalpel with its blade parallel to the longitudinal axis of the partially exposed muscle and carefully cut away the connective tissue. Be especially careful not to cut into the muscle tissue at the back of the leg, as the FDB is the most superficial muscle of the back of the foot.
8. Clamp the second hemostat to the other corner of the cut tissue and repeat **step 7**.
9. When the skin is completely cut away from the foot, the FDB should be exposed all the way to the tendons reaching the digits.
10. Turn the mouse over so that it lies on its stomach, and identify the FDB. During the next steps of the surgery, to avoid blood cell contamination of the myofiber preparation, be careful not to injure the small medial plantar artery that supplies blood to the FDB. This artery passes along the medial part of the sole of the foot and branches into the digits.
11. Carefully run the tip of the scalpel along each side of the FDB to disrupt the connective tissue holding the muscle in place.
12. When the FDB is separated from the surrounding muscles, insert the tip of the placement instrument underneath the FDB and gently lift the muscle so that the flat side of the scalpel can be inserted horizontally underneath it.

13. With the blade of the scalpel underneath the muscle, running horizontal and parallel to the muscle, cut away the underlying connective tissue. It is best to cut toward the heel and only lift that portion of the muscle directly over the scalpel.

14. Cut underneath the tendon to separate the muscle and a large portion of its tendon from the heel bone.

15. Clamp the freed tendon as far as possible from the muscle tissue with a hemostat.

16. Use the hemostatic forceps to gently lift the FDB away from the leg. Use the scalpel, running parallel to the muscle, to cut through the connective tissue while holding the FDB down.

17. Continue cutting through the connective tissue until the tendons that connect the FDB muscle to the digits have been exposed. When about half the length of the three tendons has been exposed, cut the tendons and release the entire muscle from the leg. The fourth small lateral tendon (attached to the fifth digit) and its attached myofibers can be trimmed off.

18. Retrieve from the incubator three 35-mm tissue culture dishes containing DMEM and place them close to the dissection area.

19. Place the harvested FDB in one of the 35-mm tissue culture dishes.

20. For harvesting the FDB from the other hind foot, repeat **steps 11–17** and place the muscle in a second 35-mm tissue culture dish.

21. Place the 35-mm tissue culture dishes, one at a time, under the stereo dissecting microscope.

22. Use fine-point forceps to pull the connective tissue perpendicular to the line of the muscle and use scissors to cut it off.

23. Once the muscle is clean, shorten the tendons but do not cut off all of them.

24. Use a wide-bore Pasteur pipet to transfer the cleaned muscle to another 35-mm tissue culture dish containing DMEM.

25. Repeat **steps 21–24** to clean the second FDB muscle.

3.1.3. Enzymatic Digestion

1. Transfer the two cleaned FDB muscles to a 35-mm tissue culture dish with the 0.2% collagenase I solution.

2. Place this 35-mm tissue culture dish on the low-speed agitator inside the tissue culture incubator to allow gentle and continuous collagenase digestion for 2.5 h (*see* **Notes 14** and **15**).

3. At the end of the digestion period, transfer each muscle to a 35-mm tissue culture dish containing 10% HS.

3.1.4. Separation of the Three Tendons and Release of Myofibers

All Pasteur pipets used are preflushed with 10% HS as described in **item 4** in **Subheading 2.7.1.**

1. Place one muscle at the time under the stereo dissecting microscope.

2. Identify the two grooves running between the three tendons separating the middle from the two lateral tendons.

3. Being careful not to touch the muscle, insert the tip of a forceps into one of the grooves, and, by securing the connective tissue between the tendons to the dish, hold the muscle in place.

4. Use another pair of forceps to gently pull the connective tissue that holds the tendons and their attached muscle tissue together.

5. Continue removing the connective tissue until the lateral tendons are separated from the middle tendon and its attached myofibers.

6. Holding the muscle only at its tendons, transfer the muscle preparation to a 35-mm dish containing 3 mL of 10% HS.

7. While grasping one end of the middle tendon with a pair of forceps, use a second pair of forceps to grip its surrounding connective tissue sheath and pull gently. If the sheath does not come off easily, use fine-point forceps to pull the connective tissue perpendicular to the line of the muscle and cut it off.

8. Repeat **steps 1–7** with the second FDB muscle until all six tendons and their attached myofibers are in the 35-mm tissue culture dish containing 10% HS.

9. For one tendon at a time: hold one end of the tendon with a pair of forceps and with the tip of a second pair gently separate the myofibers from the tendon. The liberation of the myofibers from the two lateral tendons should be easy; the middle tendon requires patience because the myofibers are attached to it more firmly.

10. Use a wide-bore Pasteur pipet to gently triturate the clumps of myofibers until they disengage into single myofibers.

11. Remaining clumps should be transferred to another 35-mm tissue culture dish containing 10% HS and further triturated until disengaged into single myofibers.

12. Set the stereo dissecting microscope magnification so that the small pieces of connective tissue floating around in the suspension are visible and use a pair of forceps to pick them out. Continue until the myofiber suspension is clean of connective tissue debris.

13. Triturate the myofiber suspension 10 more times using a 9-in. Pasteur pipet with a fire-polished tip to further separate small clumps of myofibers.

3.1.5. Further Purification of FDB Myofibers

1. Add 10 mL of 10% HS to each of the three glass Corex tubes.

2. Using the trimmed 100-µL pipet tip, transfer the myofiber suspension to the top of the 10% HS column in the first Corex tube. Allow the myofibers to settle (at 1*g*) through the HS column for 15 min at room temperature (*see* **Note 16**). This step is important for purifying the myofibers from free mononucleated cells, debris, and occasional broken myofibers.

3. As soon as the myofibers are settled, aspirate about 11 mL of the supernatant (leaving about 1–1.5 mL). Triturate the myofiber suspension gently with a 5-in. fire-polished Pasteur pipet and transfer the suspension to the next Corex tube as described in **step 2**.

4. Allow myofibers to settle and transfer the myofiber suspension to the third Corex tube as in **steps 2** and **3**.

5. Allow myofibers to settle and harvest the final myofiber suspension. Following the third purification, the residual volume of medium to be left with the myofiber suspension depends on the number of culture dishes and the desired myofiber number per dish. Typically in our studies, the volume of the final myofiber suspension is 300 μL, which is sufficient for culturing four to six dishes.

3.1.6. Preparation of Isotonic Vitrogen Collagen

Isotonic Vitrogen collagen can be prepared during the settling of myofibers. The isotonic mixture should be kept on ice. Stock Vitrogen is an acidic solution, and when made isotonic, it gels rapidly if not maintained at 4°C (*see* **Note 1**).

1. Place Vitrogen collagen stock bottle, 7X DMEM, and one 15-mL conical tube on ice.
2. On ice: Add 1 vol of 7X DMEM and 6 vol of Vitrogen to the 15-mL conical tube and mix gently. Calculate the volume of stock Vitrogen needed for the experiment based on using 120 μL isotonic Vitrogen collagen to coat each 35-mm tissue culture dish. Use pH paper strips to ensure a neutral pH of the Vitrogen collagen in DMEM solution. The pH of this solution rises slightly after coating the culture dish. If the pH remains acidic after coating a test dish, add 1–2 drops of 1 *M* NaOH to the Vitrogen collagen in DMEM solution.

3.1.7. Coating Culture Dishes
With Isotonic Vitrogen Collagen and Myofiber Culturing

1. On ice: Transfer 120 μL of isotonic Vitrogen collagen to the center of a 35-mm culture dish and immediately use the L-shaped spreader to coat the dish evenly.
2. Gently swirl the myofiber suspension (in the 15-mL tube) for even distribution of myofibers throughout the residual medium.
3. Remove one culture dish at a time from ice to allow rapid warming to room temperature.
4. Use a wide-bore 100-μL micropipet tip to dispense about 50 μL of the myofiber suspension per each culture dish.
5. Gently swirl the culture dish to allow even distribution of the myofibers.
6. Repeat **steps 2–5**, one dish at a time, for additional culture dishes.
7. Transfer dishes to the tissue culture incubator for a minimum of 20–30 min to allow the formation of Vitrogen collagen matrix and the adherence of the myofibers to the matrix.
8. Remove dishes from the incubator. Gently add 1 mL of myofiber culture medium to each dish without agitating the myofibers and return dishes to the incubator. When the effect of growth factors on satellite cell proliferation/differentiation is investigated, parallel cultures are maintained in myofiber culture medium with/without additives and the medium is replaced every 24 h to ensure that growth factors do not become rate limiting. Except for harvesting myofiber cultures for early time points, cultures should be left undisturbed for the initial 18 h to allow good adherence of myofibers to matrix.

3.2. Isolation of Single Myofibers From the EDL Muscle

3.2.1. Procedures Prior to Muscle Harvesting

3.2.1.1. PREPARATION OF MATRIGEL WORKING MIXTURE AND COATING TISSUE CULTURE DISHES WITH MATRIGEL

1. Thaw the required amount of stock Matrigel by placing frozen aliquots on ice for approx 20 min. When diluted into the final working solution, a 200-µL stock aliquot should be sufficient for coating at least four 24-well multiwell dishes.
2. Prechill a 15-mL conical tube on ice and transfer the thawed Matrigel into the tube. Add ice-cold DMEM to dilute the Matrigel to a final concentration of 1 mg/mL.
3. Place 500 µL of diluted Matrigel solution in the center of each of the 24 wells, using a glass 1-mL pipet.
4. Swirl the 24-well multiwell dish to allow even coating of the wells.
5. Allow the Matrigel-coated dish to sit at room temperature for 5–10 min in the tissue culture hood.
6. Transfer excess Matrigel solution from the wells back to the original tube with diluted Matrigel that is kept on ice. Use this Matrigel solution to coat additional dishes within the next 2 h. Do not keep diluted Matrigel for reuse on subsequent days.
7. Incubate the Matrigel-coated multiwell dishes in the tissue culture incubator until the end of the enzymatic digestion period, but for at least 30 min.

3.2.1.2. COATING GLASSWARE AND PLASTICWARE DISHES WITH HS FOR THE INITIAL STEPS OF MYOFIBER ISOLATION

1. Coat six plastic Petri dishes with undiluted filtered HS, prepared as described in **item 3** in **Subheading 2.7.1.** Transfer 1 mL of HS to each Petri dish and swirl the dish to coat evenly.
2. Allow the dishes to sit with HS solution for 5 min at room temperature; then, aspirate the HS and add 7 mL of DMEM to each Petri dish.
3. Incubate Petri dishes in the tissue culture incubator until needed following muscle digestion.
4. Coat the fire-polished Pasteur pipets, prepared as described in **items 1** and **2** in **Subheading 2.6.2.**, with HS by passing 10% HS solution through the pipets several times.

3.2.1.3. PREPARATION OF THE DIGESTING ENZYME SOLUTION

Prepare 0.2% (w/v) collagenase type I solution in 3 mL of DMEM and filter the solution into a 35-mm tissue culture dish using a 0.22-µm syringe filter (*see* **Note 14**).

3.2.2. Dissection of EDL Muscle

1. Euthanize one mouse according to institute regulations.
2. Shave the hind limbs and spray them lightly with 70% ethanol.

3. Secure the mouse, lying on its back, to the dissecting board by pinning down the hind limb to be operated on and the diagonal forelimb.
4. Use the straight rounded-tip scissors to cut through the skin, opening a small incision above the knee.
5. Holding the skin with fine forceps, insert the rounded-tip scissors beneath the incision, and carefully open the scissors to loosen the skin from the underlying muscles.
6. Extend the incision to a point just in front of the digits.
7. Loosen the skin as you go, being careful not to cut the underlying muscles or blood vessels.
8. Cut and remove the skin from the knee to the paw.
9. Identify the two tendons at the origin of the EDL.
10. Use microscissors to cut these tendons as far as possible from the muscle itself.
11. Identify the four tendons at the insertion of the EDL, each extending to one of the digits but not the toe.
12. Use the microscissors to cut all four tendons.
13. Using fine forceps, gently pull the portion of the tendon before its division (to the four tendons) until the four tendons slide from the paw up to the ankle.
14. Grasp the four tendons and carefully pull them in order to remove the EDL muscle.
15. The EDL should slide underneath the TA muscle and should pull out easily. It is very important not to apply any force; if the muscle does not slide out easily, one or both tendons at the origin of the muscle might still be attached to the bone. In that case, identify the attached tendon and cut it.
16. The muscle should only be handled by its tendons to prevent damage to the myofibers. Be careful not to injure the anterior tibial artery that supplies blood to the EDL, to avoid blood cell contamination of the myofiber preparation. In the upper third of its course, this artery lies between the TA and EDL muscles (very close to the origin of the EDL muscle); in the middle third, it lies between the TA and extensor hallucis longus. The lower third of the artery starts at the ankle, crossing from the lateral to the medial side, lying between the tendon of the extensor hallucis longus and the first tendon of the insertion of EDL muscle.

3.2.3. Enzymatic Digestion

1. Holding the muscle by its four tendons, transfer it to the 35-mm tissue culture dish containing 0.2% collagenase I solution.
2. Place the dish on the low-speed agitator located inside the incubator. Allow gentle and continuous agitation for collagenase digestion for 60 min (*see* **Notes 14** and **15**).

3.2.4. Liberation of Single Myofibers From Muscle Bulk

Use a stereo dissecting microscope (placed inside tissue culture hood) throughout the procedure.

1. Inspect the muscle under the stereo dissecting microscope to make sure that the myofibers are loosened from the muscle bulk. If the myofibers are not loosened, continue enzymatic digestion for another 10 min and check again.
2. Retrieve two Petri dishes containing 7 mL of DMEM from the incubator. Use the widest-bore Pasteur pipet to transfer the muscle from the collagenase solution to the DMEM to rinse away the collagenase.
3. Transfer the muscle to the second Petri dish for further dilution of any possible collagenase remains.
4. Use another wide-bore pipet (diameter: approx 2 mm) to triturate the muscle along its length. This orientation of the EDL muscle during triturations is critical to prevent myofiber breakage.
5. When single myofibers are liberated from the muscle, its diameter decreases. Therefore, use a narrower-bore pipet for subsequent triturations.
6. When 20–30 viable single myofibers are released, transfer the muscle bulk to another DMEM-containing Petri dish and place the dish with the single myofibers in the tissue culture incubator. The transfer of the muscle bulk to a second dish ensures that the already released myofibers do not break during subsequent muscle triturations.
7. Repeat trituration and transfer muscle bulk to new Petri dishes until the desired number of viable isolated myofibers is acquired.

3.2.5. Culturing Single Myofibers in 24-Well Multiwell Dishes

1. Transfer a Matrigel-coated 24-well multiwell dish from the incubator to the tissue culture hood and open its lid to allow moisture, generated during the incubation period, to evaporate.
2. Bring two Petri dishes containing single myofibers to the tissue culture hood.
3. Use a 9-in. glass Pasteur pipet to lift one myofiber, with minimal residual medium, from the suspension and gently release the myofiber in the center of a well. Alternate between the two myofiber-containing dishes in order to have both early and late isolated single myofibers in each 24-well multiwell dish.
4. After myofibers are dispensed to all 24 wells, look under the stereo dissecting microscope and make sure that indeed there is a myofiber in each well. This step is necessary because occasionally myofibers adhere to the Pasteur pipet and are not released to the well.
5. If needed, add a myofiber to any empty well.
6. Approximately 10 min after distributing myofibers, slowly add 500 µL of warm culturing medium to each well, avoiding myofiber agitation.
7. Transfer the 24-well multiwell dish to the tissue culture incubator for a minimum of 18 h (overnight).
8. Repeat **steps 1–7** until the required number of cultured myofibers is reached.
9. An additional 500 µL of fresh culturing medium is provided to each well 1 wk after culturing the myofibers. Then, to replenish the medium, every 3 d about 500 µL of the medium is aspirated and 500 µL of fresh medium is added.

3.3. Immunolabeling of Satellite Cells in FDB and EDL Myofiber Cultures

This subsection details current protocols used in our laboratory to fix myofiber cultures for immunofluorescence studies of satellite cells. FDB myofiber cultures are typically fixed with ice-cold methanol (the preferred fixative when working with Vitrogen-collagen-coated dishes). EDL myofiber cultures are typically fixed with paraformaldehyde warmed to 37°C. These protocols allow recovery of intact myofibers at the end of the fixation procedure. It should be noted that the ideal fixatives for FDB or EDL myofiber cultures are not necessarily the optimal fixatives for antigen detection. Thus, when analyzing single myofibers via immunofluorescence, fixatives should be optimized for both preserving the myofibers and the antigens being analyzed.

Fixation protocols described in this subsection are also appropriate for detecting proliferating satellite cells in single myofibers by autoradiography following labeling with ^{3}H-thymidine *(7,19)* or when analyzing proliferation using bromodeoxyuridine *(16,18)*.

3.3.1. Protocols for Fixing and Immunofluorescent Staining of Isolated Single Myofiber Cultures

3.3.1.1. FDB Myofiber Cultures

1. Warm DMEM in a water bath set at 37°C.
2. Rinse cultures with warm DMEM three times. Following the final rinse, add 1 mL ice-cold 100% methanol to each 35-mm tissue culture dish and transfer the dishes to 4°C for 10 min.
3. Return dishes to room temperature, aspirate the methanol, and allow the dishes to air-dry for 10–15 min in the tissue culture hood (*see* **Note 17**).
4. Add 1.5 mL of blocking solution (TBS-NGS) to each culture dish to block non-specific antibody binding.
5. Cultures are then kept at 4°C for overnight or longer.
6. Dilute the appropriate primary antibody in the blocking solution.
7. Rinse the cultures three times with TBS-TW20.
8. Aspirate the final TBS-TW20 rinse and add 100 µL of the primary antibody solution for 1 h at room temperature followed by an overnight incubation at 4°C in a humidified chamber (*see* **Notes 18** and **19**).
9. Dilute the appropriate secondary antibody in the blocking solution.
10. Rinse cultures with TBS-TW20 three times.
11. Aspirate the final TBS-TW20 rinse and add 100 µL of the diluted secondary antibody for 1–2 h at room temperature.
12. Aspirate the secondary antibody and wash three times with TBS-TW20.
13. For nuclear visualization, add 100 µL of DAPI solution (4',6-diamidino-2-phenylindole, dihydrochloride; Sigma-Aldrich, cat. no. D8417; stock concentration 10 mg/mL, working concentration 1 µg/mL diluted in TBS-NGS prior to use) for 30 min at room temperature (*see* **Note 20**).

14. Rinse the cultures twice with TBS-TW20 followed by a final rinse with TBS.
15. Aspirate the TBS and mount in Vectashield mounting medium (1 drop at the center of each culture dish or well) and cover with a cover slip. Mounting medium prevents the stained cultures from drying and retards fading of the immunofluorescent signal.

3.3.1.2. EDL Myofiber Cultures

1. Warm the needed volume of paraformaldehyde–fixative solution in a water bath set at 37°C.
2. While observing each myofiber under the stereo dissecting microscope, use a pipetman to gently, without agitating the culture or touching the myofiber, add 500 μL of the warm paraformaldehyde–fixative solution to the culturing medium of each of the 24 wells, for 10 min at room temperature.
3. Use a pipet to remove the paraformaldehyde–fixative-medium solution and rinse each well three times with TBS.
4. Add 500 μL of TBS-TRX100 for 5 min at room temperature.
5. Add 500 μL of blocking solution (TBS-NGS) to each of the 24 wells to block nonspecific antibody binding.
6. Follow **steps 5–15** as described in **Subheading 3.3.1.1.**

4. Notes

1. Vitrogen collagen in solution is a sterile solution of purified, pepsin-solubilized bovine dermal collagen type I dissolved in 0.012 N HCl and stored at 4°C until used (Cohesion Technologies, Palo Alto, CA). In our studies, Vitrogen collagen is made isotonic by mixing 6 vol of stock Vitrogen collagen with 1 vol of 7X DMEM. The isotonic solution is prepared just prior to coating dishes because it gels rapidly at room temperature. To obtain consistent coating, the culture dishes should be precooled and coated on ice. When removed from the ice, these dishes warm up rapidly and are ready for myofiber addition. Vitrogen collagen in solution is the recommended product and the use of collagen from other companies would require prescreening to ensure compatibility.
2. Matrigel is a solubilized basement membrane preparation extracted from the Engelbreth–Holm–Swarm mouse sarcoma, a tumor rich in extracellular matrix proteins. Its major component is laminin, followed by collagen IV, entactin, and heparan sulfate proteoglycan *(20)*. To ensure Matrigel stability, we follow the manufacturer's handling instructions and aliquot 200 μL each into 2-mL cryogenic vials sealed with O-rings (Corning Inc., cat. no. 430488). These aliquots are stored at –20°C.
3. For additional details about the FDB muscle anatomy, refer to http://www. bartleby.com/107/illus443.html and http://www.bartleby.com/107/131.html. For additional details about the EDL muscle anatomy, refer to http://www. bartleby.com/107/illus437.html, http://www.bartleby.com/107/illus441.html, and http://www.bartleby.com/107/129.html. We recommend these links as good

resources for anatomical description and schematic images of the muscles although they refer to human muscles.

4. Falcon Primaria (Becton Dickinson Biosciences) 24-well multiwell dishes have typically been used for single myofiber isolation; however, we find that the standard, less expensive Falcon 24-well multiwell dishes are as good.

5. Horse serum (HS) should be preselected by comparing sera from various suppliers. We select HS based on its capacity to support proliferation and differentiation of primary chicken myoblasts cultured at standard and clonal densities *(21)*.

6. The Controlled Processed Serum Replacement 2 (CPSR-2; Sigma–Aldrich) that had been routinely used in our earlier studies *(4,9,10,15,19)* has been discontinued. The source of both the discontinued CPSR-2 and the currently available CPSR-3 is dialyzed bovine plasma; the discontinued product was further processed in a manner that also reduced lipids. The alternative serum replacement product contains bovine serum albumin, insulin, and transferrin and its use for mouse myofiber cultures has been previously described *(18)*.

7. Fetal bovine serum (FBS) should be preselected by comparing sera from several suppliers. We select FBS based on the capacity of the serum to support proliferation and differentiation of mouse primary myoblasts cultured at various cell densities. Only sera able to support growth and differentiation over a wide range of cell densities are employed in our studies. Primary myogenic cultures are prepared as described in **refs. 4** and **22**.

8. We prepare chicken embryo extract (CEE) in our laboratory using 10-d-old White Leghorn embryos *(23)*. The procedure is similar to a previously described method *(24)* but uses the entire embryo. We recommend this approach over purchasing CEE if the investigator can obtain embryonated chicken eggs, as the quality is higher and the cost lower than that of purchased CEE.

9. Preparation of chicken embryo extract. All steps are performed in a sterile manner.

   a. Embryonated chicken eggs (8 dozen, White Leghorn; from Charles River) are maintained in a standard egg incubator (incubation conditions: a dry temperature of 38°C, a wet temperature of 30°C, and relative humidity of 56%). The following egg incubator is well suited for basic research use: Marsh Automatic Incubator, Model PRO-FI, cat. no. 910-028, manufactured by Lyon Electric Company Inc., Chula Vista, CA.

   b. After 10 d, batches of 15–30 eggs are removed from the incubator and transferred into the tissue culture hood.

   c. Place the eggs lengthwise in the rack and spray with 70% ethanol to sterilize. Wait for several minutes until the ethanol evaporates.

   d. Crack open one egg at a time into a 150-mm Petri dish.

   e. Remove the embryo from surrounding membranes by holding it with fine forceps. Rinse the embryo by transferring it through three 150-mm Petri dishes containing minimal essential medium (MEM; Gibco–Invitrogen, cat. no. 11095-080, MEM is supplemented with antibiotics as described for DMEM in **Subheading 2.7.1.**). Swirl embryo a few times in each dish for a good rinse.

f. Empty the egg remains from the initial 150-mm dish (described in **step d**) into a waste beaker and repeat **steps d–f** until the final rinse dish contains about 30 embryos.

g. The embryos are transferred with fine forceps into a 60-mL syringe, forced through with the syringe plunger, and the suspension is collected into a 500-mL sterile glass bottle.

h. The extract is diluted with an equal volume of MEM and gently agitated for 2 h at room temperature. To ensure good agitation, keep maximum volume to one-half bottle capacity.

i. The extract is frozen at –70°C for a minimum of 48 h. It is then thawed, dispensed to sterile glass Corex tubes, and centrifuged at 15,000g for 10 min to remove particulate material.

j. The supernatant is pooled, divided into 5-mL aliquots, and kept frozen until needed.

k. Prior to use, the CEE should again be centrifuged at about 700g for 10 min to remove aggregates, passed through a 0.45-µm filter, followed by a 0.22-µm filter (to clear remaining particles and to ensure sterility).

10. Methanol is a colorless, flammable liquid with an alcohol-like odor. Use nitrile gloves, safety goggles, and a fume hood when handling. It is important to refer to the MSDS instructions and institutional regulations for further information regarding storage, handling, and first-aid.

11. Preparation of Tris-buffered saline (TBS). To make 1 L of 10X TBS:

a. Weigh 60.5 g of Tris base into a beaker.

b. Add 700 mL deionized water to the beaker.

c. Place the beaker on top of a magnetic stirrer.

d. When the powder has dissolved, adjust the pH to 7.4.

e. Add deionized water to bring the volume up to 1 L, mix well, and store at 4°C.

To make 1 L of TBS:

a. Weigh 8.766 g NaCl in a beaker

b. Add 100 mL of 10X TB to the beaker and mix vigorously.

c. When the powder has dissolved, add deionized water to bring the volume up to 1 L; mix well and store at 4°C.

d. In a sterile environment: filter through a 0.45-µm disposable filter unit (Nalgene, cat. no. 0001530020) into a bottle.

e. Store at 4°C.

12. Paraformaldehyde is a white powder with a formaldehyde-like odor. It is a rapid fixative and a potential carcinogen. When handling paraformaldehyde, wear gloves, mask, and goggles. It is important to refer to the MSDS instructions and institutional regulations for further information regarding storage, handling and first-aid.

13. Preparation of 100 mL of 4% paraformaldehyde with 0.03 M sucrose in a fume hood:

a. Mix 4 g of paraformaldehyde powder and 80 mL of deionized water in a glass beaker; cover with parafilm.

  b. Warm the solution to 60°C with continuous stirring to dissolve the powder.
  c. Allow the solution to cool to room temperature.
  d. Add about 1–4 drops of 1 *N* NaOH, until the opaque color of the solution clears.
  e. Add 10 mL of 1 *M* sodium phosphate.
  f. Adjust the pH to 7.2–7.4 using color pH strips.
  g. Add 1.026 g of sucrose.
  h. Bring volume to 100 mL.
  i. Filter through a 0.45-μm disposable filter unit (Nalgene, cat. no. 0001530020) into a bottle.
  j. Store at 4°C in an aluminum-foil-wrapped bottle for no more than 1 mo.
14. Collagenase concentration, as well as the optimal time for enzymatic digestion, should be adjusted for younger or older mice and for different strains of mice.
15. FDB and EDL myofiber isolation protocols include gentle agitation during enzymatic digestion. However, if maintaining quiescence of satellite cells is an important aspect of the study, we recommend to avoid continuous agitation. Instead dishes should be gently swirled every 10–15 min.
16. The time required for the myofiber suspension to settle (at 1*g*) through 10 mL of 10% HS can vary between 5 and 15 min and the investigator should adjust this time. A prolonged period results in a preparation with more debris and remaining single cells released from the digested tissue. Depending on mouse age, the number of rounds of myofiber settling in the 15-mL glass Corex tubes, as well as the amount of medium in the tube, might also need to be adjusted.
17. The tissue culture dishes are dry when the bottom appears opaque white.
18. For some antibodies the cultures may be blocked for just 2–4 h at room temperature if overnight blocking is not desired.
19. For even and continuous distribution of the antibodies (both primary and secondary), it is recommended to place the dishes on a three-dimensional rotator (Labline Maxi Rotator; VWR Scientific Products, cat. no. 57018-500). It is especially important when staining myofibers in 24-well multiwell dishes because the antibody aliquots tend to rapidly accumulate at the well periphery, leading to uneven staining across the culture.
20. DAPI is potentially harmful. Avoid prolonged or repeated exposure; we typically dissolve the entire powder in its original container and generate a concentrated stock solution. A ready-made DAPI reagent is available from Molecular Probes. It is important to refer to the MSDS instructions and institutional regulations for further information regarding storage, handling, and first-aid.

Acknowledgments

We are grateful to Monika Wleklinski-Lee and Stefanie Kästner for helpful comments on the manuscript. ZYR thanks Stefanie Kästner and Anthony Rivera for their valuable contributions to the FDB myofiber studies. GS thanks Dr. Terrance Partridge and members of his research team for advice on EDL myofiber isolation during her former studies.

The studies described in this chapter have been supported by grants to ZYR from the National Institute of Health (AG13798 and AG21566), the Cooperative State Research, Education and Extension Service/US Department of Agriculture (National Research Initiative Agreement no. 99-35206-7934), and the Nathan Shock Center of Excellence in the Basic Biology of Aging, University of Washington. Earlier support from the Muscular Dystrophy Association and the USDA (NRI Agreements no. 93-37206-9301 and 95-37206-2356) has facilitated our initial studies on the isolation and culture of rat myofibers.

References

1. Mauro, A. (1961) Satellite cells of skeletal muscle fibers. *J. Biophys. Biochem. Cytol.* **9**, 493–495.
2. Hawke, T. J. and Garry, D. J. (2001) Myogenic satellite cells: physiology to molecular biology. *J. Appl. Physiol.* **91**, 534–551.
3. Yablonka-Reuveni, Z., Quinn, L. S., and Nameroff, M. (1987) Isolation and clonal analysis of satellite cells from chicken pectoralis muscle. *Dev. Biol.* **119**, 252–259.
4. Kästner, S., Elias, M. C., Rivera, A. J., and Yablonka-Reuveni, Z. (2000) Gene expression patterns of the fibroblast growth factors and their receptors during myogenesis of rat satellite cells. *J. Histochem. Cytochem.* **48**, 1079–1096.
5. Webster, C., Pavlath, G. K., Parks, D. R., Walsh, F. S., and Blau, H. M. (1988) Isolation of human myoblasts with the fluorescence-activated cell sorter. *Exp. Cell. Res.* **174**, 252–265.
6. Bekoff, A. and Betz, W. (1977) Properties of isolated adult rat muscle fibers maintained in tissue culture. *J. Physiol.* **271**, 537–547.
7. Bischoff, R. (1986) Proliferation of muscle satellite cells in intact myofibers in culture. *Dev. Biol.* **115**, 129–139.
8. Bischoff, R. (1989) Analysis of muscle regeneration using single myofibers in culture. *Med. Sci. Sports Exerc.* **21**, S164–S172.
9. Yablonka-Reuveni, Z. and Rivera, A. J. (1994) Temporal expression of regulatory and structural muscle proteins during myogenesis of satellite cells on isolated adult rat fibers. *Dev. Biol.* **164**, 588–603.
10. Yablonka-Reuveni, Z., Rudnicki, M. A., Rivera, A. J., Primig, M., Anderson, J. E., and Natanson, P. (1999) The transition from proliferation to differentiation is delayed in satellite cells from mice lacking MyoD. *Dev. Biol.* **210**, 440–455.
11. Rosenblatt, J. D., Lunt, A. I., Parry, D. J., and Partridge, T. A. (1995) Culturing satellite cells from single muscle fibre explants. *In Vitro Cell. Dev. Biol. Anim.* **31**, 773–779.
12. Rosenblatt, J. D., Parry, D. J., and Partridge, T. A. (1996) Phenotype of adult mouse muscle myoblasts reflects their fiber type of origin. *Differentiation* **60**, 39–45.
13. Shefer, G., Wleklinski-Lee, M., and Yablonka-Reuveni, Z. (2004) Skeletal muscle satellite cells can spontaneously enter an alternative mesenchymal pathway. *J. Cell Sci.*, in press.
14. Blaveri, K., Heslop, L., Yu, D. S., et al. (1999) Patterns of repair of dystrophic mouse muscle: Studies on isolated fibers. *Dev. Dyn.* **216**, 244–256.

15. Yablonka-Reuveni, Z., Seger, R., and Rivera, A. J. (1999) Fibroblast growth factor promotes recruitment of skeletal muscle satellite cells in young and old rats. *J. Histochem. Cytochem.* **47,** 23–42.

16. Shefer, G., Partridge, T. A., Heslop, L., Gross, J. G., Oron, U., and Halevy, O. (2002) Low-energy laser irradiation promotes the survival and cell cycle entry of skeletal muscle satellite cells. *J. Cell Sci.* **115,** 1461–1469.

17. Greene, E. C. (1963) *Anatomy of the Rat*, Hafner, New York.

18. Wozniak, A. C., Pilipowicz, O., Yablonka-Reuveni, Z., Greenway, S., Craven, S., and Scott, E. (2003) C-met expression and mechanical activation of satellite cells on cultured muscle fibers. *J. Histochem. Cytochem.* **51,** 1437–1445.

19. Yablonka-Reuveni, Z. and Rivera, A. J. (1997) Proliferative dynamics and the role of FGF2 during myogenesis of rat satellite cells on isolated fibers. *Basic Appl. Myol.* **7,** 189–202.

20. Kleinman, H. K., McGarvey, M. L., Liotta, L. A., Robey, P. G., Tryggvason, K., and Martin, G. R. (1982) Isolation and characterization of type IV procollagen, laminin, and heparan sulfate proteoglycan from the EHS sarcoma. *Biochemistry* **21,** 6188–6193.

21. Yablonka-Reuveni, Z. and Seifert, R. A. (1993) Proliferation of chicken myoblasts is regulated by specific isoforms of platelet-derived growth factor: evidence for differences between myoblasts from mid and late stages of embryogenesis. *Dev. Biol.* **156,** 307–318.

22. Yablonka-Reuveni, Z. (2004) Isolation and characterization of stem cells from adult skeletal muscle, in *Handbook of Stem Cells* (Lanza, R. P., Blau, H. M., Melton, D. A., et al., eds.), Elsevier, San Diego, CA, pp. 571–580.

23. Yablonka-Reuveni, Z. (1995) Myogenesis in the chicken: the onset of differentiation of adult myoblasts is influenced by tissue factors. *Basic Appl. Myol.* **5,** 33–42.

24. O'Neill, M. C. and Stockdale, F. E. (1972) A kinetic analysis of myogenesis in vitro. *J. Cell Biol.* **52,** 52–65.

20

Adult Ventricular Cardiomyocytes
Isolation and Culture

Klaus-Dieter Schlüter and Daniela Schreiber

Summary

Isolated cardiomyocytes are a prerequisite to study the biology of cardiomyocytes. Efficient isolation is difficult, as these cells adhere firmly together in the heart and do not divide. Therefore, any experiment is restricted to the amount of calcium-tolerant, rod-shaped cardiomyocytes that can be initially isolated from the heart. This chapter gives detailed instructions on how ventricular cardiomyocytes can be isolated from an intact adult heart. The method is based on the principle of calcium-free perfusion with collagenase supplementation to disrupt cell–cell contacts in the heart, isolation and purification of cardiomyocytes from other cell types, and, finally, re-establishing a physiological cellular calcium concentration. The chapter also summarizes some commonly used adaptations to isolate cardiomyocytes from species different from rat.

Key Words: Cell culture; adult ventricular cardiomyocytes; cell attachment; heart cells; cell biology.

1. Introduction

Although cardiomyocytes do not represent the majority of cells in the adult heart, they represent the main cell mass and determine the function of the ventricle. The analysis of heart function requires the ability to analyze functional characteristics of ventricular cardiomyocytes in regard to size control, contractility, and biochemical and molecular properties at the isolated cell level. This has two advantages: First, the investigator can fully control the conditions to which the cells are exposed, and, second, the investigator can analyze the functional behavior of ventricular cells irrespective of the influence of other cells (i.e., fibroblasts and endothelial cells known to influence the functional behavior of ventricular cells for mechanical reasons [fibrosis] or by the release of paracrine factors). In addition, in the whole heart, energy supply depends on

From: *Methods in Molecular Biology, vol. 290: Basic Cell Culture Protocols, Third Edition*
Edited by: C. D. Helgason and C. L. Miller © Humana Press Inc., Totowa, NJ

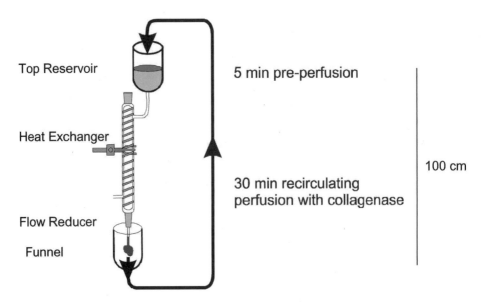

Fig. 1. Schematic of a perfusion system used for isolation of ventricular cardiomyocytes.

coronary perfusion, which might be inadequate. On the cell culture level, oxygen supply can easily be controlled. Techniques for the isolation of cardiomyocytes have been difficult to establish because heart muscle cells are firmly connected to each other by intercalated disks and the extracellular matrix network and these connections are difficult to cleave without injuring the cells. Another important issue is that ventricular cardiomyocytes are terminally differentiated and do not divide in vitro. Therefore, cells must be newly isolated for each individual experiment. This requires the establishment of highly reproducible techniques that guarantee a reproducible quality of the cardiomyocyte preparations. The isolation procedure, established for ventricular cardiomyocytes from rat ventricle, will be described in this chapter, as well as method modifications for isolating cardiomyocytes from other species.

2. Materials

All glassware and instruments are sterilized by autoclaving at 121°C or by procedures recommended by the manufacturer.

2.1. Perfusion System (Langendorff System; see Fig. 1 and Note 1)

1. Top reservoir (100 mL), double walled, temperature controlled.
2. A glass-coil heat exchanger with two cannulas fitted to its outlet; the distance between the top reservoir and cannulas is 100 cm.

3. Connection by a double-walled, temperature-controlled glass tube between the top reservoir and heat exchanger, containing a flow reducer.
4. Funnel, which can be moved below the cannulas to collect the fluid, connected with a tube leading to the top reservoir.
5. Roller pump for pumping the fluid back to the top reservoir.
6. Temperature-controlled water circulator, for 37°C temperature control of the Langendorff system.
7. Pasteur pipet for gassing the top reservoir with carbogen (95% O_2/5% CO_2).

2.2. Instruments

1. Two scissors (coarse and fine).
2. Two small forceps.
3. Two crocodile clamps.
4. Two large Petri dishes (200 mm in diameter).
5. Two scalpels and a watchglass (or a tissue chopper).
6. Nylon mesh (mesh size = 200 µm; i.e., from Neolab, cat. no. 4-1413).
7. Two 50-mL centrifuge tubes.
8. One 50-mL glass beaker.
9. One 50-mL Teflon or siliconized glass beaker.
10. 50-mL Erlenmeyer flask.
11. Two long centrifuge tubes (length = 15 cm, diameter = 1 cm).
12. Plastic Pasteur pipet (large mouth or, alternatively, glass Pasteur pipet with 90° angle tip).
13. Disposable 5-mL pipet with mouth about 2 mm in diameter.

2.3. Media

All solutions are filter-sterilized using 0.2-µm filter apparatus.

1. Perfusion buffer: 110 mM/L NaCl, 2.6 mM/L KCl, 1.2 mM/L KH_2PO_4, 1.2 mM/L $MgSO_4$, 25 mM/L $NaHCO_3$, and 11 mM/L glucose. The buffer (minus glucose) can be prepared and stored at 2–8°C for up to 1 mo and the glucose added just before use. Before use, warm the perfusion buffer to 37°C and continuously gas with 95% O_2/5% CO_2 (equilibrate to pH 7.4).
2. Ca^{2+} stock solution: 100 mM/L $CaCl_2$ in H_2O.
3. Saline: 9 g/L NaCl, ice cold.
4. Collagenase: crude collagenase, from clostridium histolyticum (*see* **Note 2**). Suitable suppliers include Worthington, Serva, and Sigma. The final concentration is about 400 mg/L in perfusion buffer. Store collagenase in aliquots sufficient for one perfusion at 4°C. These will be added to a 50-mL aliquot of the perfusion buffer and the suspension filled in the reservoir of the Langendorff perfusion system.
5. Bovine serum albumin (BSA) gradient: 4% BSA (w/v), 1% calcium stock solution (1 mM final concentration) in perfusion buffer.

6. CCT medium: Medium 199 supplemented with 5 m*M*/L creatine, 2 m*M*/L car-nitine, 5 m*M*/L taurine, 10 μ*M*/L cytosine-β-arabinofuranoside), 100 IU/mL peni-cillin, and 100 μg/mL streptomycin. CCT should be filtered to sterilize the medium and can be stored up to 3 mo at 4°C.
7. Fetal bovine serum (FBS): prescreened batch for promoting myocyte adherence (*see* **Note 3**).
8. Laminin (i.e., Roche, cat. no. 1243217)

3. Methods

The rat heart model is still the most commonly used model; therefore, the procedure for the isolation of ventricular cardiomyocytes from adult rats is described. First, the heart is perfused in a nominal Ca^{2+}-free collagenase-containing buffer to disrupt cell–cell junctions. Second, the heart tissue is fully digested to a cell suspension and then the ventricular cardiomyocytes are separated from nonmyocytes. Third, the cells are readjusted to a physiological Ca^{2+} concentration. Fourth, the cells are plated on culture dishes and allowed to adhere firmly to the dishes.

3.1. Preparation of the Heart

Before starting to remove the heart from the animal, the perfusion system must be prepared. The buffer must be filled into the perfusion system, warmed up to 37°C, and gassed with carbogen.

1. Anesthetize rat according to procedures approved by your institution (*see* **Note 4**).
2. Open the thorax and flush with ice-cold saline.
3. Quickly remove the heart from the thorax. Take the heart in your left hand, cut esophagus and trachea, tear lung and heart in your direction, and cut aorta; transfer everything to a Petri dish filled with ice-cold saline.
4. Remove adjacent tissue and cut the aorta after the first aortic arch.
5. Connect the heart to the perfusion system according to the Langendorff method as outlined in **Subheading 3.2.** It is important to perform these steps quickly.

3.2. Perfusion of the Hearts

The perfusion should be started first (1 drop per second) and then the aorta should be connected to the system.

1. While the heart is immersed in saline, open the aortic lumen with the two forceps.
2. Lift the heart to the cannula, slip aorta over cannula, and fix aorta with crocodile clamps. Do not insert the cannula too deep into the heart to ensure that the perfusion via the coronary vessels is possible.
3. Once the heart is fixed to the system with the crocodile clamps, fix it with threads (i.e., surgical tread).
4. Remove the clamps and continue to perfuse the heart with perfusion buffer for 3–5 min to remove blood from the organ.

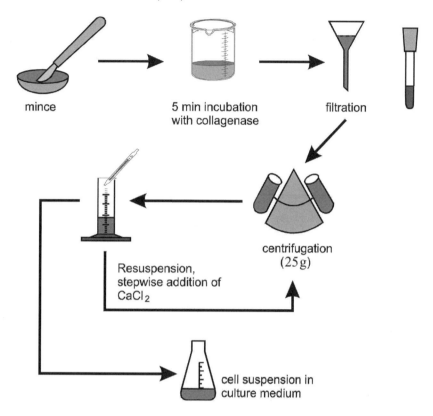

mince

5 min incubation
with collagenase

filtration

centrifugation
(25 g)

Resuspension,
stepwise addition of
$CaCl_2$

cell suspension in
culture medium

Fig. 2. Schematic drawing of the isolation steps following perfusion of the hearts.

5. Add collagenase to the buffer (to give a final concentration of 400 mg/L) in the top reservoir and continue perfusion for 25 min (*see* **Note 5**). When the number of drops per second decreases readjust perfusion to approx 10 mL/min.

3.3. Postperfusion Digestion and Separation From Nonmyocytes

The perfusion of the hearts should be stopped when the hearts are soft.

1. Carefully cut the ventricles and separate from the atrium and aorta.
2. Transfer ventricles to a separate glass dish (*see* **Fig. 2**) and mince using a scalpel.
3. Place the cell suspension into a Teflon beaker and add prewarmed (37°C) collagenase buffer. The cell suspension is gassed again for the next 5 min with carbogen. The success of digestion can be improved by moderately pipetting the solution several times.
4. Filter the cell suspension through nylon mesh (mesh size = 200 μm) into sterile 50-mL tubes and centrifuge for 3 min at 25g. The pellet contains the cardiomyocytes and the supernatant contains small nonmyocytes from the ventricles, mainly endothelial cells and fibroblasts.

5. Resuspend the cardiomyocytes pellet in the buffer medium and add 0.2% (v/v) calcium stock solution to give a final concentration of 200 µmol/L calcium.
6. Centrifuge for 3 min at 25*g* and resuspend in buffer with 0.4% (v/v) calcium stock solution (final concentration = 400 µmol/L).
7. Centrifuge again and resuspend cell pellet in buffer with 1% (v/v) calcium stock solution (final concentration = 1 mmol/l) to a total volume of about 2–3 mL.
8. Layer 2–3 mL of cell suspension on the top of 12 mL of BSA gradient and centrifuged at 15*g* for 1 min.
9. The pellet contains rod-shaped and calcium-tolerant cardiomyocytes. Suspend cell pellet in 25 mL CCT medium.
10. Count cells and estimate number of rod-shaped cardiomyocytes (*see* **Note 6**).

3.4. Culture of Cardiomyocytes

Cardiomyocytes can be cultured in their native rod-shaped form for a limited time. As they do not beat spontaneously, they will start to get atrophic within the next 2 d because of a degradation of contractile proteins (*1*). However, during this time, protein synthesis is in balance with degradation (*2*). The use of the cells requires appropriate attachment to the culture dishes. This can be achieved by preincubation of the dishes. Different attachment substrates have been described in the literature with 4% FBS or preincubation of the dishes with laminin (1 g/mL) commonly used. In both cases, the preincubation solution is removed just before plating the cells to the dishes. The ability of the cells to attach to culture dishes preincubated by FBS depends on the quality of the serum batch (*see* **Note 3**). Laminin routinely gives good and reproducible results. However, it is much more expensive than the use of FBS. *See* **Note 7** for additional information on the isolation and culture of rat cardiomyocytes.

All subsequent cultivation steps should be performed under sterile conditions in a certified biosafety cabinet.

1. Add sufficient 4% FBS in CCT medium or laminin solution (1 g/mL) to coat the tissue-culture treated dishes and incubate over night at 37°C with FBS or laminin solution.
2. Remove solution and add cardiomyocyte cell suspension (*see* **Fig. 3**). The appropriate cell density for plating your cells depends on the type of experiments you want to perform. Often, isolated ventricular cardiomyocytes are used for single-cell experiments with microscopic techniques and, therefore, a low plating density is optimal. In any case, the number of cells should not be higher than 1.4×10^5 cells per 35-mm dish to ensure that all cells have full-length contact to the culture dish.
3. After 2 h, the dishes should be washed carefully to remove non-adherent cells, and the medium replaced by fresh CCT medium.
4. Incubate at 37°C. Monitor myocyte cultures daily until use (*see* **Note 8**).

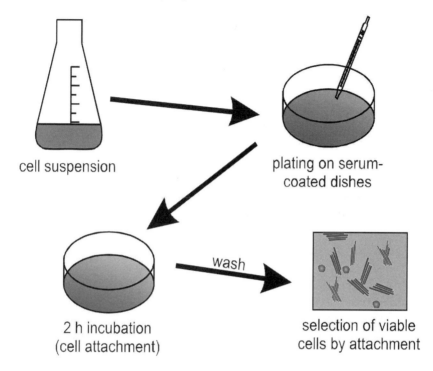

Fig. 3. Procedure to get isolated cells attached to culture dishes.

3.5. Adaptations to Other Species

The frequent use of transgenic mice strains has encouraged development of isolation procedures for mice myocytes. In general, the above-described proto-col can be used, but slight modifications are required. First, the perfusion sys-tem and volume of buffers must be reduced to the smaller size of mice heart. Second, reconstitution of a physiological calcium concentration should be per-formed more carefully. Instead of the three-step increase from nominal cal-cium-free buffer to physiological calcium concentrations described for rat myocytes, a four-step strategy leading to incremental increases to 125, 250, 500, and 1000 mmol/L is recommended. The third difference between rat and mouse myocytes relates to the choice of the attachment substrate. Whereas FBS was found to be a less expensive and attractive alternative to laminin on rat cells, mice myoyctes do not efficiently attach to FBS-pretreated culture dishes. Thus, use of laminin is strictly recommended. In addition, one has to take in account that because of the small size of the mouse, the number of cells isolated from the ventricle is about 10% of that isolated from rat ventricles. Therefore, the low cell numbers available precludes performing many biochemical experiments and limits the use of myocytes from mice mainly on single-cell experiments with regard to electrophysiology or cell contraction.

Upscaling of the method for isolation of myocytes from large animals is another goal in the cardiophysiology field. The isolation procedure for ventricular cardiomyocytes from pigs can be used as a guide for hearts from larger animals or human hearts. Ventricular cardiomyocytes are enzymatically isolated from a wedge of the left ventricular free wall, supplied by the left anterior descending coronary artery. The coronary artery is cannulated and small leakage branches at the edge of the preparation are ligated. The tissue can then be perfused through its supplying artery in a Langendorff system using a buffer system as described in **Subheading 2.3.** Collagenase (1.4 g/L) and protease (0.1 g/L) are added and perfusion is maintained for 25 min at 80 mm Hg. The enzyme solution is then washed out by perfusion with the buffer with addition of 0.18 mmol/L Ca^{2+}. Then, thin slices are cut from the wedge into small pieces, transferred through a Nylon mesh as described earlier for rat myocytes, and resuspended in a buffer containing 0.18 mmol/L Ca^{2+}. Physiological calcium concentration (1.8 mmol/L) is re-established after 20 min by slow replacement of the low-calcium buffer with a buffer containing 1.8 mmol/L calcium.

4. Notes

1. Langendorff apparatus can be purchased by many laboratory distributors (i.e., Experimetria Ltd., Budapest, Hungary).
2. Collagenase batches must be preselected. Order four to six different batches from the distributor and use them at the concentration given in **Subheading 2.3.** Use the one that gave the best results, again by increasing or decreasing the concentration briefly (±50 mg/L). Compare the results with the previous results and use the best concentration and the best batch. If inexperienced or setting up the procedure for the first time, contact a group with experience and ask for a small amount of collagenase that works quite well, so that you can compare the results with the new batches with this collagenase previously tested.
3. Fetal bovine serum must be preselected. Ask your distributor for four to six test samples. Plate dishes each with 4% serum overnight and plate cardiomyocytes after isolation. Wash the dishes after 4 h, count the number of cells per dish, and select the serum with the most cells still attached.
4. There are no specific recommendations for the procedure, as there are no reports in the literature suggesting that any of the commonly used protocols to anesthetize the animals are detrimental to the isolated cardiomyocytes. In order to avoid blood thrombosis during the time the heart is not perfused, the animals can receive heparin together with the anesthetics.
5. Crude collagenase is not a pure enzyme preparation and the amount of collagenase added to the buffer as well as the time to perfuse the heart are not fixed. They are dependent on the specific activity of the collagenase. The concentration of collagenase and the time for perfusion given here can be taken as a gross

criteria in which most of the collagenase suitable for rat heart (*see* **Note 1**) give reasonable results.

6. A successful isolation of calcium-tolerant cardiomyocytes should meet two criteria: First, the number of cardiomyocytes isolated from the ventricle should be sufficient and, second, the number of rod-shaped cardiomyocytes should be high compared to round cells. A suitable preparation has been achieved if one gets approx 4×10^6 cells per heart and a ratio between rod-shaped to round myocytes of 70:30. In addition, the cells are quite intact if they do not beat spontaneously. That means that no more than 10% of the isolated cells beat spontaneously with more than 10 beats/min.

7. Even in laboratories with long-term experience in the isolation of cardiomyocytes, from time to time the quality of cell preparations declines. Some general guidelines for troubleshooting are as follows:

   a. Avoid the use of detergents in cleaning the glassware used for cell isolation.
   b. Water quality is a limiting factor for the success of the isolation procedure. Therefore, in case of failure, solutions should be freshly prepared from a water source different from the usual one. What often happens is that the common primary ion-exchange purification system releases volatile organic impurities.
   c. It is important to keep the perfusion system clean. Therefore, the system should be washed first with clean water, then with 70% ethanol for 30 min, and, finally, dry the system using a stream of clean gas (e.g., filtered compressed air).
   d. Beginners should start to establish the method by using rats to sort out basic difficulties. The procedure is mainly adapted to this species and many laboratories have extensive experience with this system.
   e. Although you perfuse the heart for 30 min with collagenase, it is still not soft. The main reason for this problem is that the blood was not totally removed during the initiation of perfusion and has coagulated. It does not make sense to continue with the preparation. Try to reduce the time between opening the thorax to remove the heart and connecting the heart to the perfusion system, or inject heparin to the animal before starting to prepare the heart.
   f. Another problem that often occurs is that the heart is soft but the cells are all rounded up. Main reason is that the calcium gradient is too steep. View the morphology of the myocytes after each centrifugation step and increase the gradient more carefully (e.g., as indicated for mice myocytes).
   g. The number of damaged (round) cells can be reduced by trypsinization (10 mg trypsin per 50 mL buffer). Trypsin should be added to the suspension treated after perfusion. Rounded and damaged cells are more easily digested than intact cells. The amount of trypsin must be adjusted to the batch used and the combined effect to collagenase that is used in parallel. The number of damaged cells can also be reduced by gently washing the culture dishes 2 h after plating. Rounded cells are less well attached to the cell culture dish and are, therefore, dislodged more easily.

8. Cardiomyocytes can be cultured for a longer times when cultured in the CCT medium with an additional supplementation of 20% FBS. The cells then initially undergo a period of atrophy, round up, and change their morphology completely *(1)*. After 6 d, they have reached a stable situation in which the cells spread around the center. These cells have lost some of their in vivo characteristics, like the rod-shaped morphology. However, energy metabolism and coupling of specific receptors (i.e., α-adrenoceptors to the regulation of protein synthesis) are still intact *(3)*. They start to built up new cell–cell contacts *(4)*. Furthermore, they start to secrete growth factors, like transforming growth factor β, which is activated by proteases in the FBS. Therefore, although the model represents a suitable model for some questions in the cardiovascular field, one has to keep in mind the limitation of this culture procedure.

References

1. Piper, H. M., Jacobsen, S. L., and Schwartz, P. (1988) Determinants of cardiomyocyte development in long-term primary culture. *J. Mol. Cell. Cardiol.* **20,** 825–835.
2. Pinson, A., Schlüter K.-D., Zhou, X. J., Schwartz, P., Kessler-Icekson, G., and Piper, H. M. (1993) Alpha- and beta-adrenergic stimulation of protein synthesis in cultured adult ventricular cardiomyocytes. *J. Mol. Cell. Cardiol.* **25,** 477–490.
3. Schlüter, K.-D., Goldberg, Y., Taimor, G., Schäfer, M., and Piper, H. M. (1998) Role of phosphatidyl 3-kinase activation in the hypertrophic growth of adult ventricular cardiomyocytes. *Cardiovasc. Res.* **40,** 174–181.
4. Schwartz, P., Piper, H. M., Spahr, R., Hütter, F. J., and Spieckermann, P. G. (1985) Development of new intracellular contacts between adult cardiac myocytes in culture. *Basic Res. Cardiol.* **80(Suppl. 1),** 75–78.

21

Isolation and Culture of Primary Endothelial Cells

Bruno Larrivée and Aly Karsan

Summary

The purpose of this chapter is to describe the isolation techniques that result in pure cultures of human vascular endothelial cells from the umbilical vein and umbilical cord blood. We first describe the isolation of human umbilical vein endothelial cells (HUVECs). Additional protocols describe the isolation of umbilical-cord-blood-derived endothelial cells, the basic procedures of endothelial cell culture (including cryopreservation) and the methods used to characterize the phenotype of endothelial cells.

Key Words: Endothelium; endothelial progenitors; HUVEC; cell isolation; cell culture.

1. Introduction

Once viewed as a passive layer of cells forming a barrier between the blood and tissues, the endothelium is now considered to act as a distributed organ, whose functions are critical for normal homeostasis. Endothelial cells line the lumina of all blood vessels. Because of their unique position, they are the only cells in the body that form an interface among a fluid moving under relatively high pressure, the blood, and a solid substrate, the vessel wall *(1)*. The endothelium is involved in many critical processes such as maintaining vascular tone, acting as a selectively permeable barrier, regulating coagulation and thrombosis, and directing the passage of leukocytes into areas of inflammation *(2)*. To gain insight into the role of endothelial cells in normal and pathological states, investigators have isolated microvascular and macrovascular endothelial cells from a wide range of both animal and human vessels, including the human umbilical vein *(3–6)*. Human umbilical vein endothelial cells (HUVECs) have been used extensively to study the biology and pathobiology of the human endothelial cell, and most of our knowledge of human endothelial cells is derived from experiments with cultured HUVECs. The main advantage of

From: *Methods in Molecular Biology, vol. 290: Basic Cell Culture Protocols, Third Edition*
Edited by: C. D. Helgason and C. L. Miller © Humana Press Inc., Totowa, NJ

using HUVECs to study endothelial biology is the wide availability of the umbilical cord, a relatively simple method of isolation, and the general purity of the cell population obtained. Although macrovascular endothelial cells have been isolated from several human vessels, including the femoral artery, portal vein, and pulmonary artery, these vessels are, in general, not as easily obtained as umbilical cords, therefore making them not as convenient to isolate. This chapter is restricted to the isolation and culture of HUVECs and umbilical-cord-blood-derived endothelial cells.

2. Materials (see Note 1)

1. Phosphate-buffered saline (PBS).
2. Culture medium MCDB 131 (Sigma).
3. Culture medium Dulbecco's modified Eagle's medium (DMEM) (Sigma).
4. Culture medium Iscove modified Dulbecco's medium (IMDM) (Sigma)
5. Penicillin G (5000 U/mL) + streptomycin (5000 µg/mL) (Gibco).
6. Endothelial cell growth supplement (BD Biosciences).
7. Heparin (Sigma).
8. Heat-inactivated fetal bovine serum (FBS). To heat inactivate the complement present in the serum, incubate FBS at 56°C for 30 min (*see* **Note 1**).
9. HUVEC culture medium. Preparation of this medium is carried out as outlined in **Table 1**. HUVEC medium should be warmed up at room temperature or 37°C before use. It is advisable to aliquot HUVEC medium in 50-mL tubes that are kept at 4°C until use to avoid repeated warming and cooling of medium.
10. Collagenase A (Sigma). 13 mg/100 mL dissolved in serum-free DMEM containing penicillin and streptomycin (50 U/mL and 50 µg/mL, respectively). Filter through a 0.2-µm filter. Collagenase solution should be prepared fresh prior to HUVEC isolation.
11. Trypsin EDTA. A 5X stock solution of trypsin EDTA is prepared as outlined in **Table 2**. The solution is sterilized by filtration through a 0.2-µm filter. Dispense 10-mL aliquots into 50-mL tubes and store at –20°C. Before use, add 40 mL of sterile deionized water to dilute to 1X. Keep at 4°C for up to 4 wk.
12. Cotton gauze (autoclaved).
13. Cannulae.
14. Two-way stopcocks.
15. Hemostats.
16. Cable ties.
17. Scalpel blades.
18. Beakers.
19. Ethanol.
20. 0.2-µm Filters.
21. Bleach.
22. 30-mL Syringes
23. 0.2% Gelatin. For 10 mL of a 0.2% gelatin working solution, dilute 1 mL of a stock solution of 2% gelatin (Sigma) in 9 mL of PBS. Prepare fresh before each use.

Table 1
Preparation of HUVEC Medium

Component	Volume added for 500 mL	Final concentration
Fetal bovine serum	100 mL	20%
Penicillin (5000 U/mL) + streptomycin (5000 µg/mL)	5 mL	50 U/mL penicillin 50 µg/mL streptomycin
Endothelial cell growth supplement (10 mg/mL)	1 mL	20 µg/mL
Heparin (8000 U/mL)	1 mL	16 U/mL
MCDB 131	To final volume of 500 mL	

Table 2
Preparation of Trypsin EDTA (5X)

Component	Quantity added for 200 mL	Final concentration (1X)
Trypsin	2.5 g	0.25%
EDTA (250 mM)	4 mL	1 mM
PBS (10X)	100 mL	1X
Deionized water	To final volume of 200 mL	

24. 50-mL Tubes.
25. 100-mm Tissue culture plates (Falcon).
26. Ficoll–Paque (Amersham).
27. Human vascular endothelial growth factor (VEGF) (R&D).
28. Human basic fibroblast growth factor (FGF)-2 (R&D).
29. HUVEC freezing medium. Prepare fresh as needed by mixing the necessary volumes of 45% FBS, 45% HUVEC medium, and 10% dimethyl sulfoxide (DMSO). The solution must be ice cold prior to use.
30. Antibodies: Rabbit anti human von Willebrand factor (vWF) (Dako), goat anti-human VE-cadherin (Santa Cruz).
31. 4% Paraformaldehyde (Fisher). Dissolve 4 g of paraformaldehyde per 100 mL of PBS in a fume hood by heating to 65°C. Add 1 M NaOH dropwise until the solution becomes clear. Adjust to pH 7.0 with 0.5 M HCl. Store at 4°C.
32. DiI-conjugated acetylated low-density lipoproteins (AcLDL) (Molecular Probes). (**Light sensitive**). Prepare a 10-µg/mL solution in serum-free MCDB.
33. DAPI (4',6-diamidino-2-phenyindole) (Sigma) (**Light sensitive**). Prepare a stock solution (2 mg/mL) in water. The stock solution should be aliquoted and stored at −20°C. To prepare a working solution (1 µg/mL), dissolve 5 µL of the stock solution per 10 mL of PBS. The working solution can be stored in the dark at 4°C for up to 2 wk.

3. Methods

The methods described outline (1) the isolation of primary HUVECs, (2) the isolation of umbilical-cord-blood-derived endothelial cells, (3) the culture conditions used for primary endothelial cells, and (4) the characterization of the phenotype of primary endothelial cells.

3.1. Isolation of Human Umbilical Vein Endothelial Cells

Using the following procedure should result in confluent monolayers of endothelial cells in 7–10 d for a 100-mm tissue culture dish (*see* **Note 2**).

1. Fresh umbilical cords (less than 24 h old) are obtained in a sterile container and stored at 4°C until use (*see* **Note 3**).
2. Turn on biosafety hood at least 15 min before starting procedure. Spray with 70% ethanol and air-dry.
3. Prepare collagenase solution and prewarm to 37°C. Approximately 25 mL of collagenase solution is required for a cord with a length between 12 and 20 cm.
4. Remove cord from the container in a biosafety hood. Lightly soak cotton gauze in 70% ethanol and wrap around one end of the cord. Pinch the gauze gently and draw the cord through. This removes the blood off the external surface of the cord and removes some of the clotted blood from within the vessels (*see* **Note 4**).
5. Cannulate one end of the umbilical vein; umbilical cords have two arteries: (smaller and thicker walled) and one vein, usually with a large lumen. Unlike the two arteries, the vein can be easily cannulated. Often, when the end of the cord is trimmed and cleaned, the umbilical vein can be identified by the small amount of blood that continues to leak out. A cross-section of an umbilical cord displaying the two arteries and vein are shown in **Fig. 1**.
6. Use a cable tie to secure the cannulae in place.
7. Attach the stopcock to the cannulae.
8. Close stopcock valve. Remove the plunger from a syringe and attach the syringe to the stopcock. Fill the syringe with 25 mL of PBS. Insert the plunger. Open the stopcock valve and position the open end of the cord over a large waste beaker. Slowly flush the cord with PBS. Red blood cells and small clots will be flushed out. Wash the cord with PBS until most of blood is flushed through (the flowthrough should be clear) (*see* **Note 5**).
9. Close stopcock valve and detach syringe, remove plunger, and reattach syringe to stopcock. Do not try to remove plunger while it is still attached to the syringe, as the vacuum generated will pull pieces of tissue and blood into the cannula. Fill the syringe with the prewarmed collagenase solution (*see* **Note 6**). Reinsert the plunger and open the stopcock valve. Flush the cord with collagenase solution to get rid of the PBS inside the cord. Once the PBS is flushed out, clamp the end of the cord with a hemostat. Slowly inflate the vein with collagenase. Once the cord becomes turgid, close the stopcock valve. Gently massage the cord to ensure even distribution of the collagenase. Leave at room temperature for 30 min.

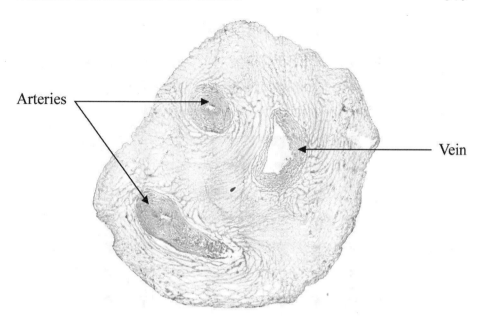

Fig. 1. Cross-section of an umbilical cord displaying the two arteries (left) and vein (right), which has a larger lumen. Note that the lower artery is sectioned tangentially.

10. Because optimal growth of endothelial cells can be achieved by culturing on gelatin-coated dishes, the necessary plates can be prepared while the collagenase digestion is in progress. Dispense sufficient gelatin solution into a culture vessel so that it completely covers the bottom. Suggested volumes are 3 mL for a 60-mm tissue culture dish or 7 mL for a 100-mm tissue culture dish. Let gelatin solution sit in contact with the plastic for 20 min at room temperature in a biosafety hood. Aspirate the gelatin solution and allow the remainder to evaporate by leaving the dish sitting open in the biosafety hood until no trace of liquid remains. Replace lid once the surface is dry (*see* **Note 7**).
11. Gently massage the cord to dislodge endothelial cells (*see* **Note 8**).
12. Position the clamped end of the cord over a 50-mL tube. Open the valve and gently flush out the endothelial cells with the remaining collagenase solution.
13. Perfuse cord with a volume of PBS equal to three times the volume of collagenase contained in the cord. Collect flowthrough in the same 50-mL tube.
14. Centrifuge the cell suspension at 4°C in a swinging-bucket rotor at 300g for 5 min.
15. Aspirate the supernatant with a sterile Pasteur pipet, in a biosafety cabinet, leaving the small cell pellet in the tube.
16. Resuspend the cell pellet in 15 mL of PBS and centrifuge for 5 min.
17. Gently resuspend the pellet in HUVEC culture medium. The resuspension volume depends on cell pellet size and choice of culture dish (pellets from different cords can be pooled and plated in the same dish). For example, with a cord length

of 20 cm, plate the cells in a 5-mL volume in a 60-mm dish. For two cords of 20-cm length, plate them in 10 mL of HUVEC medium in a 100-mm dish.

18. Plate cells onto appropriate tissue culture dish previously coated with 0.2% gelatin. Incubate the dishes at 37°C in 5% CO_2 + 95% air.

19. After incubating cells overnight, remove medium and wash with PBS to remove floating red blood cells. Add fresh HUVEC medium. Clusters of adherent cells should be visible on the culture dish.

20. Continue incubation in HUVEC medium, replacing with fresh medium every 3 d until a confluent monolayer is formed.

3.2. Isolation of Umbilical-Cord-Blood-Derived Endothelial Cells

Emerging data suggest that a subset of circulating human progenitor cells have the potential to differentiate to mature endothelial cells. Such progenitors, or angioblasts, have been successfully isolated from human bone marrow, peripheral blood, and umbilical cord blood *(7–11)*. When cultured under the right conditions, in the presence of VEGF and FGF-2, these angioblasts, defined by the expression of CD34, CD133, and VEGFR-2, can acquire the phenotype of mature endothelial cells (i.e., expression of von Willebrand factor, VE-cadherin, VEGFR-2, CD31) and uptake of acetylated low-density lipoproteins (Ac-LDL). Because these cells are derived from progenitor cells, they also exhibit a greater proliferative rate than that of mature endothelial cells.

1. Pour umbilical cord blood in 50-mL centrifuge tubes (*see* **Note 9**).
2. Centrifuge tubes containing the blood for 20 min at 300*g*.
3. Remove the platelet-rich plasma from the top of the tube. Do not disturb the buffy coat that lies on top of the red-blood-cell-rich plasma.
4. Divide the content of the tubes that contain the buffy coat and the red-blood-cell-rich plasma into two 50-mL centrifuge tubes and fill up to 25 mL with PBS. Mix gently by inversion several times.
5. Prepare new 50-mL centrifuge tubes and, to these tubes, add 10 mL of *cold* (4°C) Ficoll–Paque.
6. Carefully, using a pipet, layer the 25 mL of blood/PBS solution (prepared in **step 4**) on top of the Ficoll–Paque layer. Pipet the blood slowly along the edge of the tube. This should result in a clear Ficoll–plasma interface. Be careful not to mix the two phases.
7. Centrifuge the tubes for 30 min at 300*g* at room temperature (*see* **Note 10**).
8. Centrifugation should result in four distinct layers (from top to bottom):

   A small transparent yellowish plasma layer;
   A buffy coat layer containing the mononuclear cells;
   A clear layer of Ficoll;
   A red layer containing red blood cells, eosinophils, and polymorphonuclear cells.

An example of the isolation of human mononuclear cells is shown in **Fig. 2**.

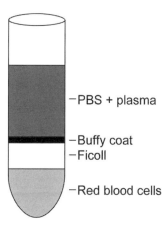

Fig. 2. Isolation of human mononuclear cells on a Ficoll–Paque gradient showing the discrete cell layers that are observed.

9. Pipet the buffy coat layer to a new 50-mL centrifuge tube. Be careful not to disturb the layer containing the red blood cells (*see* **Note 11**).
10. Fill up to 50 mL with PBS. Centrifuge tube for 7 min at 300g at room temperature (with brake on). This step removes the Ficoll.
11. Resuspend mononuclear cells in 3 mL of HUVEC medium supplemented with 5 ng/mL FGF-2 and 30 ng/mL VEGF. Plate onto a 30-mm tissue culture dish and incubate at 37°C, 5% CO_2 overnight.
12. For 3 consecutive days, replate nonadherent cells onto a new tissue culture dish. This step will remove any contaminating adherent cells, including mature endothelial cells.
13. After 3 d, harvest the nonadherent cells by centrifugation (10 min, 300g), resuspend in fresh HUVEC medium supplemented with VEGF and FGF-2, and plate onto a gelatin-coated 30-mm tissue culture dish.
14. Grow cells for 3 wk, replacing half of the medium if the color turns orange. This incubation period should result in attachment and proliferation of colonies of mature endothelial cells. Monocytes could also be adherent, but usually die off within the first 2–3 wk of culture.

3.3. Subculture of Primary Endothelial Cells

3.3.1. Passaging Primary Endothelial Cells

This procedure allows efficient amplification of endothelial cells from primary cultures of human umbilical vein at a relatively low passage number. Continued passage will eventually result in senescent changes and loss of useful replicative potential. Although there is variability among batches of cells, HUVECs can routinely be used for up to 8–10 passages. Cord-blood-derived endothelial cells can be passaged many more times (for up to 6 wk). However, there might be a loss of expression of endothelial markers in the first two to three passages. Primary endothelial cells should be passaged before the cells reach confluence and prior to the growth medium becoming acidic (*see* **Note 12**). PBS, trypsin–EDTA, and HUVEC medium should be warmed up at room temperature or 37°C before use.

1. Aspirate the medium from a dish of primary HUVECs (isolated from one or more umbilical cords) at or near confluence.
2. Wash the monolayer with 5 mL of PBS for a 100-mm dish (use 2 mL for a 60-mm dish) (*see* **Note 13**).
3. Aspirate.
4. Add 2 mL of trypsin–EDTA in PBS, allowing the surface to be covered.
5. Incubate the cells at 37°C for 1–3 min.
6. Examine the dish under a microscope to determine that the cells have detached. If cells are rounded but not detached, they can be dislodged by gently tapping the plate.
7. Add 5 mL of complete HUVEC medium directly to the dish. The serum in the medium will quench the activity of the trypsin.
8. Distribute the diluted cell suspension into the gelatin-precoated tissue culture dishes.
9. Bring the final volume in each dish up to 8–10 mL with complete HUVEC medium (*see* **Note 14**).
10. Incubate the dishes at 37°C in 5% CO_2 + 95% air. Refeed with complete HUVEC medium every 2–3 d.

3.3.2. Cryopreservation of Primary Endothelial Cells

It is advisable to freeze HUVECs that have been passaged as few times as possible in order to ensure an adequate stock of low-passage cells for subsequent experiments.

1. Label an appropriate number of cryovials, including the freezing date and the passage number of the cells, and prechill them on ice.
2. Choosing a dish that is not quite confluent (try to freeze HUVECs by the second passage), trypsinize as described in **Subheading 3.3.1.** Resuspend the cells in 10 mL of fresh HUVEC medium and transfer to a 15-mL tube.
3. Centrifuge for 5 min at 300*g* and then resuspend the cell pellet in 1 mL of HUVEC medium.
4. To prepare for a cell count, pipet 50 μL of the cell suspension in a microcentrifuge tube and add 50 μL of 0.4% trypan blue solution.
5. Mix and pipet into both chambers of a hemacytometer, count, and determine cell concentration.
6. Fill tube with 5 mL of HUVEC medium and centrifuge for 5 min at 300*g*.
7. Aspirate the majority of the media and briefly chill on ice (1–2 min).

8. Resuspend the cell pellet in ice-cold freezing medium. The ideal concentration is between 5×10^5 and 1×10^6 cells/mL.
9. Dispense the cells in 1-mL aliquots into the prechilled labeled cryovials.
10. Place the cryovials in a styrofoam rack, place this into a styrofoam box, and freeze at $-80°C$ for 24 h (*see* **Note 15**).
11. Transfer the cryovials to liquid nitrogen for long-term storage.

3.3.3. Thawing Cells

1. Warm HUVEC medium to room temperature or 37°C.
2. Remove cryovial from liquid nitrogen and immediately place in a rack in a 37°C water bath.
3. Transfer cells to a 15-mL centrifuge tube. Add 10 mL of HUVEC medium, drop by drop (*see* **Note 15**).
4. Centrifuge for 5 min at 300g and then discard the supernatant.
5. Add 8 mL of HUVEC medium, pipet gently up and down to a single-cell suspension, and plate into a 100-mm dish.
6. Incubate at 37°C, changing the medium as required.

3.4. Characterization of Primary Endothelial Cells

The most effective means of characterizing endothelial cultures is to examine various properties. Characteristic endothelial markers include von Willebrand factor, CD31, CD34, VE-cadherin, VEGF receptors 1 and 2 (flt-1, KDR), strong uptake of AcLDL, staining with Ulex europaeus lectin type 1, and morphology. The steps described in this subsection outline the procedure used to characterize the endothelial phenotype of primary endothelial cells, such as HUVECs.

3.4.1. Assessment of Cell Morphology

The monolayer of endothelial cells is characterized by a "cobblestone" appearance at confluence (*see* **Fig. 3** and **Note 16**).

3.4.2. Immunofluorescent Staining of Endothelial Cells

This procedure is used to determine the expression of markers to establish that cells recovered are endothelial. Commercially available antibodies against endothelial markers include von Willebrand factor, CD31 (PECAM-1), CD34, VE-cadherin, and the VEGF receptors 1 and 2 (flt-1, KDR). It is important to note that most of these markers are not specifically expressed on endothelial cells and that endothelial cell populations show heterogeneous expression of various markers (*12*).

1. Trypsinize cells (as described in **Subheading 3.1.1.**) and plate onto sterile multiwell glass slides (usually 75,000 cells on a 4-well glass slide 1 d prior to staining).

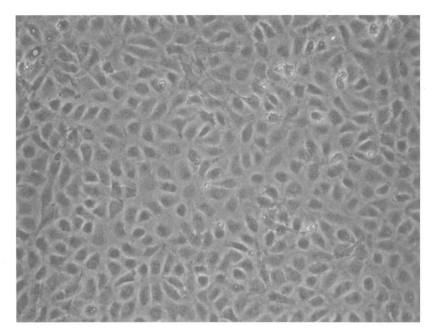

Fig. 3. Phase-contrast micrograph of a confluent HUVEC monolayer grown on gelatin-coated tissue culture dishes demonstrating typical endothelial cobblestone morphology.

2. Incubate overnight at 37°C in 5% CO_2 + 95% air. Cells should be at approx 80% confluence prior to staining.
3. Wash cells once with PBS (1 mL per chamber for this and all subsequent washing steps).
4. Fix cells in 4% paraformaldehyde in PBS for 5 min at room temperature.
5. Wash once with PBS.
6. Fix with methanol (precooled at –20°C) for 1 min at room temperature to help permeabilization.
7. Wash twice with PBS.
8. Block and permeabilize with PBS containing 4% serum (blocking solution) (*see* **Note 17**) and 0.1% Triton X-100 for 10 min at room temperature.
9. Prepare dilution of the primary antibody in PBS containing 4% serum and 0.1% Triton X-100 at the concentration specified by the manufacturer. Isotype immunoglobulin should be used as the negative control.
10. Incubate cells with primary antibody for 1 h at room temperature (*see* **Note 18**).
11. Wash twice with PBS containing 4% blocking serum and 0.1% TritonX-100.
12. Prepare dilution of the secondary antibody (conjugated with fluorescent dye) in PBS containing 4% serum and 0.1% Triton X-100 (1:64 to 1:100 is usual).
13. Incubate in the dark with secondary antibody for 30 min (*see* **Note 19**).

14. Wash once with PBS containing 4% serum and 0.1% Triton X-100.
15. Wash twice with PBS.
16. Counterstain nuclei by incubating with 1 µg/mL DAPI in PBS for 1 min.
17. Wash twice with PBS.
18. Mount with antifading solution on a cover slip and seal edges of the slide with nail polish.
19. Visualization of stained cells is performed using fluorescein (for fluorescein isothiocyanate, Alexa 488) or rodhamine (for phycoerythrin [PE], Alexa 594, Texas red) standard excitation/emission filters (*see* **Notes 20** and **21**).

3.4.3. Uptake of Acetylated LDL

Chemically modified LDL, such as acetylated LDL, are rapidly taken up by macrophages and cultured endothelial cells, and this assay is widely used to isolate and identify endothelial cells in culture *(13)*. The receptor involved in this pathway is called a scavenger receptor. However, studies have shown that the receptor present on endothelial cells is distinct from the scavenger receptor expressed on macrophages and has been termed scavenger receptor expressed by endothelial cells (SREC) *(14,15)*. Uptake of acetylated LDL labeled with the fluorescent dye 1,1'-dioctadecyl-3,3,3',3'-tetramethylindocarbocyanine (DiI) is a fast and convenient way to identify endothelial cells in culture. However, it is important to note that macrophages will also be labeled by this method.

1. Trypsinize cells (as described in **Subheading 3.3.1.**) and plate onto sterile multiwell glass slides (usually 75,000 cells on a 4-well glass slide in 1 mL of medium 1 d prior to staining).
2. Incubate overnight at 37°C in 5% CO_2 + 95% air.
3. Incubate the cells with 10 µg/mL of DiI-conjugated acetylated LDL from human plasma (Molecular Probes) for 4 h in serum-free MCDB.
4. Following labeling, wash cells once with 1 mL probe-free MCDB, then twice in 1 mL of PBS.
5. Fix in formalin for 5 min.
6. Wash twice with 1 mL of PBS.
7. Counterstain nuclei by incubating with 1 µg/mL DAPI in PBS for 1 min.
8. Wash twice with PBS
9. Mount with antifading solution and cover slip and seal the edges of the slide with nail polish.
10. Visualization of cells that incorporate AcLDL can be performed using a rodhamine standard excitation/emission filter (*see* **Note 20**).

4. Notes

1. It is very important that all material that comes in contact with cells be sterile and endotoxin-free because endotoxin interferes with normal cell proliferation and growth. The effects of endotoxin are not completely predictable and can actually

artificially enhance cell culture growth and activate endothelial cells. Because
there can be variability between batches of FBS, it is advisable to test batches of
FBS for the ability to support the viability and proliferation of HUVECs.
The most sensitive indicators of FBS quality are provided by plating efficiency
assays, whereas multiple passage tests are more reliable and correlate with long-
term culture performance. The plating efficiency assay consists of inoculating
cells at a density that will yield 100–300 discrete colonies per 100-mm tissue
culture dish in HUVEC medium (made with test FBS). The dishes are incubated
for 10–14 d until colonies of cells are visible. To count the colonies, the dishes
are fixed (2% formaldehyde, 5 min), rinsed with water, stained with Coomassie
brilliant blue dye (0.1% Coomassie in 10% acetic acid:50% methanol:40% water)
and rinsed with wash solution (10% acetic acid:50% methanol:40% water). The
total number of colonies is counted and the relative plating efficiency of the test
FBS lot is compared against reference FBS. The FBS lot qualification testing
should be quantifiable and statistically relevant to ensure interpretative objectivity.
2. Two common problems exist in the establishment of HUVEC primary cell cultures. First, bacterial and/or fungal contamination is a major cause of failure.
The second problem relates to the freshness of the umbilical cord at the com-
mencement of processing. Cords should be handled within 24 h after delivery.
3. It is important to note that, as with isolation of any cells from human tissue, there
is a potential risk of infection. Precautions for working with human tissues, such
as wearing gloves, a laboratory coat, and safety goggles, must be used at all times.
4. Discard cord if length is less than 12 cm or if the cord has crushed areas or pierced
segments, as it might result in a low HUVEC yield. Occasionally, a cord may be
severely clogged by clotted blood. If possible, cut off the clotted section. Trim
the ends with a scalpel blade to get even edges and proceed following the outlined protocol.
5. If the cord is clogged and the blood clot is not dislodged by flushing, discard the
cord. Do not force PBS through a clotted umbilical cord.
6. Use only freshly prepared collagenase solution, as it tends to precipitate, and
precipitates of collagenase can be cytotoxic.
7. If gelatin-coated dishes are not being used immediately, they can be stored at 4°C
for up to 2 wk, provided that they remain sterile by keeping them in a sealed bag.
8. Do not overmassage, as it can increase contamination with fibroblasts. Massag-
ing the whole length of the cord twice is usually sufficient.
9. Umbilical cord blood is to be harvested in 200-mL plastic bottles containing
40 mL of IMDM medium containing 800 U/mL heparin. The volume of cord
blood collected per bottle should not exceed 120 mL, as the final concentration
of heparin should not decrease below 200 U/mL.
10. It is important to turn off the brake of the centrifuge at this point, as breaking
might disrupt the Ficoll interface.
11. When harvesting the buffy coat interface band, avoid disturbing the red blood
cell pellet to reduce red blood cell contamination. Moreover, do not collect more
than 5 mL of Ficoll, as the Ficoll might make it difficult to pellet the cells afterward.

12. It is important that HUVEC passage is done regularly because overgrowth of the cultures can induce cell cycle arrest of the cells and will have deleterious effects on the subsequent ability of the cells to proliferate. We usually passage HUVECs in 100-mm tissue culture dishes. Working with three dishes allows two plates to be used for experimental studies and the remaining dish to be split 1:3 to replace the original number. Usually, a typical HUVEC preparation can be split for up to 3–4 wk in this fashion before becoming senescent.

13. When passaging the cells, it is important to remove the serum from the medium, as it inhibits trypsin activity. During trypsinization, it is important to monitor the cells carefully to avoid excessive exposure to trypsin, as this might result in lower cell viability.

14. HUVECs do not grow well if they are set up at too low a density. It is best to aim for a situation in which cells cover approx 40 to 50% of the surface area of the culture vessel at 24 h after plating. Usually, plating 5×10^5 cells per 100-mm dish is a good starting point. Watch the cultures and adjust cell numbers accordingly.

15. Cell viability can be severely compromised if the procedures for freezing and thawing are not carried out carefully. Placing the cryovials in a styrofoam rack ensures that freezing occurs slowly and reduces cell death. Similarly, when thawing the cells, it is important to add the medium slowly, as sudden dilution of DMSO can cause severe osmotic damage.

16. Morphological identification (**Subheading 3.4.1.**) is not sufficient for the determination of the endothelial phenotype, as endothelial cells can change their morphology depending on the growth supplements in the medium or the matrix onto which the cells are seeded. Indeed, studies have shown that endothelial cells isolated from different organs or different-sized vessels can differ in their antigens, expression of cellular adhesion molecules, metabolism, and growth requirements in culture. It is therefore necessary to use a combination of markers, visual identification, and/or functional assays to confirm the endothelial phenotype.

17. Blocking serum from another species, which is not recognized by the secondary antibodies, should be chosen. Alternatively, another blocking agent such as immunohistochemical-grade bovine serum albumin (BSA) can be substituted.

18. Both the incubation time and the required dilution of the primary antibody can vary with different antibodies. Whenever a new primary antibody is used, a dilution series should be performed to assay the optimal concentration for that antibody. A good starting point is to use 1:10, 1:100, and 1:1000 dilutions. Another series of dilutions (e.g., 1:50, 1:100, and 1:200) should be used once the appropriate range is found.

19. It is important to cover the slides with foil to prevent exposure to light, as fluorescent dyes are light sensitive.

20. The choice of filter to visualize the fluorescence is dependent on the fluorochrome conjugated with the secondary antibody. Some of the most commonly used fluorochromes are fluorescein isothiocyanate (FITC), tetramethyl rodhamine isothiocyanate (TRITC), Texas red, and DiI. The ideal filter to visualize FITC is excitation 450–490 nm and barrier 520–560 nm. For TRITC, Texas red, and DiI, the ideal filters combination are excitation 510–560 nm and barrier 590 nm.

21. Phenotypical characterization of endothelial cells can also be assessed by standard immunohistochemistry rather than by immunofluorescence if access to a fluorescent microscope is limited. In this case, the secondary antibody will be conjugated with horseradish peroxidase, and the antibody will be visualized with DAB staining.

Acknowledgments

The authors thank Fred Wong, Michela Noseda, and Ingrid Pollet for assistance with the manuscript.

References

1. Fishman, A. P. (1982) Endothelium: a distributed organ of diverse capabilities. *Ann. NY Acad. Sci.* **401,** 1–8.
2. Jaffe, E. A. (1987) Cell biology of endothelial cells. *Hum. Pathol.* **18,** 234–239.
3. Jaffe, E. A., Nachman, R. L., Becker, C. G., and Minick, C. R. (1973) Culture of human endothelial cells derived from umbilical veins. Identification by morphologic and immunologic criteria. *J. Clin. Invest.* **52,** 2745–2756.
4. Folkman, J., Haudenschild, C. C., and Zetter, B. R. (1979) Long-term culture of capillary endothelial cells. *Proc. Natl. Acad. Sci. USA* **76,** 5217–5221.
5. Davison, P. M., Bensch, K., and Karasek, M. A. (1980) Isolation and growth of endothelial cells from the microvessels of the newborn human foreskin in cell culture. *J. Invest. Dermatol.* **75,** 316–321.
6. Schutz, M. and Friedl, P. (1996) Isolation and cultivation of endothelial cells derived from human placenta. *Eur. J. Cell Biol.* **71,** 395–401.
7. Gehling, U. M., Ergun, S., Schumacher, U., et al. (2000) In vitro differentiation of endothelial cells from AC133-positive progenitor cells. *Blood* **95,** 3106–3112.
8. Lin, Y., Weisdorf, D. J., Solovey, A., and Hebbel, R. P. (2000) Origins of circulating endothelial cells and endothelial outgrowth from blood. *J. Clin. Invest.* **105,** 71–77.
9. Peichev, M., Naiyer, A. J., Pereira, D., et al. (2000) Expression of VEGFR-2 and AC133 by circulating human CD34(+) cells identifies a population of functional endothelial precursors. *Blood* **95,** 952–958.
10. Quirici, N., Soligo, D., Caneva, L., Servida, F., Bossolasco, P., and Deliliers, G. L. (2001) Differentiation and expansion of endothelial cells from human bone marrow CD133(+) cells. *Br. J. Haematol.* **115,** 186–194.
11. Reyes, M., Dudek, A., Jahagirdar, B., Koodie, L., Marker, P. H., and Verfaillie, C. M. (2002) Origin of endothelial progenitors in human postnatal bone marrow. *J. Clin. Invest.* **109,** 337–346.
12. Garlanda, C. and Dejana, E. (1997) Heterogeneity of endothelial cells. Specific markers. *Arterioscl. Thromb. Vasc. Biol.* **17,** 1193–1202.
13. Voyta, J. C., Via, D. P., Butterfield, C. E., and Zetter, B. R. (1984) Identification and isolation of endothelial cells based on their increased uptake of acetylated-low density lipoprotein. *J. Cell. Biol.* **99,** 2034–2040.

14. Adachi, H. and Tsujimoto, M. (2002) Characterization of the human gene encoding the scavenger receptor expressed by endothelial cell and its regulation by a novel transcription factor, endothelial zinc finger protein-2. *J. Biol. Chem.* **277,** 24,014–24,021.

15. Adachi, H., Tsujimoto, M., Arai, H., and Inoue, K. (1997) Expression cloning of a novel scavenger receptor from human endothelial cells. *J. Biol. Chem.* **272,** 31,217–31,220.

22

Studying Leukocyte Rolling and Adhesion In Vitro Under Flow Conditions

Susan L. Cuvelier and Kamala D. Patel

Summary

Leukocyte recruitment from the vasculature occurs under conditions of hemodynamic shear stress. The parallel-plate flow chamber apparatus is an in vitro system that is widely used to study leukocyte recruitment under shear conditions. The flow chamber is a versatile tool for examining adhesive interactions, as it can be used to study a variety of adhesive substrates, ranging from monolayers of primary cells to isolated adhesion molecules, and a variety of adhesive particles, ranging from leukocytes in whole blood to antibody-coated Latex beads. We describe methods for studying leukocyte recruitment to cytokine-stimulated endothelial cells using both whole blood and isolated leukocyte suspensions. These methods enable multiple parameters to be measured, including the total number of recruited leukocytes, the percentage of leukocytes that are rolling or firmly adherent, and the percentage of leukocytes that have transmigrated. Although these methods are described for interactions between leukocytes and endothelial cells, they are broadly applicable to the study of interactions between many combinations of adhesive substrate and adhesive particles.

Key Words: Parallel-plate flow chamber; recruitment; tethering; rolling; adhesion; detachment; transmigration; shear stress; whole blood; leukocyte; PBMC; PMN; lymphocyte; monocyte; neutrophil; eosinophil.

1. Introduction

The recruitment of leukocytes from the vasculature is a process that is essential for the normal function of the immune system. When inappropriately regulated, leukocyte recruitment can contribute to inflammatory diseases such as asthma, sepsis, and multiple sclerosis. The importance of leukocyte recruitment to both physiological and pathological processes has led to extensive research in this field (*1*). Critical to this field is the parallel-plate flow chamber

From: *Methods in Molecular Biology, vol. 290: Basic Cell Culture Protocols, Third Edition*
Edited by: C. D. Helgason and C. L. Miller © Humana Press Inc., Totowa, NJ

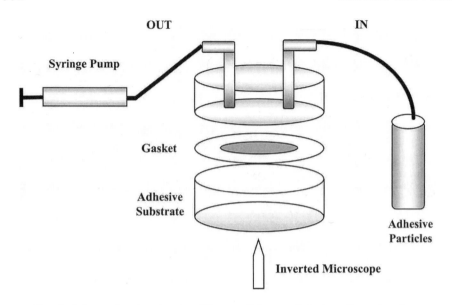

Fig. 1. Schematic of a circular, 35-mm-dish, parallel-plate flow chamber.

apparatus, which is widely used as a tool for studying leukocyte recruitment in vitro *(2–4)*.

Parallel-plate flow chambers are used to generate shear conditions that are similar to those that are found in the vasculature. Although the design of flow chambers is quite varied, all have a number of features in common (*see* **Fig. 1**). All flow chambers have two surfaces arranged in parallel: one lower plate onto which an adhesive substrate is immobilized and one upper plate that serves as the top of the chamber. All flow chambers have a device such as a gasket that maintains a defined distance between the lower and upper plates. Finally, all flow chambers are attached to a device such as a syringe pump that moves liquid over the adhesive substrate on the lower plate at a defined rate. These elements of the flow chamber apparatus enable defined shear rates and shear stresses to be generated in vitro, thereby facilitating the study of leukocyte recruitment under physiologically relevant conditions.

The two primary requirements for a flow chamber experiment are an adhesive substrate, which is immobilized on the lower plate of the flow chamber, and a suspension of adhesive particles, which is perfused over the adhesive substrate. Many different choices exist for both the adhesive substrate and the adhesive particles (*see* **Table 1**). Adhesive substrates that can be used in the flow chamber range from monolayers of primary cells to isolated adhesion molecules, and adhesive particles range from leukocytes in whole blood to antibody-coated Latex beads. The wide variety of adhesive substrates and

Table 1
Examples of Adhesive Surfaces and Adhesive "Particles"
That Can Be Used in a Parallel-Plate Flow Chamber Apparatus

Adhesive surface	Adhesive particles
Monolayer of isolated primary cells	Whole blood
Monolayer of adherent cell line	Isolated primary cells
Lipid bilayer containing isolated adhesion molecule	Suspension cell line
Isolated adhesion molecule	Ligand-coated Latex beads
Fragment of isolated adhesion molecule	Antibody-coated Latex beads

adhesive particles that can be used in the flow chamber system makes it a versatile tool for addressing multiple questions regarding leukocyte recruitment.

In these protocols, we detail methods for examining leukocyte recruitment to cytokine-stimulated endothelial cells. Methods for measuring recruitment from both whole blood and isolated leukocyte suspensions are described. Although a specific adhesive substrate and specific adhesive particles are used here for illustrative purposes, these protocols should be applicable to any combination of substrate and particles.

2. Materials

2.1. Endothelial Cell Stimulation

1. Cultured human endothelial cells in 35-mm culture dishes (Corning Inc. Life Sciences, Acton, MA).
2. 25% Human serum albumin (HSA) (Bayer Corp., Elkhart, IN). Supplied sterile; store at 4°C.
3. Sterile-filtered M199 (Sigma Chemicals, St. Louis, MO). Store at 4°C; heat to 37°C before use.
4. Hank's balanced salt solution; with calcium chloride, magnesium chloride, and magnesium sulfate; without sodium bicarbonate and phenol red (HBSS) (Gibco–BRL, Grand Island, NY). Store at 4°C; heat to 37°C before use.
5. Recombinant human cytokine(s) (R&D Systems, Minneapolis, MN). Store in aliquots at –20°C and transfer to storage at 4°C as needed.

2.2. Whole Blood Preparation and Leukocyte Isolation

1. HBSS (as described in **Subheading 2.1.**).
2. Heparin LEO (10,000 IU/mL), preservative-free (LEO Pharma, Ballerup, Copenhagen).
3. Human blood donors.
4. Hemacolor stain set (VWR International, West Chester, PA). Store at 20°C.

2.3. Flow Chamber Setup

1. HBSS (as described in **Subheading 2.1.**).
2. Phase-contrast microscope with ×10, ×20, and ×40 objectives and equipped with a stage that holds 35-mm dishes (Carl Zeiss, Inc., Thornwood, NY).
3. Charge-coupled device (CCD) camera (Hitachi Denshi, Ltd., Tokyo, Japan).
4. Videocassette recorder (VCR) (Panasonic, Secaucus, NJ).
5. Time/date/title generator (GE Interlogix, Corvalis, OR).
6. Flow chamber and gasket (Glycotech, Rockville, MD).
7. Infuse/refill syringe pump (Harvard Apparatus, Inc, Holliston, MA).
8. Pharmed 65, 1/16-in.-internal diameter, 1/16-in.-external diameter tubing (Saint-Gobain Performance Plastics Co., Akron, OH).
9. Vacuum pump and vacuum flask (VWR International).
10. 37°C Water bath (VWR International).

2.4. Whole Blood Recruitment Experiment

1. HBSS and hemacolor stain set (as described in **Subheadings 2.1.** and **2.2.**).
2. Stimulated human endothelial cells (from **Subheading 2.1.**).
3. Heparinized (10 IU/mL) whole blood from human donors (from **Subheading 2.2.**).
4. Flow chamber setup (from **Subheading 2.3.**).

2.5. Whole Blood Recruitment Analysis

VCR recording (from **Subheading 2.3.**).

2.6. Isolated Leukocyte Recruitment Experiment

1. HBSS (as described in **Subheading 2.1.**).
2. Stimulated human endothelial cells (from **Subheading 2.1.**).
3. Isolated leukocytes from human donors (from **Subheading 2.2.**).
4. Flow chamber setup (from **Subheading 2.3.**).

2.7. Isolated Leukocyte Resistance to Detachment Experiment

1. HBSS (as described in **Subheading 2.1.**).
2. Stimulated human endothelial cells (from **Subheading 2.1.**).
3. Isolated leukocytes from human donors (from **Subheading 2.2.**).
4. Flow chamber setup (from **Subheading 2.3.**).

2.8. Isolated Leukocyte Recruitment Analysis

VCR recording (from **Subheading 2.3.**).

3. Methods

3.1. Endothelial Cell Stimulation

1. Isolate endothelial cells and grow in 35-mm culture dishes as described for human umbilical vein endothelial cells (HUVECs; refer also to Chapter 21) *(5,6)*, human pulmonary microvascular endothelial cells (HPMECs) *(7)*, and human dermal

microvascular endothelial cells (HDMECs) *(8–10)*. Endothelial cells are also available commercially from companies such as Cambrex (East Rutherford, NJ) (*see* **Note 1**). Grow to tight confluence (*see* **Note 2**).

2. Remove the medium and wash cells once with approx 2 mL of M199.
3. Stimulate cells in 1.5 mL of M199 with 0.5% HSA containing cytokine(s) of interest (*see* **Note 3**) at 37°C and 5% CO_2. If the stimulation time is 4 h or less, cells can be stimulated in HBSS with 0.5% HSA instead of M199 with 0.5% HSA.

3.2. Whole Blood Preparation and Leukocyte Isolation

1. For whole blood recruitment experiments, draw blood from human donors into a syringe containing heparin (10 U heparin/mL blood). Gently invert the tube two to three times to ensure that the blood and heparin are mixed. Prepare peripheral blood smears and stain with Wright Giemsa stain according to manufacturer's instructions. Whole blood should be used within 90 min of blood draw.
2. For isolated leukocyte recruitment experiments, isolate leukocytes as previously described for T-lymphocytes *(11)*, B lymphocytes *(12)*, monocytes *(13)*, neutrophils *(5)*, and eosinophils *(14)*. Resuspend isolated leukocytes in 37°C HBSS. In previous studies, isolated leukocytes have been used at concentrations ranging from 0.5 to 1.5×10^6 cells/mL *(11,15,16)*.

3.3. Flow Chamber Setup (see Note 4)

1. Connect a phase-contrast microscope to a CCD camera and a VCR according to the manufacturer's instructions. A time/date/title generator can be connected to the VCR if desired.
2. Clean the flow chamber and gasket with isopropanol swabs. Assemble the flow chamber according to the manufacturer's instructions, using an empty 35-mm culture dish as the bottom plate. Place the flow chamber in the microscope stage.
3. Pull back on the plunger of a 30-mL syringe to break the seal. Attach a two-way stopcock to the syringe. Load the syringe into an infuse/refill syringe pump. Program the syringe pump for the desired refill rate (*see* **Note 5**) according to the manufacturer's instructions.
4. Connect the two-way stopcock to the outlet fitting of the flow chamber using a piece of Pharmed tubing (*see* **Note 6**). Connect a vacuum flask and vacuum pump to the vacuum fitting of the flow chamber using a second piece of Pharmed tubing. Attach a third piece of Pharmed tubing to the inlet fitting of the flow chamber and place the free end of this piece into a 50-mL Falcon tube containing approx 40 mL of HBSS. Replenish this HBSS as needed during experiments. Place the Falcon tube containing HBSS into a 37°C water bath. Place a second 50-mL Falcon tube containing the whole blood or isolated leukocyte suspension into the same water bath.
5. Turn on the vacuum pump. Manually pull on the pusher block of the syringe pump to fill the inlet line and flow chamber with HBSS. Ensure that there are no air bubbles in the inlet line or flow chamber. Close the two-way stopcock and clamp the inlet line with a hemostat.

6. Remove the empty culture dish and replace it with a culture dish containing endothelial cells (*see* **Note 7**). To do this, invert the flow chamber, cover the open section of the gasket with HBSS, and lower the new culture dish onto the flow chamber. Alternatively, replace the culture dish by lowering the flow chamber onto the new culture dish.

3.4. Whole Blood Recruitment Experiment

1. Visualize the endothelial cell monolayer at ×200 magnification using brightfield optics.
2. Place the inlet line into the whole blood. Open the two-way stopcock and unclamp the hemostat from the inlet line. Start the syringe pump and pull blood into the flow chamber for 5 min (*see* **Note 8**).
3. Switch the inlet line into the HBSS (*see* **Note 9**). Buffer will begin to enter the chamber soon after the inlet line is switched into the HBSS; the time required for this to happen will depend on the length of the inlet line and the flow rate used.
4. After buffer has entered the chamber and enough blood has cleared from the field to permit visualization of accumulated cells, begin recording fields using the CCD camera and VCR (*see* **Note 10**). Record four random fields for 15 s each.
5. Hold the flow chamber at a 45°–90° angle relative to the stage and remove the inlet line from the flow chamber. Allow air to flow over the monolayer until it has displaced all of the buffer in the flow chamber. Remove the culture dish from the flow chamber, holding the dish at a 45°–90° angle relative to the stage to prevent any residual buffer from flowing back over the monolayer.
6. Wright Giemsa stain the culture dish according to manufacturer's instructions.

3.5. Whole Blood Recruitment Analysis (see Notes 11 and 12)

1. The four fields that were recorded in **step 4** of **Subheading 3.4.** can be used to measure total leukocyte recruitment, the percentage of firmly adherent or rolling leukocytes, and the rolling velocities of leukocytes.
2. To measure total leukocyte recruitment, count the total number of cells in all of the recorded fields. Divide the total by the number of recorded fields to give the average number of cells per field. Divide the average number of cells per field by the area of the field to give the average number of cells/mm^2 (*see* **Note 13**).
3. To measure the percentage of leukocytes that are firmly adherent, count the number of firmly adherent cells in all of the four fields. We define firmly adherent cells as those that moved less than one cell diameter in a 10-s period. Divide the number of firmly adherent cells by the total number of cells (measured in **step 2**) and multiply by 100%.
4. To measure the percentage of leukocytes that are rolling, count the number of rolling cells in all of the four fields. We define rolling cells as those that moved one cell diameter or more in a 10-s period. Divide the number of rolling cells by the total number of cells (measured in **step 2**) and multiply by 100%. The percentage of leukocytes that are firmly adherent (measured in **step 3**) and the percentage of leukocytes that are rolling should add up to 100%.

5. To measure the rolling velocity of a leukocyte, select a leukocyte and measure the distance that it traveled in a 10-s period. Divide this distance by 10 to give the distance traveled per second. We usually measure all of the cells that are present on the endothelial cell monolayer and represent the data using a histogram of rolling velocities.

6. Perform a 200-cell differential on the culture dish that was stained in **step 6** of **Subheading 3.4.** (*see* **Note 14**).

7. Perform a 200-cell whole-blood differential on the peripheral blood smears that were prepared in **step 1** of **Subheading 3.2.**

8. Calculate a recruitment factor (R-factor) for each leukocyte subclass. The R-factor for a given leukocyte subclass is calculated by dividing the percentage of leukocytes on the plate that are of that subclass by the percentage of leukocytes in the whole-blood differential that are of that subclass (*see* **Note 15**).

3.6. Isolated Leukocyte Recruitment Experiment

1. Visualize the endothelial cell monolayer at ×100 magnification using phase-contrast optics.

2. Place the inlet line into the isolated leukocytes. Open the two-way stopcock and unclamp the hemostat from the inlet line. Start the syringe pump.

3. When leukocytes enter the chamber, begin recording a single field using the CCD camera and VCR. Record this field for 1 min. Wait 3 min.

4. Switch the inlet line into the HBSS (*see* **Note 9**). Record six random fields for 10 s each (*see* **Note 16**).

5. Switch to ×400 magnification and wait 1 min. Scan the monolayer to find groups of cells. Record multiple groups of cells while focusing up and down on the monolayer for a total of 1 min.

6. Analysis of the data is carried out as described in **Subheading 3.8.**

3.7. Isolated Leukocyte Resistance to Detachment

1. Visualize the endothelial cell monolayer at ×100 magnification using phase-contrast optics.

2. Place the inlet line into the isolated leukocytes. Open the two-way stopcock and unclamp the hemostat from the inlet line. Start the syringe pump at 0.5 dynes/cm^2. Wait 1 min. Switch the inlet line into the HBSS (*see* **Note 9**).

3. Wait until 1 min after leukocytes enter the chamber. Stop the syringe pump and wait 5 min to allow the leukocytes in the chamber to settle and adhere to the endothelial cell monolayer.

4. Start the syringe pump at 0.5 dynes/cm^2. Wait 30 s. Record three random fields for 10 s each (*see* **Note 16**).

5. Increase the shear stress by a chosen increment. Record three random fields for 10 s each.

6. Repeat **step 5** until all of the leukocytes have detached from the endothelial cell monolayer or until only transmigrated leukocytes remain on the endothelial cell monolayer.

7. Analysis of the data is described in **Subheading 3.8.**

3.8. Isolated Leukocyte Recruitment Analysis (see Note 11)

1. The six fields that were recorded in **step 4** of **Subheading 3.6.** can be used to measure total leukocyte recruitment, the number of firmly adherent or rolling leukocytes, and the rolling velocity of leukocytes. The methods for measuring these parameters are described in **Subheading 3.5.**

2. The single field that was recorded in **step 3** of **Subheading 3.6.** can be used to measure leukocyte tethering. Analyze this recording frame by frame for primary tethers, secondary tethers, and leukocyte–leukocyte interactions. We define primary tethers as direct tethers between a flowing leukocyte and the endothelial cell monolayer, secondary tethers as tethers in which a flowing leukocyte makes contact with an adherent leukocyte prior to attaching to the endothelial cell monolayer, and leukocyte–leukocyte interactions as direct tethers between a flowing leukocyte and an adherent leukocyte. We express the data as the number of each of primary tethers, secondary tethers, and leukocyte–leukocyte interactions per square millimeter per minute.

3. The fields that were recorded in **step 5** of **Subheading 3.6.** can be used to measure the percentage of leukocytes that have transmigrated (*see* **Note 17**). Count the total number of each of transmigrated and nontransmigrated cells in all of the recorded fields. Divide the number of transmigrated cells by the total number of cells (transmigrated and nontransmigrated) and multiply by 100%.

4. The fields that were recorded in **steps 4–6** of **Subheading 3.7.** can be used to measure the leukocyte resistance to detachment. Determine the total leukocyte recruitment in the three fields recorded in **step 4** of **Subheading 3.7.** as described in **Subheading 3.5.** Set this value as 100% accumulation. Determine the total leukocyte recruitment in each set of three fields recorded in **steps 5** and **6** of **Subheading 3.7.** as described in **Subheading 3.5.** Convert each of these values to % accumulation by dividing the value by the total leukocyte recruitment in the field recorded in **step 4** of **Subheading 3.7.** and multiplying by 100%. We represent leukocyte resistance to detachment as a line graph of % accumulation versus shear stress.

4. Notes

1. Although endothelial cells can be purchased from commercial sources, we isolate our own endothelial cells for use in our laboratory and recommend freshly isolated endothelial cells over commercially available endothelial cells, especially when experiments require primary or first-passage cells.

2. Many protocols for culturing endothelial cells recommend using fetal bovine serum (FBS) in the culture medium; however, our laboratory has found that endothelial cells often grow better in medium containing human serum than in medium containing FBS. Human serum can be isolated from normal donors and pooled as described *(5)*.

3. The cytokines tumor necrosis factor (TNF), interleukin-1β (IL-1β), and IL-4 are frequently used to stimulate human endothelial cells for leukocyte recruitment assays using parallel-plate flow chambers. The concentrations and incubation periods that are used for these cytokines vary greatly. We present here some

examples of concentrations and incubation periods that have been used for these cytokines. These examples are by no means exhaustive; we recommend that you consult the relevant literature to determine the concentration and incubation period that should be used in your experimental setup. TNF has been used at concentrations between 10 and 20 ng/mL *(14,17,18)* and for incubation periods between 6 and 24 h *(14,17–20)*. IL-1β has been used at concentrations between 0.1 and 10 ng/mL *(21,22)* and for incubation periods between 4 and 24 h *(21–23)*. IL-4 had been used at concentrations between 10 and 20 ng/mL and for incubation periods between 24 and 48 h *(15,18,20,24)*.

4. These protocols are written for use with circular flow chambers (for 35-mm culture dishes), which are commercially available through GlycoTech (www.glyco tech.com).

5. The relationship between the wall shear stress in a flow chamber setup and the flow rate through the chamber is given by the equation $\tau_W = \mu\gamma = 6\mu Q/a^2 b$, where τ_W is the wall shear stress (dynes/cm^2), μ is the viscosity of the medium (P), γ is the shear rate (s^{-1}), Q is the volumetric flow rate (mL/s), a is the channel height (gasket thickness) (cm), and b is the channel width (gasket width) (cm). The viscosity of whole blood varies between donors; thus, we use shear rates rather than wall shear stresses for whole blood recruitment experiments. In previous studies, leukocyte recruitment from whole blood has been examined at shear rates ranging from 50 to 400 s^{-1} *(18)*. For isolated leukocyte experiments, the shear stress can be calculated using 1 cP (0.01 P) as the viscosity of the cell suspension. In previous studies, isolated leukocyte recruitment has been examined at shear stresses ranging from 0.5 to 4 dynes/cm^2 *(15,16)*.

6. It is not necessary to use a specific length of tubing for the outlet, vacuum, or inlet line; instead, these lengths should be chosen to facilitate movement of the flow chamber in a particular experimental setup. The length of tubing used for the inlet line, however, should be kept constant both within and between experiments. A fresh inlet line should be used for each new experiment or donor and the inlet line should be switched after using blood or a leukocyte suspension that has been treated with an antibody or inhibitor.

7. It is very important to prevent the introduction of air into the flow chamber setup at any stage in an experiment, as endothelial cells can become activated or damaged if air is perfused over them *(25)*. Before using the flow chamber for the first time, we recommend practicing placing empty culture dishes onto the flow chamber until you can do this without introducing air into the system.

8. The specific perfusion times used in whole blood and isolated leukocyte recruitment assays vary between laboratories. The times that are given in these protocols are the times that we use in our laboratory.

9. When switching the inlet line during an experiment, close the two-way stopcock to prevent air from being pulled into the inlet line and work quickly to prevent pressure from building up in the system. This should take less than 5 s.

10. The amount of time required for the field to clear after the HBSS has entered the chamber can vary, but usually ranges from 30 s to 2 min in our system. For consis-

tency, we recommend that you select an amount of time to wait between the time that the HBSS enters the chamber and the time that you begin recording accumulated cells and use this in all experiments.

11. There are a number of different computer programs that can facilitate the analysis of whole blood recruitment and leukocyte recruitment assays. We analyze our experiments using NIH Image and Adobe Premiere. NIH Image works on Macintosh computers and is available for download at rsb.info.nih.gov/nih-image/Default.html. A similar program called Image J can be used with Windows, Linux, Unix, and OS-2 and is available for download at rsbweb.nih.gov/ij/.

12. The analysis of whole-blood rolling experiments is difficult because of the presence of red blood cells, which can obscure the field of view. This analysis can be particularly difficult for experiments in which endothelial cells are used as a substrate. As a result, you might be unable to perform some of the analysis that is described here, such as the measurement of rolling velocities, on whole-blood rolling experiments in which endothelial cells are used as a substrate.

13. An object of known dimension, such as a hemocytometer, can be used to measure the dimensions of a field.

14. It might be difficult to differentiate between lymphocytes and monocytes that have interacted with endothelial cells using only Wright Giemsa staining. We do not attempt to differentiate between these two leukocyte subclasses in whole blood recruitment experiments, but, instead, classify them all as peripheral blood mononuclear cells (PBMCs).

15. For a given leukocyte subclass, an R-factor of 1 indicates that there is no selective or preferential recruitment of that subclass. An R-factor of less than 1 indicates that there is selectivity against that subclass, whereas an R-factor of greater than 1 indicates that there is selectivity for that subclass.

16. Leukocytes can become activated and transmigrate across the endothelium, making them difficult to identify at ×100 magnification. If this occurs, the 10-s fields can be recorded at ×200 magnification.

17. Transmigrated cells will generally appear as flattened, phase-dark cells with irregular edges and will be below the focal plane of the endothelial cell monolayer. When measuring transmigration for the first time, we recommend that you carefully examine the interactions between leukocytes and endothelial cells at ×800 to ×1000 magnification while focusing up and down on the monolayer so as to learn to differentiate between activated and transmigrated cells. Photographs and supplementary videos of transmigrating leukocytes can be found in articles on lymphocyte and eosinophil transmigration *(14,26)*; these images can assist with the identification of transmigrated leukocytes.

References

1. Kubes, P. (2002) The complexities of leukocyte recruitment. *Semin. Immunol.* **14,** 65–72.
2. Lawrence, M. B. and Springer, T. A. (1991) Leukocytes roll on a selectin at physiologic flow rates: distinction from and prerequisite for adhesion through integrins. *Cell* **65,** 859–873.

3. Lawrence, M. B., McIntire, L. V., and Eskin, S. G. (1987) Effect of flow on polymorphonuclear leukocyte/endothelial cell adhesion. *Blood* **70,** 1284–1290.

4. Forrester, J. V. and Lackie, J. M. (1984) Adhesion of neutrophil leucocytes under conditions of flow. *J. Cell Sci.* **70,** 93–110.

5. Zimmerman, G. A., McIntyre, T. M., and Prescott, S. M. (1985) Thrombin stimulates the adherence of neutrophils to human endothelial cells in vitro. *J. Clin. Invest.* **76,** 2235–2246.

6. Jaffe, E. A., Nachman, R. L., Becker, C. G., and Minick, C. R. (1973) Culture of human endothelial cells derived from umbilical veins. Identification by morphologic and immunologic criteria. *J. Clin. Invest.* **52,** 2745–2756.

7. Hewett, P. W. and Murray, J. C. (1993) Human lung microvessel endothelial cells: isolation, culture, and characterization. *Microvasc. Res.* **46,** 89–102.

8. Richard, L., Velasco, P., and Detmar, M. (1998) A simple immunomagnetic protocol for the selective isolation and long-term culture of human dermal microvascular endothelial cells. *Exp. Cell Res.* **240,** 1–6.

9. Marks, R. M., Czerniecki, M., and Penny, R. (1985) Human dermal microvascular endothelial cells: an improved method for tissue culture and a description of some singular properties in culture. *In Vitro Cell Dev. Biol.* **21,** 627–635.

10. Gupta, K., Ramakrishnan, S., Browne, P. V., Solovey, A., and Hebbel, R. P. (1997) A novel technique for culture of human dermal microvascular endothelial cells under either serum-free or serum-supplemented conditions: isolation by panning and stimulation with vascular endothelial growth factor. *Exp. Cell Res.* **230,** 244–251.

11. Chan, J. R., Hyduk, S. J., and Cybulsky, M. I. (2000) Alpha 4 beta 1 integrin/VCAM-1 interaction activates alpha L beta 2 integrin-mediated adhesion to ICAM-1 in human T cells. *J. Immunol.* **164,** 746–753.

12. Yago, T., Tsukuda, M., Tajima, H., et al. (1997) Analysis of initial attachment of B cells to endothelial cells under flow conditions. *J. Immunol.* **158,** 707–714.

13. Luscinskas, F. W., Kansas, G. S., Ding, H., et al. (1994) Monocyte rolling, arrest and spreading on IL-4-activated vascular endothelium under flow is mediated via sequential action of L-selectin, beta 1-integrins, and beta 2-integrins. *J. Cell. Biol.* **125,** 1417–1427.

14. Cuvelier, S. L. and Patel, K. D. (2001) Shear-dependent eosinophil transmigration on interleukin 4-stimulated endothelial cells: a role for endothelium-associated eotaxin-3. *J. Exp. Med.* **194,** 1699–1709.

15. Patel, K. D. (1998) Eosinophil tethering to interleukin-4-activated endothelial cells requires both P-selectin and vascular cell adhesion molecule-1. *Blood* **92,** 3904–3911.

16. Ostrovsky, L., King, A. J., Bond, S., et al. (1998) A juxtacrine mechanism for neutrophil adhesion on platelets involves platelet-activating factor and a selectin-dependent activation process. *Blood* **91,** 3028–3036.

17. Fitzhugh, D. J., Naik, S., Caughman, S. W., and Hwang, S. T. (2000) Cutting edge: C–C chemokine receptor 6 is essential for arrest of a subset of memory T cells on activated dermal microvascular endothelial cells under physiologic flow conditions in vitro. *J. Immunol.* **165,** 6677–6681.

18. Patel, K. D. (1999) Mechanisms of selective leukocyte recruitment from whole blood on cytokine-activated endothelial cells under flow conditions. *J. Immunol.* **162,** 6209–6216.
19. Ulfman, L. H., Joosten, D. P., van der Linden, J. A., Lammers, J. W., Zwaginga, J. J., and Koenderman, L. (2001) IL-8 induces a transient arrest of rolling eosinophils on human endothelial cells. *J. Immunol.* **166,** 588–595.
20. Kitayama, J., Mackay, C. R., Ponath, P. D., and Springer, T. A. (1998) The C–C chemokine receptor CCR3 participates in stimulation of eosinophil arrest on inflammatory endothelium in shear flow. *J. Clin. Invest.* **101,** 2017–2024.
21. Kukreti, S., Konstantopoulos, K., Smith, C. W., and McIntire, L. V. (1997) Molecular mechanisms of monocyte adhesion to interleukin-1beta-stimulated endothelial cells under physiologic flow conditions. *Blood* **89,** 4104–4111.
22. von Hundelshausen, P., Weber, K. S., Huo, Y., et al. (2001) RANTES deposition by platelets triggers monocyte arrest on inflamed and atherosclerotic endothelium. *Circulation* **103,** 1772–1777.
23. Abe, Y., El-Masri, B., Kimball, K. T., et al. (1998) Soluble cell adhesion molecules in hypertriglyceridemia and potential significance on monocyte adhesion. *Arteriosc. Thromb. Vasc. Biol.* **18,** 723–731.
24. Woltmann, G., McNulty, C. A., Dewson, G., Symon, F. A., and Wardlaw, A. J. (2000) Interleukin-13 induces PSGL-1/P-selectin-dependent adhesion of eosinophils, but not neutrophils, to human umbilical vein endothelial cells under flow. *Blood* **95,** 3146–3152.
25. Patel, K. D. Unpublished observations, 1997–1998.
26. Cinamon, G., Shinder, V., and Alon, R. (2001) Shear forces promote lymphocyte migration across vascular endothelium bearing apical chemokines. *Nature Immunol.* **2,** 515–522.

23

Isolation and Characterization of Side Population Cells

Margaret A. Goodell, Shannon McKinney-Freeman, and Fernando D. Camargo

Summary

The protocol for isolation of side population (SP) cells was originally established for murine bone marrow hematopoietic stem cells (HSCs), but it has also been adapted for other species and tissues. This purification strategy offers a simple and reproducible strategy to obtain a highly homogeneous population of HSCs. The method is based on the differential efflux of the fluorescent DNA-binding dye Hoechst 33342 from stem cells relative to nonstem cells. The protocols outlined in this chapter describe the isolation of murine SP cells from both bone marrow and skeletal muscle using the fluorescent DNA-binding dye Hoechst 33342. In these tissues, the SP cells that are isolated are HSCs.

Key Words: Hematopoietic stem cells; skeletal muscle; bone marrow; murine; SP cells; stem cell purification; Hoechst 33342.

1. Introduction

The protocols outlined in this chapter describe the isolation of murine side population (SP) cells from both bone marrow and skeletal muscle using the fluorescent DNA-binding dye Hoechst 33342 *(1)*.

1.1. Isolation of Bone Marrow SP Cells

Side population cells isolated from murine bone marrow are hematopoietic stem cells (HSCs), and most of those isolated from murine skeletal muscle are also HSCs *(2)*. The method is based on the differential efflux of Hoechst dye relative to other bone marrow cells *(1)*. The protocol was originally established for murine HSCs, but modifications for other species and nonhematopoietic tissues have also been described *(3–8)*. In **Subheading 3.1.**, we describe the isolation of SP cells from normal mouse bone marrow.

From: *Methods in Molecular Biology, vol. 290: Basic Cell Culture Protocols, Third Edition*
Edited by: C. D. Helgason and C. L. Miller © Humana Press Inc., Totowa, NJ

1.2. Isolation of Skeletal Muscle SP Cells

This protocol is employed for the purification of a population of muscle-derived cells from the skeletal muscle of C57Bl/6 mice. The resulting preparation is a mixture of many cell types including satellite cells (aka muscle progenitor cells) and hematopoietically active muscle-derived cells. This protocol is based on that described by Yablonka-Reuveni et al., with slight modifications *(9)*. In **Subheading 3.2.**, we describe the isolation of skeletal muscle SP cells.

2. Materials

2.1. Isolation of Bone Marrow SP Cells

1. C57Bl/6 mice (*see* **Note 1**).
2. 10-cm Tissue culture dishes
3. Two 10-cm^3 Syringes.
4. 18G and 27G needles.
5. 70 μM Nylon mesh (Becton Dickinson "Cell Strainers").
6. DME+: Dulbecco's modified Eagle's medium, high glucose (Gibco, cat. no. 11965-092) supplemented with 2% fetal calf serum (FCS) and 10 mM HEPES buffer.
7. Hoechst 33342 at 1 mg/mL in water. We obtain the dye from Sigma (called Bis-Benzimide, cat. no. B2261) as a powder and resuspend at 1 mg/mL in water, filter-sterilize, and freeze in 250-μL aliquots. Aliquots are not refrozen.
8. Verapamil: dissolve in 95% ethanol as a 5 mM 100X stock. Stored at –20°C in 100-μL aliquots.
9. Propidium iodide (PI) in PBS at 200 μg/mL. PI powder is obtained from Sigma and a stock solution (10 mg/mL) is dissolved in water and stored at –20°C in 100-μL aliquots. The "working" stock (covered with aluminum foil and kept in the refrigerator) is at 200 μg/mL in PBS. The final concentration of PI in your sample should be 2 μg/mL.
10. Hank's buffered salt solution+: HBSS+ is prepared by supplementing HBSS with 2% FCS and 10 mM HEPES buffer. It is stored and used at 4°C
11. Circulating water bath at *exactly* 37°C.
12. Flow cytometer with an ultraviolet (UV) laser (usually high-power argon tuned to 350 nm emission).

2.2. Isolation of Skeletal Muscle SP Cells

1. C57Bl/6J mice, 6–8 wk of age (*see* **Note 1**).
2. HBSS+ (as in **Subheading 2.1.**)
3. 0.2% Type II collagenase (Worthington Biochemicals). This solution is prepared fresh as needed by dissolving the entire vial in HBSS. As long as sterility is maintained, the solution can be kept at 4°C and used for up to 2 d.
4. DMEM/HS: DMEM (Gibco, cat. no. 11965-092) supplemented with 10% horse serum, 100 U/mL penicillin, and 100 mg/mL streptomycin

5. 70-μm Cell filter (Becton Dickinson, "cell strainer").
6. Percoll™ (Amersham Pharmacia Biotech). First, make a 90% Percoll solution by diluting 9 parts Percoll with one part 10X PBS. You can prepare the 90% Percoll solution ahead of time and keep it refrigerated for 1 wk. Next, prepare 40% and 70% Percoll solutions by diluting the 90% Percoll solution with 1X PBS to the appropriate concentration. The 70% and 40% solutions should be made fresh as needed.
7. 10X Phosphate-buffered saline (PBS) (Gibco).

3. Methods

3.1. Isolation of Bone Marrow SP Cells

The ability to discriminate Hoechst SP cells is based on the *differential efflux* of Hoechst 33342 by a multidrug-like transporter *(4,10)*. Because this is an active biological process, the optimal SP profile is obtained with great attention to the staining conditions. The method described will yield highly purified HSCs if performed carefully (*see* **Note 2**).

3.1.1. Extraction of Bone Marrow Cells

1. Using C57Bl/6 mice 5–8 wk of age, excise femurs and tibias, removing as much muscle as possible, and place in tissue culture dish containing HBSS+. If sterility of the stem cells is important (i.e., for subsequent culture), this procedure should be performed in a biological safety cabinet.
2. Using a 10-cm^3 syringe tipped with a 27G needle and filled with HBSS+, flush bone marrow into a fresh sterile dish. We typically prepare marrow from 10 animals at a time and use a total of 10–20 mL of medium to flush the bone marrow from all these bones. Flush from both ends while turning the bone to ensure all the marrow is removed. Bones (without muscle) should be very pale after all marrow is removed.
3. Change the needle to 18G and draw the medium-marrow mixture up and down into the needle several times to break the marrow into a single-cell suspension.
4. Filter through a 70 μ*M* mesh and place the cells in a 50-mL polypropylene tube.
5. Count the nucleated cells accurately (*see* **Note 3**). We find an average of 5×10^7 nucleated cells per C57Bl/6 mouse when collecting marrow from two femurs and tibias.
6. Spin down the cells in a clinical centrifuge (3000 rpm [1700*g*], 5 min).
7. Resuspend cells at 10^6 cells/mL in prewarmed (37°) DME+. The 250-mL polypropylene tubes (Corning) are convenient for staining large volumes of marrow.

3.1.2. Hoechst Staining of Bone Marrow Cells

1. Ensure that the water bath is at precisely 37°C (check this with a thermometer). Prewarm DMEM+ to this temperature while preparing the bone marrow.
2. Add Hoechst to a final concentration of 5 μg/mL (a 200X dilution of the stock). Mix the cells by gentle inversion and place in the 37°C water bath for exactly 90 min (*see* **Note 4**).

3. After 90 min, spin the cells down in a centrifuge at 3000 rpm (1700*g*) at 4°C and resuspend in ice-cold HBSS+. If samples are to be used directly for fluorescent-activated cell sorting (FACS) analysis, use HBSS+ containing 2 μg/mL PI for dead cell discrimination because dead cells can contribute to a poor profile (*see* **Note 5**).

3.1.3. Antibody Staining of Hoechst-Stained Cells

In order to confirm identification of HSC, cells should be co-stained with antibodies such as Sca-1, c-Kit, and lineage markers (*see* **Note 6**).

1. Hoechst-stained bone marrow cells are aliquotted into staining tubes at 10^7 cells per tube in 100 μL of HBSS+.
2. Add appropriately titered antibodies (usually a 1/100 dilution of anti-mouse antibodies obtained from Pharmingen).
3. Incubate the cells with the antibodies on ice for 15 min.
4. Wash cells once by centrifugation with a 10- to 50-fold excess of HBSS+.
5. Resuspend cells in PI solution as described in **Subheading 3.1.2.** for analysis by flow cytometry.

3.1.4. Flow Cytometry Analysis for SP Cells

Side population cells can be analyzed on cytometers from either BD (Facstar-plus and Vantage) or Cytomation (MoFlow). An ultraviolet laser is used to excite the Hoechst dye and PI. A second laser (e.g., argon at 488, or a HeNe) can be used to excite additional fluorochromes. The Hoechst dye is excited with the UV laser at 350 nm and its fluorescence is measured in the "blue" with a 450/20 band pass (BP) filter and in the "red" with a 675 edge filter long pass (EFLP; Omega Optical, Brattleboro VT). A 610 dichroic mirror short pass (DMSP) is used to separate the emission wavelengths. PI fluorescence is also measured through the 675 EFLP (having been excited at 350 nm). Note that PI is read off the UV laser, and dead cells are much brighter red than the Hoechst red signal (*see* **Fig. 1**). Hoechst blue is the standard wavelength for Hoechst 33342 DNA content analysis. Although some other filter sets work sufficiently, we have found these to give the best results.

1. Hoechst-stained cells are placed on the cytometer and preferably kept cold by the use of a chilling apparatus.
2. First, the Hoechst blue versus red profile is displayed, with blue (450 BP filter) on the vertical axis and red (675 LP) on the horizontal axis.
3. With the detectors in linear mode, the voltages are adjusted so that the red blood cells are seen in the lower left corner and dead cells are seen against the far right (very bright PI, thus dead cells) (*see* **Fig. 1**). The major G0–G1 population with S-G2M cells is oriented to the upper right corner.
4. After a profile similar to that shown in **Fig. 1** is obtained, a live gate is drawn to exclude the red and dead cells and 100,000 events should be collected within this

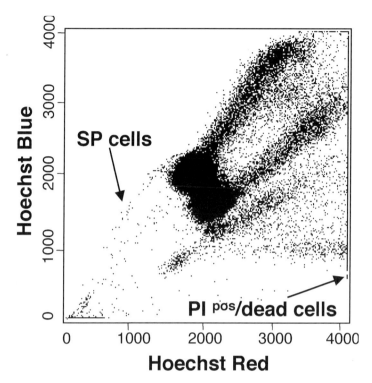

Fig. 1. Hoechst-stained profile of murine bone marrow cells. The flow cytometric profile obtained after staining cells with 5 µg/mL of Hoechst 33342 is shown. Note the very distinct and small subset of cells at the left side of the plot (SP cells). Also, note that PI-positive cells (dead events) are much brighter than Hoechst in the red channel and easily discernible.

live gate. The SP region should appear as shown in **Fig. 1**. The prevalence is around 0.05% of whole bone marrow in the mouse. In human samples, the prevalence is lower (0.01–0.03% of ficolled marrow) (*see* **Notes 7–9**).

3.2. Isolation of Skeletal Muscle SP Cells

The procedures required for the isolation of skeletal muscle SP cells are outlined in **Fig. 2**.

3.2.1. Extraction of Muscle Cells

1. Excise the muscle from the lower limbs of 6- to 8-wk-old mice and place in a 10-cm tissue culture dish containing approx 10 mL HBSS+ (*see* **Note 10**). Remove any bones and tendons from the muscle.
2. In a minimal amount (approx 2 mL) of HBSS+, thoroughly mince the muscle into a fine slurry of 1-mm^2 or smaller particles (*see* **Note 11**).

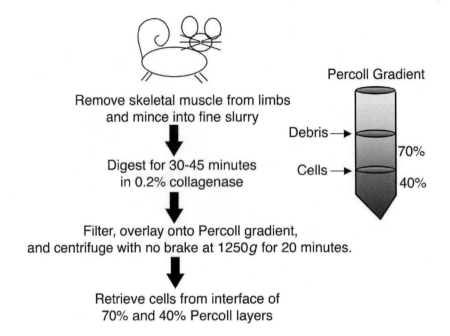

Remove skeletal muscle from limbs
and mince into fine slurry

Digest for 30-45 minutes
in 0.2% collagenase

Filter, overlay onto Percoll gradient,
and centrifuge with no brake at 1250*g* for 20 minutes.

Retrieve cells from interface of
70% and 40% Percoll layers

Fig. 2. Schematic of the isolation of skeletal-muscle-derived cells. Summary of purification of skeletal muscle-derived cells as described in **Subheading 3.2.**

3. Transfer the minced muscle to a 50-mL conical tube and spin down at 2000 rpm (1250*g*) for 3 min in a clinical centrifuge. If extracting muscle from more than one animal, transfer muscle from two animals into one 50-mL tube.

4. After centrifugation, discard the supernatant. Add an equivalent volume of 0.2% type II collagenase to the tube. Mix well and incubate mixture in a circulating water bath for 30 min. If you still note large pieces of undigested muscle, the digestion should be extended another 15 min. This step should result in a very liquidy suspension in which very few large muscle pieces are apparent.

5. Fill the tube to the brim with HBSS+ and spin down at 3000 rpm (1700*g*) for 5 min. Discard supernatant.

6. Resuspend in 10 mL of warm DMEM/HS. Triturate the sample five times with 10 mL of DMEM/HS using a 10-mL plastic pipet. This is done by drawing the mixture up and down inside the pipet repeatedly in order to break up any remaining pieces of muscle tissue and release muscle progenitors from beneath the basal lamina of the intact fibers.

7. After trituration, transfer the medium containing the cells through a 70-μm filter into a fresh 50-mL tube.

8. Collect cells by centrifugation at 3000 rpm (1700*g*) for 5 min and resuspend in 3 mL HBSS+.

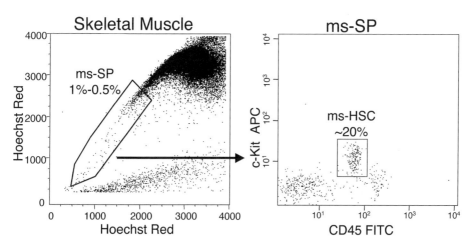

Fig. 3. Hoechst distribution of skeletal muscle and expression of c-kit and CD45 by muscle SP cells. **(A)** Typical FACS plot of skeletal muscle that has been stained with Hoechst. The SP regularly comprises 0.5 to 1% of all muscle-derived cells. **(B)** Distribution of c-kit and CD45 on muscle SP cells. CD45^{pos}c-kit^{dim} cells, the functionally active muscle HSCs, usually comprise about 20% of the muscle SP cells.

3.2.2. Percoll Gradient With Muscle Cells

1. A Percoll gradient is prepared by gently overlaying 3 mL of 70% Percoll with 3 mL of 40% Percoll in a 15-mL conical tube.
2. Gently overlay cell suspension onto Percoll gradient. Wash 50-mL tube that cells were transferred from with 3 mL HBSS+ and overlay this onto same Percoll gradient.
3. Centrifuge at 1250g for 20 min at 25°C with the brake off.
4. Remove cells from 70% Percoll–40% Percoll interface and transfer to fresh 50-mL tube (*see* **Fig. 2**).
5. Fill tube to brim with HBSS+ in order to wash away Percoll and collect cells via centrifugation.
6. Count cells via hemocytometer. Typical cell yields range from 5×10^5 to 1×10^6 cells/mouse.

3.2.3. Staining Muscle-Derived Cells With Hoechst

1. Ensure that the water bath is at precisely 37°C (check this with a thermometer). Prewarm DMEM+ while preparing the muscle-derived cells.
2. Suspend muscle-derived cells at 5×10^5 cells/mL in prewarmed DMEM+ and add Hoechst to a final concentration of 7.5 µg/mL (a 133X dilution of the stock). Place cells in the 37°C water bath for exactly 90 min (*see* **Note 4**).
3. After 90 min, spin the cells down in a centrifuge at 1700g for 5 min at 4°C and resuspend in ice-cold HBSS+ containing 2 µg/mL PI for dead cell discrimination if samples are to be used directly for FACS analysis because dead cells can con-

tribute to a poor profile (*see* **Note 5**). HSCs present in muscle SP are CD45^pos c-kit^dim (*see* **Fig. 3**). This population comprises approx 20% of muscle SP cells.

4. The FACS analysis of muscle SP is performed according to the protocols described for bone marrow in **Subheadings 3.1.3.** and **3.1.4.**

4. Notes

1. C57Bl/6J mice, 6–8 wk of age are recommended for these protocols because this is the gold-standard strain on which this procedure was developed; therefore, comparisons with published literature will be facilitated. Other strains will also work, but we recommend starting with this strain to first establish the method before applying it to other strains.

2. Nonadherence to precise staining conditions can result in low-quality Hoechst stain and potentially a lower purity of stem cells after sorting. Initial experiments should be performed using murine bone marrow as described, in order to definitively identify the SP. The Hoechst concentration, cell concentration, staining time, and staining temperature can all affect the profile. Likewise, following staining, cells must be maintained at 4°C in order to prohibit further dye efflux. If a Ficoll separation, or other lengthy procedures, is to be applied to cells, this should be done *prior* to Hoechst staining.

3. Nucleated cell counts must be performed accurately to ensure that the correct concentration of nucleated cells is set up in the staining medium. As indicated in **Note 2**, variance from the precise staining conditions could affect purity. Counts should be performed to exclude nonnucleated erythrocytes. This can be done by the eye of an experienced investigator or with the aid of one of many red blood cell lysis protocols or commercially available agents.

4. The staining tubes must be well submerged in the bath water to ensure that the temperature of the cells is maintained at 37°C. The tubes should be mixed several times during incubation to ensure equal exposure of the cells to the dye.

5. At this point, samples can be run directly on the FACS or further stained with antibodies as described in **Subheading 3.1.3.** All further manipulations must be performed at 4°C to prohibit efflux of Hoechst dye from the cells. Magnetic enrichments performed at 4°C can be employed at this stage (or, alternatively, prior to Hoechst staining).

6. Mouse SP cells are highly homogeneous with respect to cell surface markers: About 85% of SP cells will be Sca-1+, c-Kit+, CD45+, and lineage marker-negative/low. We recommend staining with at least two antibodies, one which positively stains most SP cells (Sca-1 or c-kit) and one which does not stain SP cells but stains a large fraction of the bone marrow (e.g., Gr-1). All antibodies suggested are available from Pharmingen.

7. In order to confirm the identity of the SP cells, the population can be blocked with verapamil or costained with antibodies as described in **Note 6**. Verapamil is used at 50 μM (Sigma, make a 100X stock in 95% ethanol) and is included during the entire Hoechst-staining procedure. Absence of the SP in the presence of verapamil confirms the identity of the SP cells.

8. Because analysis of the Hoechst dye is performed on a linear scale, optimal setup of the flow cytometer is critical. Good CVs (coefficients of variation) are important. In keeping with having good CVs, the sample differential pressure must be as low as possible. A relatively high power on the UV laser gives the best CVs. We find 50–100 mW to give the best Hoechst signal. Less power will suffice, but the populations might not be as clearly resolved. Likewise, using sensitive red detectors (photomultiplier tubes [PMTs]) is helpful in detecting the best signal from Hoechst red.

9. Hoechst staining should be performed nearly identically for all species, with potentially small variations in staining time. We found 90 min to be optimal for mouse SP cells, whereas 120 min is optimal for human, rhesus, and swine cells *(3,4)*.

10. Younger or female mice generally result in a higher yield. If your goal is to purify myogenic satellite cells, then the diaphragm is also useful to excise at this point, as it contains high numbers of satellite cells with low fibroblast contamination.

11. It is critical that the muscle be thoroughly minced for efficient collagenase digestion. We recommend mincing with one pair of curved mincing scissors, two pairs of forceps (one large and one small), and a pair of sharp surgical scissors.

References

1. Goodell, M. A., Brose, K., Paradis, G., Conner, A. S., and Mulligan, R. C. (1996) Isolation and functional properties of murine hematopoietic stem cells that are replicating in vivo. *J. Exp. Med.* **183,** 1797–1806.

2. McKinney-Freeman, S. L., Jackson, K. A., Camargo, F. D., Ferrari, G., Mavilio, F., and Goodell, M. A. (2002) Muscle-derived hematopoietic stem cells are hematopoietic in origin. *Proc. Natl. Acad. Sci. USA* **99,** 1341–1346.

3. Heinz, M., Huang, C. A., Emery, D. W., et al. (2002) Use of CD9 expression to enrich for porcine hematopoietic progenitors. *Exp. Hematol.* **30,** 809–815.

4. Goodell, M. A., Rosenzweig, M., Kim, H., et al. (1997) Dye efflux studies suggest that hematopoietic stem cells expressing low or undetectable levels of CD34 antigen exist in multiple species. *Nature Med.* **3,** 1337–1345.

5. Gussoni, E., Soneoka, Y., Strickland, C. D., et al. (1999) Dystrophin expression in the mdx mouse restored by stem cell transplantation. *Nature* **401,** 390–394.

6. Welm, B. E., Tepera, S. B., Venezia, T., Graubert, T. A., Rosen, J. M., and Goodell, M. A. (2002) Sca-1(pos) cells in the mouse mammary gland represent an enriched progenitor cell population. *Dev. Biol.* **245,** 42–56.

7. Wulf, G. G., Luo, K. L., Jackson, K. A., Brenner, M. K., and Goodell, M. A. (2003) Cells of the hepatic side population contribute to liver regeneration and can be replenished with bone marrow stem cells. *Haematologica* **88,** 368–378.

8. Asakura, A. and Rudnicki, M. A. (2002) Side population cells from diverse adult tissues are capable of in vitro hematopoietic differentiation. *Exp. Hematol.* **30,** 1339–1345.

9. Yablonka-Reuveni, Z. and Nameroff, M. (1987) Skeletal muscle cell populations. Separation and partial characterization of fibroblast-like cells from embryonic tissue using density centrifugation. *Histochemistry* **87,** 27–38.
10. Zhou, S., Schuetz, J. D., Bunting, K. D., et al. (2001) The ABC transporter Bcrp1/ABCG2 is expressed in a wide variety of stem cells and is a molecular determinant of the side-population phenotype. *Nature Med.* **7,** 1028–1034.

24

Scalable Production of Embryonic Stem Cell-Derived Cells

Stephen M. Dang and Peter W. Zandstra

Summary

Embryonic stem (ES) cells have the ability to self-renew as well as differentiate into any cell type in the body. These traits make ES cells an attractive "raw material" for a variety of cell-based technologies. However, uncontrolled cell aggregation in ES cell differentiation culture inhibits cell proliferation and differentiation and thwarts the use of stirred suspension bioreactors. Encapsulation of ES cells in agarose microdrops prevents physical interaction between developing embryoid bodies (EBs) that, in turn, prevents EB agglomeration. This enables use of stirred suspension bioreactors that can generate large numbers of ES-derived cells under controlled conditions.

Key Words: Embryonic stem cell; differentiation; embryoid body; EB agglomeration; cell aggregation; cell encapsulation; agarose; bioreactor; stirred suspension; cell culture.

1. Introduction

Embryonic stem (ES) cells are a renewable source for many cell types because they have the ability for long-term self-renewal while maintaining the ability to differentiate into any cell type in the body. Whereas ES-derived cells have tremendous potential in many experimental and therapeutic applications, their utility is dependent on the capacity to generate relevant cell numbers under controlled in vitro culture conditions.

Removal of antidifferentiation agents such as leukemia inhibitory factor (LIF) and/or mouse embryonic fibroblasts (MEFs) permits ES cells to differentiate. ES cells can either differentiate in adherent or suspension culture. In suspension, differentiating ES cells spontaneously form tissuelike spheroids called embryoid bodies (EBs). The EB system recapitulates aspects of early embryogenesis by creating a complex microenvironment that supports the development of many different cell lineages *(1)*. Therefore, EB differentiation is a robust method of generating target cell types, particularly when knowledge

From: *Methods in Molecular Biology, vol. 290: Basic Cell Culture Protocols, Third Edition*
Edited by: C. D. Helgason and C. L. Miller © Humana Press Inc., Totowa, NJ

of important differentiation cues is lacking. The EB system has been used to generate potentially therapeutically useful cells, including cardiomyocytes *(2)*, insulin-secreting cells *(3)*, dopaminergic neurons *(4)*, and hematopoietic progenitors *(5,6)*.

Almost all mouse and human ES cells require aggregation of multiple ES cells to initiate EB formation; however, the tendency of EBs to agglomerate prevents their direct addition to stirred suspension culture *(7)*. EB agglomeration is mediated primarily by the cell–cell adhesion molecule E-cadherin *(8)*, whose expression is downregulated as ES cells differentiate *(9)*. Established EB differentiation systems balance the competing requirements of allowing ES cell aggregation while preventing EB agglomeration. Encapsulation of ES cells within agarose microcapsules allows control of both these processes and thus enables culture of EBs in stirred suspension bioreactors. Importantly, stirred suspension bioreactors allow for scalable cell production as well as control of important culture conditions that can affect cell growth and differentiation *(10)*.

In this chapter, we describe methods for the preparation of encapsulation reagents, generation of mouse and human ES cell aggregates, encapsulation of mouse and human ES cell aggregates, and formation and differentiation of mouse and human EBs in stirred suspension culture.

2. Materials

1. Mouse ES cell line (R1, CCE, and D3 have all been tested).
2. Mouse primary embryonic fibroblasts (MEFs).
3. Human ES cell line (H9, H9.2, and I6 have all been tested).
4. Phosphate-buffered saline (PBS) (Gibco–BRL, Rockville, MD): arrives in aqueous form; store in the dark at room temperature.
5. Hank's balanced saline solution (HBSS) (Gibco–BRL): arrives in aqueous form; store in the dark at room temperature.
6. Dulbecco's modified Eagle's medium (DMEM) (Gibco–BRL): arrives in aqueous form; store in the dark at 4°C.
7. Knockout DMEM (Gibco–BRL): arrives in aqueous form; store in the dark at 4°C.
8. ES-qualified fetal bovine serum (FBS) (Hyclone, Logan, UT): arrives frozen; thaw and prepare aliquots at working volumes and store at –20°C.
9. Knockout serum replacement (Gibco–BRL): arrives frozen; thaw and prepare aliquots at working volumes and store at –20°C.
10. Bovine serum albumin (BSA) (Sigma, St. Louis, MO): arrives as a dry powder; store at 4°C.
11. Penicillin and streptomycin (Gibco-BRL): arrives in aqueous form; prepare aliquots at working volumes and store at –20°C.
12. L-Glutamine (Gibco-BRL): arrives in aqueous form; prepare aliquots at working volumes and store at –20°C.

13. Nonessential amino acids (Gibco-BRL): arrives in aqueous form; store in the dark at 4°C.

14. 2-Mercaptoethanol (Sigma): arrives in liquid form; dilute 2-mercaptoethanol in PBS to stock concentration of 10 mM (100X working concentration), prepare aliquots at working volumes, and store at –20°C.

15. Leukemia inhibitory factor (LIF) (Chemicon, Temecula, CA): arrives in aqueous form; prepare aliquots at working volumes and store at –20°C.

16. Human basic fibroblast growth factor (bFGF) (Gibco-BRL): arrives as a lyophilized powder; resuspend powder in 0.2% BSA in PBS to stock concentration of 40 ng/μL (10,000X working concentration), prepare aliquots at working volumes, and store at –20°C.

17. Mouse ES cell media. DMEM supplemented with 15% ES-qualified FBS, 50 U/mL penicillin, 50 μg streptomycin, 2 mM L-glutamine, 0.1 mM of 2-mercaptoethanol, and 500 pM LIF. Prepared media should be stored in the dark at 4°C.

18. Mouse ES cell differentiation media. Same as mouse ES cell media without LIF; Prepared media should be stored in the dark at 4°C.

19. Human ES cell media. Knockout DMEM supplemented with 20% knockout serum replacement, 2 mM L-glutamine, 1 mM nonessential amino acids, 0.1 mM of 2-mercaptoethanol, and 4 ng/mL human bFGF. Prepared media should be stored in the dark at 4°C.

20. Human ES cell differentiation media. DMEM (Gibco-BRL) supplemented with 20% ES-qualified FBS (Hyclone), 2 mM L-glutamine (Gibco-BRL), 0.1 mM of 2-mercaptoethanol. Prepared media should be stored in the dark at 4°C.

21. 0.25% Trypsin–ethylenediaminetetraacetic acid (EDTA) (Sigma, St. Louis, MO): arrives in aqueous form; store at –20°C.

22. Collagenase B (Sigma): arrives as a lyophilized powder. Resuspend collagenase B at 2 mg/mL in 2% FBS in PBS. Prepare aliquots at working volumes and store at –20°C.

23. 15-cm Petri dishes (Fisher, Nepean, ON).

24. Low-gelling temperature agarose (type VII, Sigma or SeaPlaque, FMC, Rockland, ME), (*see* **Note 1**): arrives as a powder; store at room temperature.

25. Dimethylpolysiloxane, 200 cs viscosity (DMPS) (Sigma): arrives in liquid form; store at room temperature.

26. Pluronic F-68 (Sigma): arrives in aqueous form; store at room temperature.

27. Glass scintillation vials, 20 mL (Kimble Glass, Vineland, NJ).

28. Heat/stir plate (Cimerac 1, Barnstead/Thermolyne, Dubuque, IA).

29. CellSys Microdrop Maker (One Cell Systems, Cambridge, MA).

30. Spinner flasks (Bellco Glass, Vineland, NJ).

   Optional:

31. pH sensor (DasGip, Juelich, Germany).

32. Oxygen sensor (DasGip).

33. Gasmix controller unit (DasGip).

34. Air, nitrogen, carbon dioxide, and oxygen gas cylinder tanks (Boc Gases, Mississauga, ON).

35. Computer and data acquisition software (DasGip).

3. Methods

The methods described outline (1) the preparation of reagents, (2) formation of mouse and human ES cell aggregates, (3) encapsulation of mouse and human ES cell aggregates, (4) setup of a stirred suspension culture, and (5) differentiation culture of encapsulated mouse and human ES cell aggregates in a stirred suspension culture.

3.1. Preparation of a 2% (w/v) Agarose Solution in PBS

1. Using a heat/stir plate, bring 10 mL of PBS to a boil in a scintillation vial (*see* **Note 2**).
2. Add 0.2 g agarose powder to the boiling PBS. Use a magnetic stir bar to help the agarose dissolve.
3. Once agarose has fully dissolved, remove the stir bar, cap the scintillation vial, and autoclave (20 min at 120°C, 0.15 MPa).
4. If the agarose solution has cooled (<28°C) and gelled after autoclaving, heat the mixture until molten again (>60°C; *see* **Note 1**). Aliquot 0.4 mL of agarose solution into individual sterile Eppendorf tubes and store at 4°C.

3.2. Formation of Mouse ES Cell Aggregates

Aggregation of approx 40 or more mouse ES cells will efficiently induce the formation of an EB *(11)*. ES cell aggregates can be formed in static liquid suspension culture, hanging-drop culture, or by partial dissociation of attached ES cell colonies. However, static liquid suspension culture is the simplest method of quickly generating large numbers of similar-sized ES cell aggregates (*see* **Note 3**). Encapsulation prevents further contact between cell aggregates and in this way prevents further decrease in aggregate number (*see* **Fig. 1**).

1. Maintain mouse ES cells on gelatin in mouse ES cell medium for a minimum of two passages to remove unwanted MEF feeder cells (*see* **Note 4**).
2. Generate a single-cell suspension by incubating ES cells with 0.25% trypsin–EDTA for 2 min at 37°C, followed by mechanical dissociation using a pipet.
3. Inactivate trypsin–EDTA by adding ES cell media at a ratio of 5:1 (media to trypsin).
4. Transfer the mixture to a centrifuge tube and pellet the ES cells by centrifugation at 1000 rpm (200g) for 5 min.
5. Aspirate the supernatant.
6. Prepare 15 mL of ES cell suspension at a cell density of 3×10^5 ES cells/mL in ES cell media (*see* **Note 5**).
7. Transfer 15 mL of ES cell suspension into a 15-cm Petri dish and culture overnight (16–24 h) at 37°C in humidified air with 5% CO_2
8. Harvest ES cell aggregates by transferring the cell suspension to a 15-mL conical centrifuge tube. Cell aggregates can be enumerated at this point by taking a small

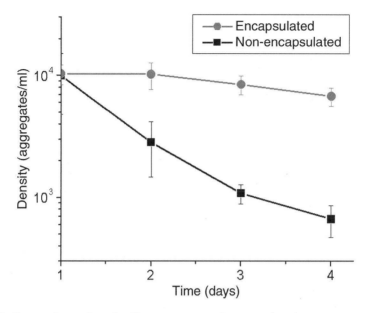

Fig. 1. Comparison of total cell aggregate number over time between encapsulated and nonencapsulated liquid suspension culture. Encapsulation prevents decline in cell aggregate density.

aliquot of the sample (e.g., 0.1 mL), transferring it to a gridded 35-mm Petri dish containing 2 mL PBS, and using a microscope to visually inspect and count the number of cell aggregates.

9. Centrifuge at 500 rpm ($50g$) for 3 min. Alternatively, ES cell aggregates can sediment out of the media after standing for 20 min.

10. Aspirate supernatant. The expected yield is 3×10^4 ES cell aggregates with an average size of 40 ES cells/aggregate. Mouse ES cell aggregates are now ready for encapsulation.

3.3. Formation of Human ES Cell Aggregates

Human ES cell aggregates are prepared by chemically and mechanically detaching whole human ES cell colonies from culture plates and MEF feeder cells. For efficient EB formation, it is recommended that individual human ES cell colonies are kept intact and not further dissociated to smaller cell aggregates.

1. Allow freshly passaged human ES cell colonies to grow for 3–4 d on gelatin-coated, MEF-covered, six-well plates before harvesting (*see* **Note 4**). Individual human ES cell colonies should contain $(1–5) \times 10^3$ cells. Prepare at least two wells of human ES cells.

2. Aspirate media and add 2 mL (or 1 mL/10 cm²) of 2 mg/mL collagenase B to each well. Incubate cells for 30 min at 37°C.

3. Using a 5-mL pipet, gently wash the sides of the culture plate until most human ES cell colonies have visibly detached from the culture plate surface. Transfer liquid mixture to a 15-cm conical centrifuge tube.

4. Add 2 mL (or 1 mL/10 cm²) of human ES cell media to the culture plate and again wash the surface to detach and suspend any remaining human ES colonies. Transfer liquid mixture to the centrifuge tube. Cell aggregates can be enumerated at this point, as previously described in **Subheading 3.2., step 8**.

5. Centrifuge cells at 500 rpm ($50g$) for 3 min.

6. Aspirate supernatant. The expected yield from harvesting two wells is approx 2×10^3 human ES cell aggregates. Human ES cell aggregates are now ready for encapsulation.

3.4. Encapsulation of Mouse and Human ES Cell Aggregates

Cell encapsulation is necessary to control ES cell aggregation. Agarose, a naturally derived polysaccharide molecule, was selected for this purpose because it has been widely used in various cell-encapsulation applications. Agarose gels are highly porous, allowing rapid diffusional exchange of high-molecular-weight molecules (up to 500 kDa) and they do not significantly alter cell physiology *(12)*. ES cells within a particular capsule are permitted to aggregate and initiate EB formation; however, contact between cells in different capsules is physically prevented by the encapsulating agarose.

As encapsulated ES cell aggregates form EBs and grow in size, they degrade the surrounding agarose matrix. Capsules are designed to encapsulate differentiating ES cells for as long as E-cadherin expression remains high. This corresponds to the first 4 d and first 8 d of differentiation for mouse and human ES cells, respectively (*see* **Fig. 2**). Mouse and human ES cells are therefore encapsulated in 2% agarose capsules with a diameter range of 80–120 µm and 150–200 µm, respectively.

An emulsification technique is described to encapsulate ES cell aggregates although alternative techniques are available (*see* **Note 6**). Cells partition into the aqueous agarose solution that is immiscible with the nonpolar DMPS solvent. A rapidly spinning impeller is then used to shear agarose droplets into the appropriate size distribution. The mixture is then cooled to allow the agarose droplets to gel.

1. Aliquot 15 mL of DMPS into an autoclaved scintillation vial and place in a 37°C water bath for a minimum of 10 min.

2. Prepare molten agarose by microwaving a 0.4-mL agarose aliquot for 25 s on high power or place in a 70°C water bath until molten.

3. Add 25 µL Pluronic F-68 to molten agarose. This protects cells from shear forces during encapsulation. Place molten agarose in a 37°C water bath for 1 min to allow temperatures to equilibrate (*see* **Note 7**).

Encapsulated Mouse EB

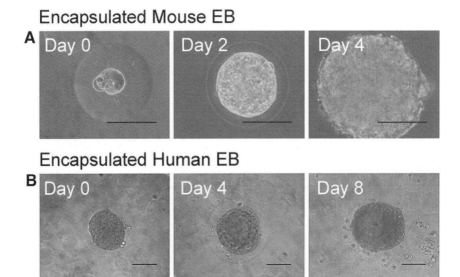

Encapsulated Human EB

Fig. 2. Encapsulated mouse (**A**) and human (**B**) ES cells form EBs that grow and degrade the encapsulating agarose matrix over time. Mouse EBs (**A**) emerge after 4 d of differentiation culture; human EBs (**B**) emerge after 8 d. Scale bars are 100 μm.

4. Resuspend mouse or human ES cell aggregates (generated in **Subheading 3.2.** or **3.3.**, respectively) in 100 μL of HBSS, dispense mixture into molten agarose, and gently mix by pipetting.
5. Using a 1-mL pipet, dispense the agarose mixture into the previously prepared scintillation vial containing 15 mL DMPS at 37°C. The pipet tip should be flicked or rotated rapidly by hand as the agarose is dispensed to create small immiscible agarose droplets in DMPS.
6. Secure the scintillation vial to the CellSys Microdrop Maker.
7. Stir the mixture at 850 rpm for 2 min at room temperature.
8. Immerse the scintillation vial in an ice-water bath and continue stirring at 850 rpm for an additional 4 min.
9. First wash. Divide by pipetting the emulsion mixture evenly between two 15-mL conical centrifuge tubes and gently overlay with 5 mL HBSS (per tube).
10. Centrifuge the mixture at 4°C and 1500 rpm (400g) for 5 min. Centrifugation will partition the mixture into three phases, from bottom to top: (1) encapsulated ES cell aggregate pellet, (2) HBSS, and (3) DMPS. Transfer the top layer of DMPS to the scintillation vial for disposal. Aspirate the remaining aqueous phase, leaving the encapsulated ES cell aggregate pellet.
11. Second wash. Prepare two new 15-mL centrifuge tubes with 10 mL of HBSS each. Resuspend encapsulated ES cell aggregates in 1 mL PBS and overlay onto the prepared 15-mL centrifuge tubes containing 10 mL HBSS.

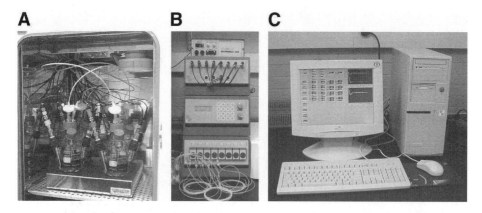

Fig. 3. Bioreactor setup: (**A**) spinner flasks with gas lines, oxygen sensors, and pH sensors; (**B**) gas mix control unit; (**C**) computer and data acquisition software.

12. Centrifuge the mixture at 4°C and 1500 rpm (400*g*) for 5 min.
13. Aspirate the aqueous phase and resuspend encapsulated ES cell aggregates in culture media according to desired experimental protocol. Encapsulated cell aggregates can be enumerated at this point as previously described in **Subheading 3.2., step 8**. The expected yield is 1.5×10^4 encapsulated mouse ES cell aggregates or 1×10^3 encapsulated human ES cell aggregates. If a controlled bioreactor setup will not be used, continue to **Subheading 3.6.**

3.5. Bioreactor Setup

Differentiation culture of encapsulated ES cell aggregates can be performed in most vessel types and configurations. Here, we describe the assembly of the bioreactor for the generation of differentiated ES-cell-derived progenitors in stirred, controlled bioreactors (*see* **Fig. 3**).

1. Calibrate the pH sensors following the manufacturer's instructions. If these are not available, perform a two-point calibration by immersing sensors in standardized pH-buffered solutions (e.g., pH 7 and pH 4).
2. Assemble the bioreactor according to the manufacturer's instructions. **Figure 3A** shows assembled DasGip 500-mL bioreactors with glass ball stirrer, pH sensor, oxygen sensor, and gas inlet and outlet ports.
3. Fill the assembled vessel with 150 mL of PBS and autoclave (25 min at 120°C, 0.15 MPa).
4. In a tissue culture hood, empty PBS and replace with 100–200 mL of ES-cell-differentiation media.
5. Attach the electrical leads to the pH and oxygen sensors and connect the gas line to the inlet port.
6. Set impeller stir speed at 50 rpm.
7. Flow 100% air through the gas line for 6 h.

8. Calibrate oxygen electrodes according to the manufacturer's instructions. If these are not available, perform a single-point calibration of oxygen electrode for 100% dissolved oxygen (in equilibrium with air). Optionally, a second calibration point for 0% dissolved oxygen can be obtained after flowing 100% nitrogen gas through the gas line for 6 h.
9. Input control options and setpoint values for pH and dissolved oxygen. Sample conditions are pH 7.4 and 100% dissolved oxygen (20% oxygen tension).

3.6. Encapsulated Mouse ES Cell Stirred Suspension Differentiation Culture

Different media conditions, such as those described for hematopoietic development *(13,14)*, can be used to encourage differentiation along specific pathways. After selecting the desired ES cell differentiation media, perform the following:

1. Inoculate each sterile 500-mL bioreactor containing 200 mL of mouse ES-cell-differentiation media with 1.5×10^4 encapsulated mouse ES cell aggregates generated in **Subheading 3.4.** to achieve a cell density of 2.5×10^3 ES cells/mL (*see* **Notes 8** and **9**).
2. Cells are cultured at 37°C in humidified air. Default gas mixture in the headspace should be set to 21% $O_{2(g)}$ and 5% $CO_{2(g)}$; however, on-line gas mix controllers will adjust $O_{2(g)}$ and $CO_{2(g)}$ levels to maintain 21% oxygen tension and pH 7.4 media conditions. Impeller stir speed can be set between 40 and 60 rpm.
3. Harvest cells after desired time in differentiation culture. Expected yield for differentiating mouse EBs after 7 d is approx 30 million cells (60 times cell fold expansion) per bioreactor.

3.7. Encapsulated Human ES Cell Stirred Suspension Differentiation Culture

1. Inoculate each sterile 500-mL bioreactor containing 100 mL ES-cell-differentiation media with 1×10^3 encapsulated human ES cell aggregates generated in **Subheading 3.4.** to achieve a cell density of 5×10^3 ES cells/mL (*see* **Note 9**).
2. Cells are cultured at 37°C in humidified air. Default gas mixture in the headspace should be set to 21% $O_{2(g)}$ and 5% $CO_{2(g)}$; however, on-line gas mix controllers will adjust $O_{2(g)}$ and $CO_{2(g)}$ levels to maintain 21% oxygen tension and pH 7.4 media conditions. Impeller stir speed can be set between 40 and 60 rpm.
3. Harvest cells after desired time in differentiation culture. Expected yield for differentiating human EBs after 15 d is approx 2 million cells (four times cell fold expansion) per bioreactor.

4. Notes

1. Agarose solutions exhibit an upper critical and lower critical solution temperature: The low-gelling-temperature agarose (gel state) liquefies at 60°C and aqueous (liquid state) agarose gels at 28°C.
2. Glass scintillation vials are convenient for preparing agarose solutions because they fit inside standard sterilization pouches for autoclaving.

3. Single ES cells will rapidly aggregate with neighboring ES cells. ES cell aggregate size can be controlled by input ES cell density and culture time. A minimum culture period of 16 h is necessary for the formation of tightly adhered cell aggregates that can maintain their structure when sheared during the encapsulation process.

4. The methodology for routine maintenance and passage of undifferentiated mouse and human ES cells is beyond the scope of this chapter. The reader is referred to *Embryonic Stem Cells: Methods and Protocols (15)* for appropriate culture conditions for murine ES cells and to *Human Embryonic Stem Cells (16)* for appropriate culture conditions for human ES cells.

5. If mouse ES cell aggregates fail to form in suspension culture, increase the ES cell density and/or time in static liquid suspension culture. Alternatively, the partial dissociation method described for human ES cell aggregate formation in **Subheading 3.3.** is also applicable for mouse ES cell aggregate formation.

6. Alternative methods for encapsulating ES cells are readily available. Encapsulation of ES cells in alginate beads by polyelectrolyte complexation was previously described *(17)*. Other agarose encapsulation protocols can be readily adapted for encapsulating ES cells *(18,19)*. Other encapsulation techniques include interfacial phase inversion *(20)*, *in situ* polymerization *(21)*, and conformal coating *(22)*.

7. It is important that molten agarose is cooled to 37°C in a water bath before cells are introduced. Once cells have been transferred to the agarose (**Subheading 3.4., step 3**), the encapsulation **steps 5–7** should be performed as rapidly as possible to prevent mixture from cooling and gelling prematurely.

8. An input ES cell density of 2.5×10^3 mouse ES cells/mL was selected to permit batch-style culture; that is, media exchange was not required over the 7-d culture period (based on glucose consumption). Higher input cell densities can be realized with media perfusion or exchange.

9. An acceptable pH range is 7.2–7.6, glucose concentration >5 mM, oxygen tension >80%, and cell density <5 × 10^5 cells/mL (unless other setpoint values are desired). The input ES cell density can be increased above suggested values provided that culture conditions remain within the accepted ranges.

Acknowledgments

This work was funded by the Natural Sciences and Engineering Research Council (NSERC) of Canada. StemCell Technologies Inc., One Cell Systems Inc., and DasGip AG are acknowledged for reagent and equipment support. Stephen Dang is supported by a NSERC postgraduate award. PWZ is a Canada Research Chair in Stem Cell Bioengineering.

References

1. Keller, G. M. (1995) In vitro differentiation of embryonic stem cells. *Curr. Opin. Cell Biol.* **7,** 862–869.
2. Boheler, K. R., Czyz, J., Tweedie, D., Yang, H. T., Anisimov, S. V., and Wobus, A. M. (2002) Differentiation of pluripotent embryonic stem cells into cardiomyocytes. *Circ. Res.* **91,** 189–201.

3. Soria, B., Roche, E., Berna, G., Leon-Quinto, T., Reig, J. A., and Martin, F. (2000) Insulin-secreting cells derived from embryonic stem cells normalize glycemia in streptozotocin-induced diabetic mice. *Diabetes* **49,** 157–162.
4. Lee, S. H., Lumelsky, N., Studer, L., Auerbach, J. M., and McKay, R. D. (2000) Efficient generation of midbrain and hindbrain neurons from mouse embryonic stem cells. *Nature Biotechnol.* **18,** 675–679.
5. Kaufman, D. S., Hanson, E. T., Lewis, R. L., Auerbach, R., and Thomson, J. A. (2001) Hematopoietic colony-forming cells derived from human embryonic stem cells. *Proc. Natl. Acad. Sci. USA* **98,** 10,716–10,721.
6. Keller, G., Kennedy, M., Papayannopoulou, T., and Wiles, M. V. (1993) Hematopoietic commitment during embryonic stem cell differentiation in culture. *Mol. Cell. Biol.* **13,** 473–486.
7. Dang, S. M., Kyba, M., Perlingeiro, R., Daley, G. Q., and Zandstra, P. W. (2002) Efficiency of embryoid body formation and hematopoietic development from embryonic stem cells in different culture systems. *Biotechnol. Bioeng.* **78,** 442–453.
8. Larue, L., Antos, C., Butz, S., et al. (1996) A role for cadherins in tissue formation. *Development* **122,** 3185–3194.
9. Viswanathan, S., Benatar, T., Rose-John, S., Lauffenburger, D. A., and Zandstra, P. W. (2002) Ligand/receptor signaling threshold (LIST) model accounts for gp130-mediated embryonic stem cell self-renewal responses to LIF and HIL-6. *Stem Cells* **20,** 119–138.
10. Zandstra, P. W. and Nagy, A. (2001) Stem cell bioengineering. *Annu. Rev. Biomed. Eng.* **3,** 275–305.
11. Dang, S. M., Gerecht-Nir, S., Chen, J., Itskovitz-Eldor, J., and Zandstra, P. W. (2004) Controlled, scalable embryonic cell differentiation culture. *Stem Cells* **22,** 275–282.
12. Weaver, J. C., McGrath, P., and Adams, S. (1997) Gel microdrop technology for rapid isolation of rare and high producer cells. *Nature Med.* **3,** 583–585.
13. Adelman, C. A., Chattopadhyay, S., and Bieker, J. J. (2002) The BMP/BMPR/Smad pathway directs expression of the erythroid-specific EKLF and GATA1 transcription factors during embryoid body differentiation in serum-free media. *Development* **129,** 539–549.
14. Chadwick, K., Wang, L., Li, L., et al. (2003) Cytokines and BMP-4 promote hematopoietic differentiation of human embryonic stem cells. *Blood* **102,** 906–915.
15. Turksen, K. (ed.) (2002) *Embryonic Stem Cells: Methods and Protocols*, Humana, Totowa, NJ.
16. Chiu, A. and Rao, M. S. (eds.) (2003) *Human Embryonic Stem Cells*, Humana, Totowa, NJ.
17. Magyar, J. P., Nemir, M., Ehler, E., Suter, N., Perriard, J. C., and Eppenberger, H. M. (2001) Mass production of embryoid bodies in microbeads. *Ann. NY Acad. Sci.* **944,** 135–143.
18. Gin, H., Dupuy, B., Baquey, C., Ducassou, D., and Aubertin, J. (1987) Agarose encapsulation of islets of Langerhans: reduced toxicity in vitro. *J. Microencapsul.* **4,** 239–242.

19. Tashiro, H., Iwata, H., Tanigawa, M., et al. (1998) Microencapsulation improves viability of islets from CSK miniature swine. *Transplant. Proc.* **30,** 491.
20. Stevenson, W. T., Evangelista, R. A., Sugamori, M. E., and Sefton, M. V. (1988) Microencapsulation of mammalian cells in a hydroxyethyl methacrylate-methyl methacrylate copolymer: preliminary development. *Biomater. Artif. Cells Artif. Organs* **16,** 747–769.
21. Dupuy, B., Gin, H., Baquey, C., and Ducassou, D. (1988) In situ polymerization of a microencapsulating medium round living cells. *J. Biomed. Mater. Res.* **22,** 1061–1070.
22. Zekorn, T., Siebers, U., Horcher, A., et al. (1992) Alginate coating of islets of Langerhans: in vitro studies on a new method for microencapsulation for immuno-isolated transplantation. *Acta Diabetol.* **29,** 41–45.

Index